Progress in Sensory Physiology

Editors: H. Autrum, D. Ottoson,
E.R. Perl, R.F. Schmidt, H. Shimazu, W.D. Willis
Editor-in-Chief: D. Ottoson

Volume 3
W.D. Willis
Control of Nociceptive Transmission in the Spinal Cord
1982. 51 figures. VI, 159 pages
ISBN 3-540-11510-2

Contents: Introduction: Centrifugal Control of Sensory Pathways. – Behavioral Evidence for Descending Control of Nociceptive Transmission. – Pharmacology of Analgesia Due to Descending Control Systems. – Descending Control of the Flexion Reflex. – Descending Control of Spinal Cord Nociceptive Neurons. – Correlations Between the Descending Control of Spinal Cord Nociceptive Pathways and the Operation of the Analgesia Systems.

Volume 4
1983. 41 figures. V, 118 pages
ISBN 3-540-12498-5

Contents: *N. Mei:* Sensory Structures in the Viscera. – *G.R. Martin:* Schematic Eye Models in Vertebrates. – *A. Gallego:* Organization of the Outer Plexiform Layer of the Tetrapoda Retina.

Volume 5
1985. 83 figures. V, 174 pages
ISBN 3-540-15339-X

Contents: *R.C. Hardie:* Functional Organization of the Fly Retina. – *H. Machemer, J.W. Deitmer:* Mechanoreception in Ciliates. – *M. Lindauer, H. Martin:* The Biological Significance of the Earth's Magnetic Field. – *H. Bleckmann:* Perception of Water Surface Waves: How Surface Waves are Used for Prey Identification Communications.

Volume 6
1986. 86 figures. V, 227 pages
ISBN 3-540-15340-3

Contents: *T. Sato:* Receptor Potential in Rat Taste Cells. – *K.B. Døving:* Functional Properties of the Fish Olfactory System. – *J.A. Coles:* Homeostasis of Extracellular Fluid in Retinas of Invertebrates and Vertebrates. – *S. Mense:* Slowly Conducting Afferent Fibers from Deep Tissues: Neurobiological Properties and Central Nervous Actions.

Progress in

Sensory Physiology 7

Editors: H. Autrum D. Ottoson
E.R. Perl R.F. Schmidt H. Shimazu W.D. Willis

Editor-in-Chief: D. Ottoson

D.R.F. Irvine: The Auditory Brainstem

A Review of the Structure and Function of Auditory Brainstem Processing Mechanisms

With 60 Figures and 5 Tables

Springer-Verlag
Berlin Heidelberg New York Tokyo

Editor-in-Chief

Professor Dr. David Ottoson
Karolinska Institutet, Fysiologiska Institutionen II
Solnavägen 1, 10401 Stockholm 60, Sweden

Editors

Professor Dr. Hansjochem Autrum
Zoologisches Institut der Universität München
Luisenstrasse 14, 8000 München 2, Germany

Professor Dr. Edward R. Perl
Department of Physiology
University of North Carolina at Chapel Hill
Chapel Hill, NC 27514, USA

Professor Dr. Robert F. Schmidt
Physiologisches Institut der Universität
Röntgenring 9, 8700 Würzburg, Germany

Professor Dr. Hiroshi Shimazu
Department of Neurophysiology
University of Tokyo, Institute of Brain Research
7.3.1. Hongo, Bunkyo Ku, Tokyo, Japan

Professor Dr. William D. Willis
The Marine Biomedical Institute
University of Texas Medical Branch
Galveston, TX 77550, USA

Author

Dr. Dexter R.F. Irvine
Department of Psychology, Monash University
Clayton, Victoria 3168, Australia

ISBN-13: 978-3-642-71059-9　　e-ISBN-13: 978-3-642-71057-5

DOI: 10.1007/978-3-642-71057-5

Progress in sensory physiology. – 1- – Berlin; New York: Springer-Verlag, 1981– v.: ill.; 25 cm. Irregular. Editor: D. Ottoson. ISSN 0721-9156 = Progress in sensory physiology. 1. Senses and sensation-Collected works. 2. Sense-organs-Collected works. 3. Neurophysiology-Collected works. I. Ottoson, David, 1918– .
[DNLM: 1. Neurophysiology-periodicals. 2. Sensation-physiology-periodicals. W1 PR681G] QP431.P73 599'.01'8-dc19 85-647743 AACR 2 MARC-S [8509]

Softcover reprint of the hardcover 1st edition 1986

Typesetting: K+V Fotosatz GmbH, Beerfelden

2121/3140-543210

Contents

Common Abbreviations

AI	primary auditory cortex
AA	anterior part of anterior division of anteroventral cochlear nucleus
ABI	average binaural intensity
AChE	acetylcholinesterase
ALSR	average localized synchronized rate
AM	amplitude-modulated
AN	auditory nerve
AP	posterior part of anterior division of anteroventral cochlear nucleus
APD	posterodorsal part of anterior division of anteroventral cochlear nucleus
AVCN	anteroventral cochlear nucleus
BIC	brachium of inferior colliculus
CD	characteristic delay
CF	characteristic frequency
CLL	commissure of lateral lemniscus
CN	cochlear nucleus
CNS	central nervous system
CV	consonant-vowel
DAS	dorsal acoustic stria (stria of Monakow)
DCN	dorsal cochlear nucleus
DMPO	dorsomedial periolivary nucleus
DNLL	dorsal nucleus of lateral lemniscus
EM	electron-microscopic
EMI	excitatory monaural intensity
EPSP	excitatory postsynaptic potential
FCL	fusiform cell layer (of dorsal cochlear nucleus)
FM	frequency-modulated
FTC	frequency tuning curve
HRP	horseradish peroxidase
IAS	intermediate acoustic stria (stria of Held)
IC	inferior colliculus
ICC	central nucleus of inferior colliculus
ICO	commissure of inferior colliculus
ICP	pericentral nucleus of inferior colliculus
ICX	external nucleus of inferior colliculus
IHC	inner hair cell
IID	interaural intensity difference

INLL	intermediate nucleus of lateral lemniscus
IPD	interaural phase difference
IPSP	inhibitory postsynaptic potential
ISIH	interspike interval histogram
ITD	interaural time difference
JND	just-noticeable difference
LF	limited-field
LL	lateral lemniscus
LM	light-microscopic
LNTB	lateral nucleus of trapezoid body
LSO	lateral superior olivary nucleus
MAA	minimum audible angle
MGB	medial geniculate body
ML	molecular layer (of dorsal cochlear nucleus)
MLD	nucleus *mesencephalicus lateralis dorsalis*
MNTB	medial nucleus of trapezoid body
MSO	medial superior olivary nucleus
NBIC	nucleus of brachium of inferior colliculus
NLL	nuclei of the lateral lemniscus
OCA	octopus cell area
OCB	olivo-cochlear bundle
OHC	outer hair cell
OTD	ongoing time difference
PD	dorsal part of posterior division of anteroventral cochlear nucleus
PL	polymorphic layer (of dorsal cochlear nucleus)
PSTH	peristimulus-time histogram
PV	ventral part of posterior division of anteroventral cochlear nucleus
PVCN	posteroventral cochlear nucleus
RF	receptive field
SC	superior colliculus
SCD	deep layers of superior colliculus
SCS	superficial layers of superior colliculus
SOC	superior olivary complex
SPL	sound pressure level
SR	spontaneous discharge rate
TB	trapezoid body
TTD	transient time difference
VCN	ventral cochlear nucleus
VNLL	ventral nucleus of lateral lemniscus
VNTB	ventral nucleus of trapezoid body
2-DG	^{14}C-2-deoxyglucose

Orientation of Anatomical Sections

C caudal
D dorsal
L lateral
M medial
R rostral
V ventral

Binaural Input Categories

For listing of binaural input categories see description in Sect. 5.2.1.

1 Introduction and Plan of the Review

The auditory system differs from the other major sensory systems in the complexity of its subcortical pathways. In the auditory system four major processing levels are interposed between the primary afferents and the thalamorecipient laminae of the primary cortical field, whereas there is only one such intermediate processing level in the visual system and two in the somatosensory system. Associated with this complexity is a high degree of divergence and convergence in the auditory brainstem pathways.

The major nuclear divisions of the brainstem auditory pathway are shown in Fig. 1, and the ascending projections are summarized schematically in Figs. 26, 34, and 37. The fibers of the eighth nerve innervate three major divisions of the cochlear nucleus (CN), each of which contains further subdivisions defined by morphologically different cell types receiving different patterns of primary innervation. The efferent projections of the various subdivisions form parallel pathways, proceeding either directly, or indirectly via other brainstem nuclei in the superior olivary complex (SOC) and the lateral lemniscus (NLL), to the auditory midbrain nucleus, the inferior colliculus (IC). Neurons in some divisions of the SOC and NLL receive afferents from the CN on both sides of the brain, and some divisions project bilaterally to IC. The central nucleus of the inferior colliculus (ICC) consequently receives major projections from approximately ten auditory brainstem structures, and less substantial projections from many more (see Fig. 37). All of these projections converge in a highly organized fashion onto a single representation of the cochlea. At different levels within this system of divergent and convergent projections the information conveyed to the central nervous system (CNS) by each auditory nerve undergoes major transformations, information from the two ears is integrated, and significant monaural and binaural features are extracted from the input. It is the purpose of this review to examine current knowledge of the mechanisms by which such processing is carried out in these brainstem pathways.

The preceding paragraph and Fig. 1 present a conventional account of the auditory brainstem pathways in terms of gross nuclei and nuclear subdivisions. Much of the recent progress in our understanding of auditory brainstem processing has come with the realization that the major nuclear subdivisions of classical anatomy contain further subdivisions defined by cells of distinctive neuronal morphology and patterns of afferent and efferent connections, and with the establishment of correlations between morphologically and physiologically defined cell "types." The complexity of the convergence onto IC is even more remarkable in these terms, some 24 reasonably well-defined classes of cell from each side of the brainstem giving rise to ascending projections to the central nucleus (Adams 1979). A review of brainstem processing therefore requires as much attention to the evidence on neuronal architecture and connections as to that on physiological mechanisms.

In most natural environments, the sound reaching our ears arises from a number of sources at different points in space and includes reflections from nearby sur-

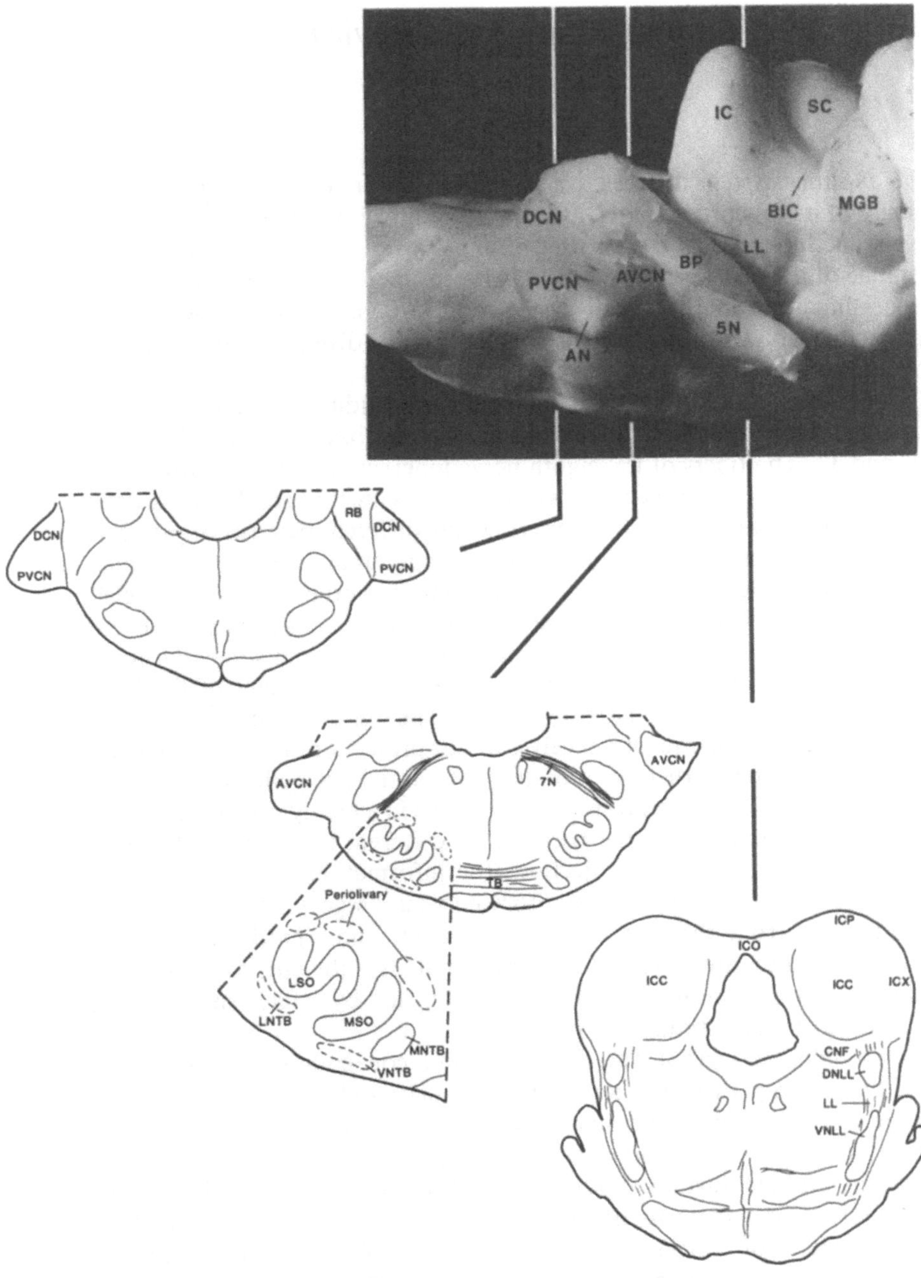

Fig. 1. Major structures of auditory brainstem shown on a photograph of the cat's brainstem in lateral view and on a set of transverse sections, the approximate levels of which are indicated by the *lines* joining them to the brainstem photograph. The auditory nerve (*AN*) enters the cochlear nucleus (*CN*), which has traditionally been divided into three: the anteroventral, posteroventral, and dorsal cochlear nuclei (*AVCN, PVCN,* and *DCN*). Neurons in CN project via the trapezoid body (*TB*) and other striae to the superior olivary complex (SOC) and to higher centers. The major nuclei of SOC shown here are the lateral and medial superior olivary nuclei (*LSO* and *MSO*) and the medial nucleus of the trape-

faces. Yet we are able to segregate the auditory input, identifying different sources on the basis of their spectral and temporal properties and their spatial location. This review will focus on the mechanisms by which spectral and spatial information is processed and, for the most part, will consider spectral and spatial processing in *simple* environments (i.e., single sources in anechoic space) and as *independent* processes. This restriction reflects the fact that some of the higher-order processes on which the perceptual "parsing" of acoustic input in complex environments depends are beyond the scope of current auditory neuroscience (see, e.g., Bregman 1978; Moore 1982). It should be emphasized, however, that the separate treatment of spectral and spatial processing is to a large extent artificial. Not only do some aspects of spatial coding rest upon spectral information, but the properties of many lower brainstem neurons indicate that they are involved in both forms of processing, and both ultimately involve similar pattern recognition processes in the CNS.

Any discussion of the processing of auditory information by the CNS must begin with some specification of the tasks performed by the central processor and the level of performance achieved by it. In the case of processing by auditory brainstem centers, a further requirement is specification of the information conveyed to the central processor by the auditory nerve. In the initial sections of the review (Sects. 2 and 3) evidence on these issues will be examined as a foundation for the subsequent account of brainstem structure and mechanisms. The performance data to be reviewed are, of course, those of the central processor as a whole and undoubtedly reflect the activity not only of the brainstem, but also of thalamic and cortical mechanisms. In general, however, the precision with which the features of the acoustic stimulus are represented in the lower levels of the auditory pathway establishes limits on the performance of the entire system. Furthermore, as we shall see, fundamental transformations of many features of the information provided by the auditory nerve are carried out in highly specialized brainstem centers.

Most of our detailed knowledge of the performance of the auditory central processor is derived from psychophysical studies with conscious human subjects, and a fundamental aim of auditory neuroscience is to understand the mechanisms underlying human perception. For obvious reasons, however, most of the anatomical and physiological data concerning these mechanisms have been obtained from laboratory animals (most commonly the domestic cat), and the phys-

Fig. 1 (*continued*) zoid body (*MNTB*). The approximate locations of the lateral and ventral nuclei of the trapezoid body (*LNTB* and *VNTB*) and of the dorsal "periolivary" cell groups are indicated by *broken lines*. Axons of CN and SOC neurons ascend to the midbrain in the lateral lemniscus (*LL*), some of them terminating in the ventral and dorsal nuclei of the lateral lemniscus (*VNLL* and *DNLL*), which in turn project rostrally. Almost all ascending axons terminate in the auditory midbrain nucleus, the inferior colliculus (*IC*), which has traditionally been divided into a central, pericentral, and external nucleus (*ICC, ICP*, and *ICX*). The IC projects via the brachium of the inferior colliculus (*BIC*) to the auditory thalamic nucleus, the medial geniculate body (*MGB*). Auditory afferents also reach the deep layers of the superior colliculus (*SC*). Other abbreviations: *BP* brachium pontis; *CNF* cuneiform nucleus; *ICO* commissure of IC; *RB* restiform body; 5 *N* trigeminal nerve; 7 *N* facial nerve

iological data have usually been obtained from surgically anesthetized preparations. Attempts to relate these disparate lines of evidence are justified by the anatomical similarity of peripheral and brainstem structures across a wide range of mammals, and by similarities in the basic auditory discriminative capacities of humans and at least some mammals. Anatomical and physiological similarities at the level of the cochlea and auditory nerve have been summarized by Kiang (1984). The basic plan of the brainstem auditory system is similar in all mammals, although there are significant variations between species in the size and cellular structure of various subdivisions of brainstem nuclei (e.g., Harrison and Howe 1974a; Strominger 1978; Willard and Martin 1983; Willard and Ryugo 1983; Zook and Casseday 1982a). Evidence on spectral and spatial discrimination in humans and animals is reviewed in Sect. 2. Although some species differences of functional significance will be emphasized in this review, the basic similarities justify attempts to relate human psychophysical and animal physiological data. The problem of anesthesia is less easily dealt with. There have been relatively few studies of the responses of auditory brainstem neurons in unanesthetized animals, and any review of brainstem processing must, of necessity, be based largely on the extensive body of data obtained using anesthetized preparations. Where data are available from unanesthetized preparations, they will be examined in detail and an attempt will be made to determine the extent to which anesthetic effects qualify interpretation of the main body of data.

In many respects the auditory brainstem pathways are best discussed in a serial ascending fashion, by dealing sequentially with successive levels of the pathways from the CN to the midbrain. However, the discussion of some aspects of spatial coding, in particular, is not well served by this procedure, since information from higher centers is more detailed in some respects than that from lower levels and in a number of cases is essential to its proper interpretation. These considerations have determined the organization of the major sections of the review dealing with central auditory processing. In Sects. 4–7 the structure and physiology of the nuclei and fiber tracts comprising the pathways from CN to midbrain will be discussed in a serial order. In Sect. 8 the mechanisms by which spectral and spatial information is processed will be considered more generally and free of the restrictions imposed by serial organization.

Finally, a number of limitations on the scope of the review should be emphasized. The ascending auditory pathway is paralleled throughout its length by descending projection systems of comparable size and complexity. Our knowledge of these descending or centrifugal pathways is almost entirely structural: we know little about their physiology and virtually nothing about their contribution to processing in the auditory brainstem. For these reasons, the descending pathways will not be considered in detail in this review; the reader is referred to the detailed account of the anatomy of the descending system by Harrison and Howe (1974b). No attempt has been made to provide a comprehensive review of every aspect of current research on the auditory brainstem; evidence on a number of topics (e.g., ultrastructure, neurochemistry, development) will be considered only to the extent that it has direct bearing on processing. Attention will for the most part be restricted to the mammalian auditory system, and largely to that of the domestic cat, for which most information is available. Consideration will be given to important results in other mammals and in nonmammalian vertebrates

where necessary; when reference is made to data obtained from animals other than the cat, the species will be identified.
The area under review has been one of considerable research activity and progress in recent years and this activity is reflected in the number of reviews of similar or overlapping scope that have appeared in the last decade and to which the reader is referred (Aitkin 1986; Aitkin et al. 1984b; Altman 1978; Brugge and Geisler 1978; Erulkar 1975; Evans 1975; Goldberg 1975; Harrison 1978; Harrison and Howe 1974a; Knudsen 1983c; Masterton and Imig 1984; Merzenich et al. 1984; Phillips and Brugge 1985; Tsuchitani 1978; Warr 1982; Webster and Aitkin 1975; Yin and Kuwada 1984) and Osen's chapter in Brodal (1981).

2 The Performance of the Auditory Central Processor

2.1 Spectral Discrimination

Sustained sinusoidal stimuli (pure tones) are the spectrally simplest acoustic stimuli. Although pure tones are rare in nature, the fact that the ear carries out a sort of running Fourier analysis on the input signal has prompted extensive use of pure tones in psychophysical and physiological studies. The minimum detectable change, or just-noticeable difference (JND), in the frequency F of a pure tone (the frequency difference limen, ΔF) is an important value in this context, since it reflects the limits of frequency selectivity of the auditory central processor. In Fig. 2A, data are presented from a number of studies in which human ΔF for pure tone stimuli was determined as a function of frequency. In the 1 – 2 kHz range in which the Weber fraction $\Delta F/F$ is minimal, ΔF for practiced observers under optimal conditions is of the order of 1 – 2 Hz (Moore 1973; Wier et al. 1977).

Despite the basic similarity in the structure of mammalian peripheral and lower brainstem auditory systems, there is significant functional variation between species, notably in the high- and low-frequency limits of hearing (see Heffner and Heffner 1982; Heffner and Masterton 1980; Masterton et al. 1969 for reviews). Data summarized in Fig. 2B indicate that, below about 8 kHz, ΔF for humans is consistently smaller than for other mammals. This is true even for animals such as the chinchilla which have an audibility curve similar to that of humans (Miller 1970). Such comparisons are complicated, however, by the difficulty in equating human and animal experiments in terms of task factors and motivational variables. Whether or not the apparent superiority of the human is real, the general mammalian data indicate remarkable frequency resolution performance which must be accounted for by any proposed mechanism for the processing and coding of spectral information.

As has been emphasized by others (e.g., Konishi 1978; Worden and Galambos 1972), mammalian auditory systems have evolved for the processing of natural sounds, which are characterized by complex and time-varying spectra. Among the most important natural sounds are human speech and the species-specific vocalizations of nonhuman animals. In the former case, at least, it should be noted that the direction of evolutionary pressure is not as unequivocal as might perhaps be thought: it is likely that the discriminative properties of the auditory system have constituted one of the factors exerting pressure on the selection of the sounds used (e.g., Kuhl 1981; Stevens 1980, 1981). Evidence on the performance of the central processor with respect to spectrally and temporally complex sounds cannot be summarized as simply as that on performance with pure tones, and in the present context it is perhaps sufficient merely to invoke the existence of music and speech perception. Details of quantitative performance data will be found in other accounts of the discrimination of partials in, and perception of the pitch of, complex tones (e.g., Goldstein 1978; Moore 1982; Plomp 1964) and of discrimination along the various acoustic dimensions on which speech sounds differ (e.g., Stevens 1980, 1981; Stevens and House 1972).

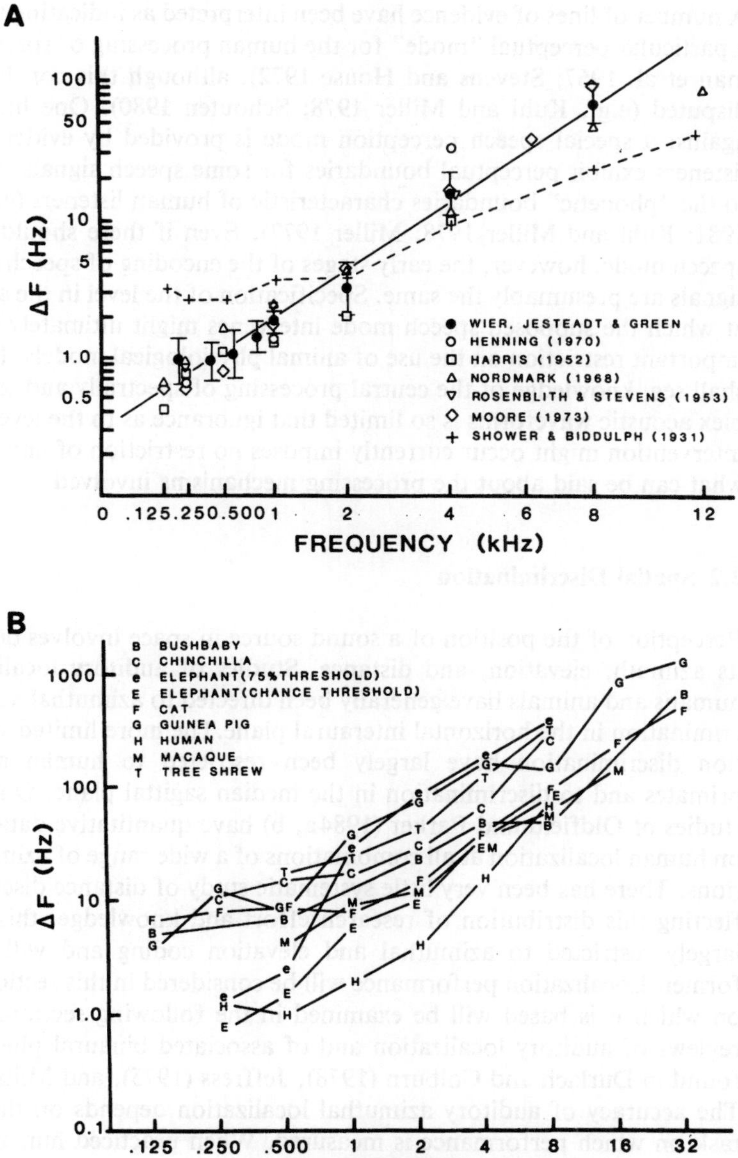

Fig. 2 A, B. Just-noticeable difference in frequency ΔF as a function of frequency for **A** humans and **B** a number of mammals. **A** is redrawn from Wier et al. (1977) and presents data from a number of studies in which ΔF was measured with pulsed sinusoids at sensation levels between 30 and 50 dB. The *full line* is the regression line fitted to the data of Wier et al., which also provides a good description of the data from the other studies. The *broken line* is data from a study using frequency-modulated stimuli. See Wier et al. (1977) for details. **B** Redrawn from Heffer and Heffer (1982) Hearing in the elfant. *Journal of Comparative and Physiological Psychology* 96:926–944. Copyright 1982 by the American Psychological Association. Adapted by permission of the author

A number of lines of evidence have been interpreted as indicating the existence of a particular perceptual "mode" for the human processing of speech (e.g., Liberman et al. 1967; Stevens and House 1972), although this conclusion has been disputed (e.g., Kuhl and Miller 1978; Schouten 1980). One line of argument against a special speech perception mode is provided by evidence that animal listeners exhibit perceptual boundaries for some speech signals that correspond to the "phonetic" boundaries characteristic of human listeners (e.g., Kuhl 1978, 1981; Kuhl and Miller 1978; Miller 1977). Even if there should prove to be a speech mode, however, the early stages of the encoding of speech and nonspeech signals are presumably the same. Specification of the level in the auditory system at which the supposed speech mode intervenes might ultimately prove to be an important restriction on the use of animal physiological models. However, as we shall see, knowledge of the central processing of spectrally and temporally complex acoustic waveforms is so limited that ignorance as to the level at which such intervention might occur currently imposes no restriction of any significance on what can be said about the processing mechanisms involved.

2.2 Spatial Discrimination

Perception of the position of a sound source in space involves determination of its azimuth, elevation, and distance. Studies of auditory localization in both humans and animals have generally been directed to azimuthal acuity and to discrimination in the horizontal interaural plane.The more limited studies of elevation discrimination have largely been restricted to human and nonhuman primates and to discrimination in the median sagittal plane. Only in the recent studies of Oldfield and Parker (1984a, b) have quantitative data been obtained on human localization at all combinations of a wide range of azimuths and elevations. There has been very little systematic study of distance discrimination. Reflecting this distribution of research effort and knowledge, this review will be largely restricted to azimuthal and elevation coding and will emphasize the former. Localization performance will be considered in this section, and the cues on which it is based will be examined in the following section. More detailed reviews of auditory localization and of associated binaural phenomena can be found in Durlach and Colburn (1978), Jeffress (1975), and Mills (1972).

The accuracy of auditory azimuthal localization depends on the nature of the task on which performance is measured. When practiced human observers are required to detect a change in the azimuthal direction of the source of a pure tone (*relative* localization), the minimum detectable change (minimum audible angle, MAA) under binaural listening conditions without head movements is of the order of 1° (e.g., Harris 1972; Mills 1958, 1972). Other methods of testing localization acuity, such as indicating the source direction either physically or verbally (*absolute* localization) result in larger thresholds, of the order of 4° – 5° or more (e.g., Casseday and Neff 1973; Stevens and Newman 1936; Terhune 1974). It is generally reported that accuracy is highest for sources near the median plane (Mills 1958; Sandel et al. 1955; Stevens and Newman 1936) and at low and high frequencies, with poorer performance at intermediate frequencies (between approximately 1500 and 2500 Hz, e.g., Mills 1958; Stevens and Newman 1936). In the localization of low-frequency pure tones, front-back reversals are common (i.e., the azimuthal displacement of the sound source from the median sagittal plane is correctly discriminated, but the sound is localized to the mirror image position behind or in front of the observer, e.g., Stevens and Newman 1936). The accuracy figures for azimuthal discrimination cited here are "corrected" for such reversals. Broad-band stimuli such as white or band-pass noise are more accurately localized than pure tones and front-back reversals are uncommon with such stimuli (Oldfield and Parker 1984a; Sandel et al. 1955; Stevens and Newman 1936; Terhune 1974).

Binaural azimuthal localization as discussed here is based largely on a number of interaural stimulus disparities, the nature of which is to be considered in the following section. In the absence of head movements, monaural azimuthal localization of pure tones is impossible or very inaccurate (Angell and Fite 1901a, b; Butler 1971). However, accurate monaural localization is possible with broad-band stimuli having high-frequency components (Angell and Fite 1901a, b; Belendiuk and Butler 1975, 1977; Butler 1975; Freedman and Fisher 1968). The major cue under these conditions is provided by spectral transformations of the waveform produced by the structures of the pinna (e.g., Butler 1975; Watkins 1982; see discussion in next section). The extent to which this cue contributes to normal binaural localization is unclear. Oldfield and Parker (1984b) reported that elimination of pinna cues (by partial occlusion of the pinnae) produced an increase in front-back reversals, but little or no change in "corrected" accuracy scores, in binaural azimuthal localization of broad-band noise signals. Conversely, Musicant and Butler (1984) reported that similar procedures produced an increase in *both* reversals and corrected error scores in localization of broad-band and high-pass noise. Although these results are in conflict with respect to the contribution of pinna effects to the *accuracy* of binaural azimuthal localization, they support other evidence that these effects are responsible for the absence of front-back reversals with broad-band signals (e.g, Blauert 1969/70).

Spectral transformations produced by the pinna are the major cue for human discrimination of elevation. Such discrimination is not possible with pure tones (e.g., Roffler and Butler 1968a, b) and requires broad-band stimuli with energy above 6–8 kHz (Butler and Planert 1976; Hebrank and Wright 1974b; Roffler and Butler 1968a). With such stimuli, discrimination thresholds of the order of 4°–5° are reported for elevation discrimination in the median plane (Oldfield and Parker 1984a; Roffler and Butler 1968a; Wettschureck 1973). Occlusion or distortion of the pinnae results in total loss or systematic disturbance, respectively, of discrimination (Gardner and Gardner 1973; Oldfield and Parker 1984b; Roffler and Butler 1968a).

The spectral transformations produced by the pinna (the nature of which is discussed in the following section) provide a monaural cue to elevation, and subjects can be trained to localize in the median plane as accurately monaurally as binaurally (Hebrank und Wright 1974a). However, the fact that less highly trained subjects are usually more accurate binaurally than monaurally (Bothe and Elfner 1972; Butler 1969; Butler and Planert 1976; Gardner 1973; Hebrank and Wright 1974a) has led to the suggestion that interaural spectral differences attributable to pinna asymmetries might function in addition to monaural spectral cues (Searle et al. 1975). Although there is physical evidence for the binaural pinna disparity cue (Searle et al. 1975; Shaw 1982), there is disagreement as to whether it is actually used (Hebrank and Wright 1974a; Searle et al. 1975, 1976).

Evidence on the physical cues to auditory distance (depth), and psychophysical data on the distance perception performance of human observers, have been reviewed by Coleman (1963) and by Mershon and King (1975). Two of the cues to distance suggested by physical acoustics, namely the decrease in sound pressure level of a given source with distance (the inverse first-power law) and change in frequency spectrum, cannot provide *absolute* information as to distance unless the source characteristics are known (i.e., unless the stimulus is familiar). In accordance with this dependence on knowledge of source characteristics, inability to localize the sources of unfamiliar sounds (e.g., Coleman 1962), and disturbance of distance judgments by manipulation of the level of a fixed source (Gardner 1969) have been reported. Strybel and Perrott (1984) have reported that discrimination of changes in the distance of familiar broad-band noise sources is in accordance with predictions from the inverse first-power equation at distances in excess of 3 m, but not at shorter distances. Physical acoustics also suggests the existence of binaural cues to distance that are independent of knowledge of the source characteristics (e.g, Coleman 1963; Hirsch 1968; Molino 1973), but psychophysical data concerning the use of these cues is sparse and unclear (Holt and Thurlow 1969; Molino 1973).

Two final points should be noted in connection with the localization performance achieved by the human auditory central processor. The first is that for the spectral transformations produced by the pinna (or by distance) to serve as cues for sound location, familiarity with the sound source is necessary (e.g, Coleman 1962; Hebrank and Wright 1974a). Central processing in these cases must involve comparison of the input spectrum with some form of stored template of the untransformed spectrum. A similar template comparison process is postulated by some current models of the processing of frequency spectrum per se (e.g., Srulovicz and Goldstein 1983; see discussion in Sects. 3.5 and 8.1). There is currently no information on the neural mechanisms by which the processes of template storage and comparison might be realized.

The second point is that attempts by human observers to localize sounds under natural conditions are often associated with head movements, and such movements can reduce localization error (e.g., Freedman and Fisher 1968; Thurlow and Runge 1967; Thurlow et al. 1967). Even when the head is stationary, auditory information must be combined with information about head position to enable localization of the sound in spatial coordinates. Neither the locus nor the neural mechanisms by which auditory and somatosensory/motor information are integrated have yet received experimental attention (see discussion in Sect. 8.2.4).

Most studies of animal azimuthal and elevational localization have reported thresholds higher than those in humans (see Gourevich 1980 and Heffner and Heffner 1982 for reviews). As noted previously in the context of frequency discrimination, these comparisons are complicated by the difficulty in equating human and animal experiments in terms of task requirements and motivational level. Although these factors preclude detailed comparisons, it is noteworthy that absolute thresholds comparable to those of humans have in fact been reported in a number of studies. Brown and his colleagues (Brown 1982; Brown et al. 1980, 1982) have reported MAAs (i.e., relative localization thresholds) as small as 3° – 4° for azimuthal and vertical localization of noise signals and species-specific vocalizations by monkeys. As in humans, azimuthal acuity is greatest in the region of the median plane. Absolute azimuthal localization thresholds of 5° or less have been reported in the cat (Casseday and Neff 1975; Moore et al. 1974; see Sect. 5.3 for a more detailed account of cat localization data). More generally, the localization ability of any animal will reflect the extent to which it depends on auditory information to locate prey or conspecifics and to avoid predators, and on the nature of its acoustic environment. The barn owl, for example, a nocturnal predator whose aerial attack can be entirely acoustically guided (Konishi 1973a, b; Payne 1971), can localize frontally located broad-band targets with errors of less than 2° in azimuth and elevation (Knudsen et al. 1979; Payne 1971). This absolute localization performance is superior to that of any terrestrial animal so far tested, including humans. Comparative aspects of auditory localization have been reviewed in detail by Erulkar (1972).

2.3 The Cues for Auditory Localization

As noted in the previous section, the spatial location of a sound must be inferred from a number of cues that are only indirectly associated with its direction. This situation is in marked contrast to that in the visual and somatosensory systems, in which the spatial location of a stimulus determines the point on the receptor epithelium at which it produces activation. Thus, in the visual system, information as to the retinal locus of activation, together with information as to the position of the eye in the orbit, specifies the location of the stimulus. In the auditory system, location must be inferred from *binaural* cues, deriving chiefly from the fact that the ears are separated by the head, and *monaural* cues, deriving chiefly from the effect of the pinna on the stimulus waveform (see Lewis 1983; Mills 1972; Searle et al. 1976 for reviews).

The major binaural cues are provided by differences in the time of arrival and sound pressure level (SPL) of the sound at the two ears. Each of these interaural differences subsumes a number of cues, each of which has implications for the task performed by the central processor; they will therefore be considered in some detail. The initial discussion of these cues will be restricted to their variation with azimuthal position in the interaural horizontal plane and in the frontal half of space; variation with azimuthal position at different elevations and behind the observer will be considered subsequently.

Interaural time differences (ITDs) arise from the greater length of the sound path to the further ear when the sound is located off the median plane. The traditional analysis of this cue was in terms of a wave traveling around a sphere (the head) at the ambient speed of sound in air, and therefore predicted ITDs that were independent of frequency (e.g., Woodworth 1938). Recent analytic and empirical investigations have shown, however, that the frequency dispersion of the diffracted wave causes ITDs to be frequency dependent. Measured "steady state" ITDs as a

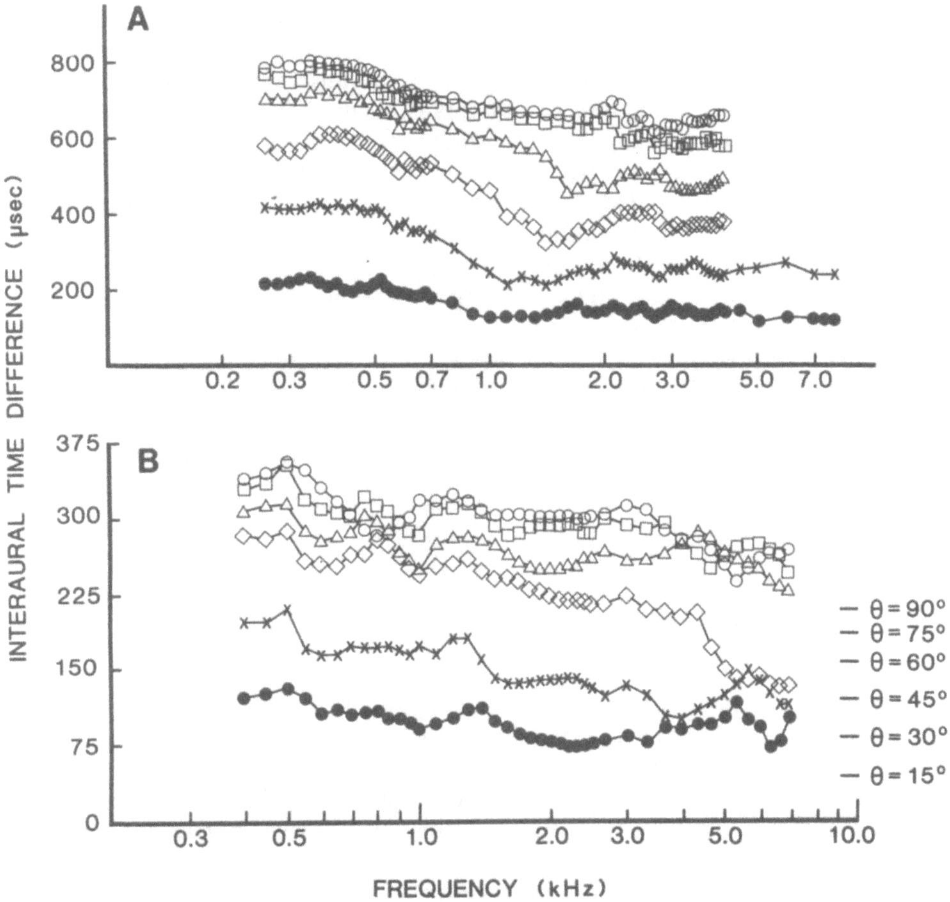

Fig. 3A,B. Measured interaural time differences (ITDs) as a function of frequency and angle of incidence (θ_{inc}) for **A** a manikin without torso and **B** an anesthetized cat. Points for each θ_{inc} on *right* of **B** indicate ITDs calculated from Woodworth's (1938) spherical model using the cat's tragus–tragus distance. Key to symbols in **A** and **B**: *full circles* θ_{inc} = 15°; *crosses* θ_{inc} = 30°; *open diamonds* θ_{inc} = 45°; *open triangles* θ_{inc} = 60°; *open squares* θ_{inc} = 75°; *open circles* θ_{inc} = 90°. (**A** Redrawn from Kuhn 1977; **B** redrawn from Roth et al. 1980)

function of frequency for an artificial human head (Kuhn 1977) and for the cat (Roth et al. 1980) are shown in Fig. 3A and Fig. 3B, respectively. In each case, ITD decreases as a function of frequency; since the ITD associated with a given azimuthal displacement varies with frequency, it follows that ITD alone does not provide a reliable cue to the azimuthal location of the source, but must be combined with frequency information (cf. comment by McFadden 1981).

Two other features of the ITD cue are also apparent in Fig. 3A and B. The first is that, at all frequencies and for both humans and cats, the increase in ITD for a given increase in azimuthal displacement is greater for low angles of incidence θ_{inc}. A fixed ITD threshold would therefore give greater acuity near the median plane than at extreme azimuths.[1] The second point is that the measured ITDs exceed those calculated from Woodworth's spherical model on the basis of tragus-tragus measurements (indicated at the right of Fig. 3B).

Two components of ITD have commonly been distinguished. The first is the difference in the time of arrival of the onset of the sound at each ear, viz., *transient* disparity. Transient disparity provides a cue for all sounds (at onset and offset), but is the *only* form of ITD for very brief stimuli such as clicks. For sustained stimuli, a second form of ITD is present in the form of *ongoing* time differences. For pure tones, ongoing time differences are manifest only in the fine structure of the signal as *phase* differences, while for complex sounds ongoing time differences in the envelope of the signal also provide a cue. The following abbreviations will be used to refer to particular forms of ITD throughout this review: transient time difference (TTD); ongoing time difference (OTD); interaural phase difference (IPD).

For any given species, phase differences are completely unambiguous only at those frequencies for which the period is greater than twice the maximum interaural delay that can occur in that species (see Brown et al. 1978; Mills 1972 for more detailed discussion). At these frequencies, IPDs cannot exceed 180° and – if we consider, for the moment, only frontal space and the interaural horizontal plane – a given IPD can therefore arise from only a single position. At frequencies for which the period is less than twice the maximum delay, but is greater than the maximum delay, *some* IPDs (those arising from larger azimuthal displacements) could be associated with either a phase lead at one ear or a phase lag at the other, and are therefore ambiguous as to the side on which the source is located. At higher frequencies, for which the period is less than the maximum delay, any given IPD can be associated with multiple source locations on either side of the head, and IPDs are therefore totally ambiguous. If one assumes maximum interaural delays of approximately 800 μs in humans and 350 μs in cats (Fig. 3), the frequencies at which IPDs become completely ambiguous are approximately 1250 and 2850 Hz, respectively. In fact, a number of lines of evidence support the view that IPD is the major cue for human azimuthal localization of pure tones at frequencies below about 1500 Hz (see Mills 1972 for review). MAAs at these frequencies correspond to ITDs of the order of 10–15 μs, which agree well with the ITD thresholds for pure tones obtained in dichotic studies (Domnitz 1973; Hershkowitz and Durlach 1969; Klump and Eady 1956; Mills 1960; Yost 1974; Zwislocki and Feldman 1956). This discriminative ability on the part of the central processor is remarkable given that the information provided to it is in the form of spatiotemporal patterns of action potentials that have rise times of the order of hundreds of microseconds.

The restriction of IPD as a cue to low frequency tones has led to the common view, as one component of the "duplex" theory of auditory localization, that

---

[1] Yost (1974) in fact reported that the difference threshold for ITDs in dichotically presented low-frequency (<2 kHz) tones increases with increases in the base ITD. However, the subjects' task in his experiments was to detect a difference in the *lateral position* of the sound image, and the variation in ITD threshold could therefore be a reflection (rather than a cause) of greater azimuthal acuity in the region of the median plane. See discussion of lateralization task on p. 19.

ITD in general serves as a cue only at low frequencies. Even for pure tones, this is true only for IPD and not for TTD, while for more complex stimuli ongoing disparities are not restricted to low frequencies. Thus, ITDs in the envelope of amplitude-modulated high-frequency carriers can be detected with sensitivity comparable to that of the detection of ITDs for pure tones at the modulation frequency (e.g., Henning 1974; Nuetzel and Hafter 1976). High-frequency noise bands similarly provide ongoing disparities in the stimulus envelope (e.g., McFadden and Pasanen 1976). The remarkably low ITD threshold of 6 μs for dichotically presented broad-band noise bursts (Tobias and Zerlin 1959) presumably reflects the use of this cue in addition to the presence of IPD in a number of low-frequency bands. These observations make it clear that ITD as a cue for azimuthal localization is not in fact restricted to low frequencies. As we shall see, however, there is only limited evidence on the central processing of ITDs for complex stimuli.

The second major binaural cue to azimuthal location is provided by interaural differences in SPL. These have commonly been termed interaural intensity differences (IIDs); although this term is acoustically incorrect,[2] it will be retained here in the interests of compatibility with the literature to be reviewed. IIDs are commonly attributed to the shadowing effect of the head, i.e., the fact that for sound sources displaced from the median plane there is a frequency-dependent attenuation of waves diffracted about the head. However, for most mammals, the pinna has a directional amplifying effect at high frequencies (e.g., Calford and Pettigrew 1984; Phillips et al. 1982; Wiener et al. 1966; humans: Shaw 1974a, b; Wiener 1947; echolocating bats: Fuzessery and Pollack 1984; Grinnell and Grinnell 1965), and IIDs are in fact the joint product of amplification at the near (or directly irradiated) ear and attenuation at the far (or shadowed) ear.[3]

Physical measurements for humans are presented in Fig. 4, in which the full lines are curves fitted by Shaw (1974a) to data from a number of different studies. In the left-hand column, sound pressure transformation from free-field to the eardrum as a function of azimuth is shown for frequencies 7.6 – 12 kHz. Note that at the directly irradiated ear (positive azimuths) there is an azimuth- and frequency-dependent amplification, while at the shadowed ear (negative azimuths) there is an azimuth- and frequency-dependent attenuation. The resultant IIDs are shown in the right-hand column and reveal that not only IID magnitude, but also the nature of the IID – azimuth relationship varies with frequency. In humans, the

---

[2] The term "intensity" is used generally in auditory neural and behavioral science to refer to stimulus magnitude, the parameter usually measured being the root-mean-square sound pressure. Measured SPLs are commonly expressed in decibels (dB) *re* 20 μPa. In acoustics, the intensity of a sound in a specified direction at a point is defined as the average rate of sound energy transmitted in the specified direction through a unit area normal to the direction of propagation, and is equal to the square of sound pressure divided by the characteristic impedance. Although the misuse of the term "intensity" has understandably provoked the ire of acousticians (e.g., Behar 1984), the term will be used in this review as a synonym for SPL, according to the practice that has become established in auditory neuroscience and psychophysics.

[3] The amplifying effect of the pinna is also reflected in the fact that behaviorally determined auditory thresholds of monaural cats with pinna intact are significantly lower at high frequencies than those of cats with the pinna removed (Flynn and Elliott 1965).

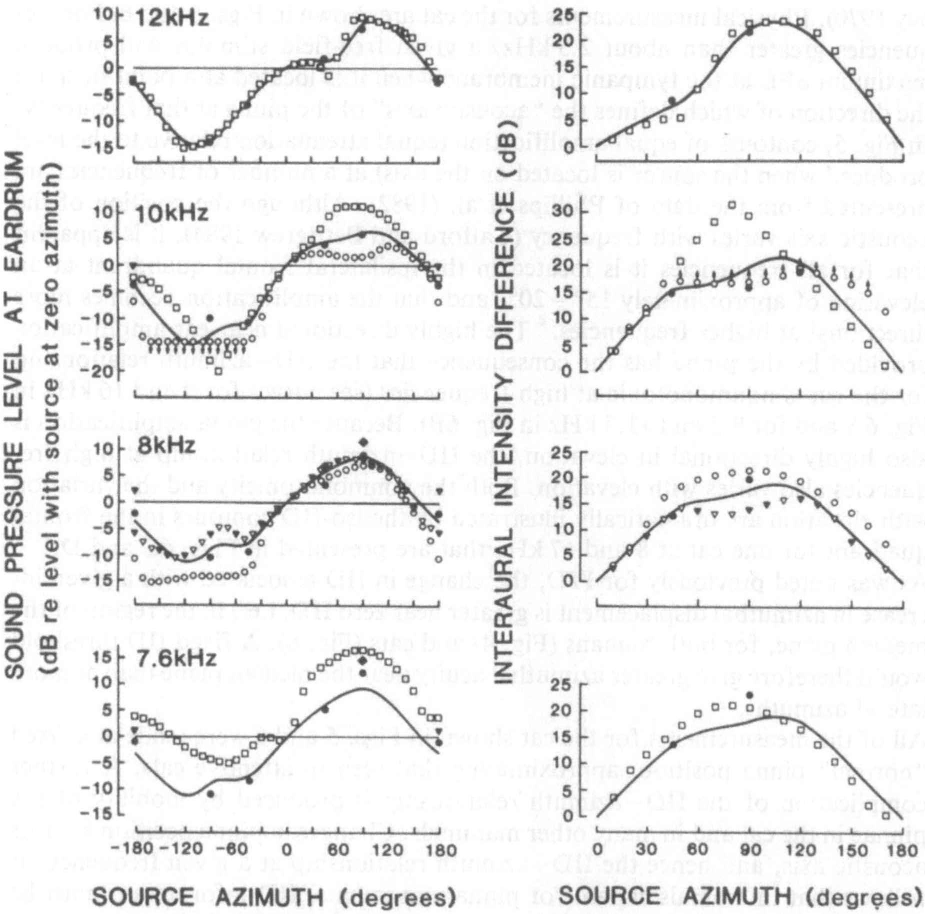

Fig. 4. Sound pressure level (SPL) at eardrum as a function of sound source azimuth (*left-hand column*) and resultant interaural intensity differences (IIDs) (*right-hand column*) at four frequencies for human. In *left-hand column*, positive azimuths indicate source locations in the horizontal plane on the same side as the ear in which measurements were made; negative azimuths indicate contralateral source locations. SPLs are expressed in dB *re* sound pressure at eardrum when source was at zero azimuth (i.e., in median plane). At each frequency, IIDs in right-hand column were derived from azimuthal dependence data in *left-hand column* by subtracting SPL change at negative azimuth from that at the corresponding positive azimuth. In both columns, *lines* are curves fitted by Shaw (1974) to points obtained from pooled data from a number of different studies (indicated by different symbols). Redrawn from Shaw (1974a), from which details of original studies and pooling procedures can be obtained

amplification and attenuation curves are roughly mirror symmetric and IID increases monotonically with azimuth in the frontal quadrant (i.e., up to 90° – 100°).

For animals in which the amplifying effect of the pinna is more directional, amplification and attenuation are usually not mirror symmetric and IID does not increase monotonically with azimuth (see Calford et al. 1986; Harrison and Downey 1970). Physical measurements for the cat are shown in Figs. 5 and 6. For frequencies greater than about 2.5 kHz, a given free-field stimulus will produce maximum SPL at the tympanic membrane when it is located at a point in space the direction of which defines the "acoustic axis" of the pinna at that frequency. In Fig. 5, contours of equal amplification (equal attenuation relative to the level produced when the source is located on the axis) at a number of frequencies are presented from the data of Phillips et al. (1982). Although the position of the acoustic axis varies with frequency (Calford and Pettigrew 1984), it is apparent that for all frequencies it is located in the ipsilateral frontal quandrant at an elevation of approximately 15° – 20°, and that the amplification becomes more directional at higher frequencies.[4] The highly directional near-ear amplification provided by the pinna has the consequence that the IID – azimuth relationship for the cat is nonmonotonic at high frequencies (see curves for 9 and 16 kHz in Fig. 6A and for 8.5 and 11.5 kHz in Fig. 6B). Because the pinna amplification is also highly directional in elevation, the IID – azimuth relationship at high frequencies also varies with elevation. Both the nonmonotonicity and the variation with elevation are dramatically illustrated by the iso-IID contours in the frontal quadrant for one cat at 8 and 17 kHz that are presented in Figs. 6C and D.

As was noted previously for ITD, the change in IID associated with a given increase in azimuthal displacement is greater near zero IID, i.e., in the region of the median plane, for both humans (Fig. 4) and cats (Fig. 6). A fixed IID threshold would therefore give greater azimuthal acuity near the median plane than at more lateral azimuths.

All of the measurements for the cat shown in Figs. 5 and 6 were made at a fixed "normal" pinna position, approximating that seen in attentive cats. A further complication of the IID – azimuth relationship is produced by mobility of the pinnae in the cat and in many other mammals. Changes in pinna position alter its acoustic axis, and hence the IID – azimuth relationship at a given frequency. It follows that in animals capable of pinna movements, IID information must be combined with pinna position information if it is to provide information as to azimuth.

The dependence of IIDs on pinna amplification and head shadow effects has the consequence that IIDs are small or absent at low frequencies (where the wavelength is long relative to the dimensions of the head and pinnae) and larger at

[4] Grinnell and Grinnell (1965) presented similar data (at much higher frequencies) for echolocating bats. Although they measured a central auditory system response (the gross evoked potential from the inferior colliculus, N_4), the fact that they determined variation in response *threshold* as a function of azimuth and elevation strongly suggests that their data reflected the directional amplification of the pinna contralateral to the recording site. Indeed, they found that changes in contralateral pinna position had marked effects on directionality whereas manipulation of the ipsilateral pinna had virtually no effect, and Grinnell (1963b) had earlier shown that the auditory nerve gross action potential (N_1) showed almost the same (ipsilateral) directionality as shown by N_4 contralaterally. The reflection of pinna amplification in central auditory system responses is discussed in detail in Sect. 7.4. More recently, Fuzessery and Pollack (1984) have used cochlear microphonic measurements of tympanic SPL in an echolocating bat and obtained pinna amplification data that are similar to those presented by Grinnell and Grinnell (1965).

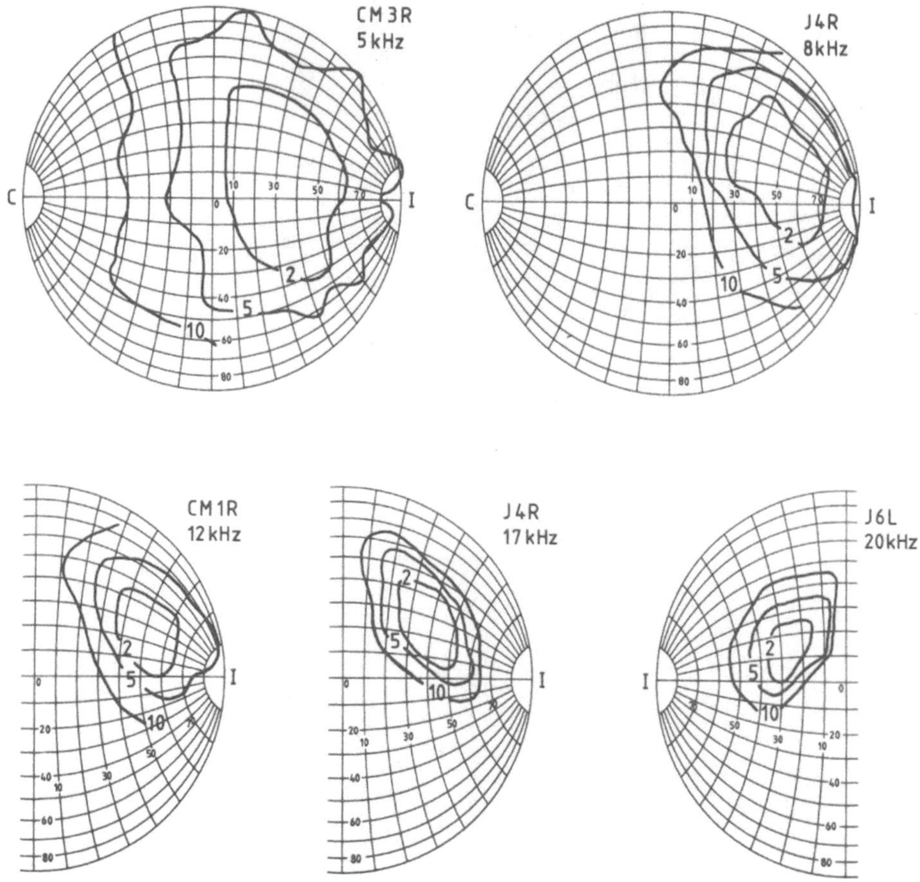

Fig. 5. Directionality of pinna amplification in cat as a function of frequency. Labeled *bold lines* are contours of equal amplification – equal attenuation of sound pressure level (SPL) at tympanic membrane relative to the level when the source is at the point generating maximum tympanic SPL, i.e., on the acoustic axis. Variation in tympanic SPL was measured using the cochlear microphonic potential. Isoamplification contours are plotted on a scale-area representation of the frontal hemifield (*upper plots*) or frontal ipsilateral quadrant (*lower plots*). In these representations, lines of equal azimuth are represented by equidistant arcs that terminate on a circular representation of speaker positions in the frontal field. The separation of contours representing arcs of equal elevation was adjusted to produce $10^0 \times 10^0$ enclosures of equal area within a given azimuthal range, preserving the areal relationships on the coordinate sphere formed by the "hoop" system used to move the speaker in space. For a description of hoop system see Sect. 7.3 and Fig. 47. *C* and *I* indicate 90° azimuths contralateral and ipsilateral, respectively, to the ear at which measurements were made. Note that contours become closer together and enclose a smaller area, indicating increased directionality, as frequency increases. From Phillips et al. (1982), where more detailed accounts of measurement procedures and coordinate system can be obtained

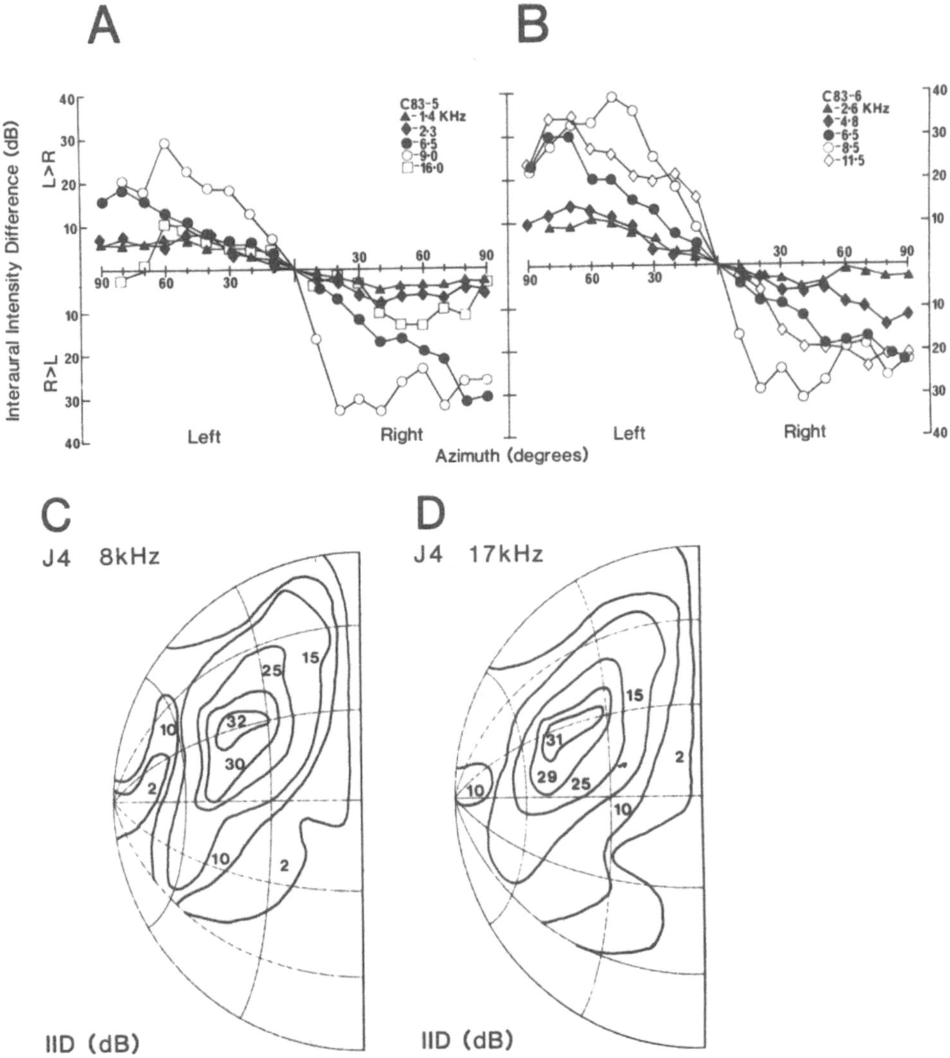

Fig. 6. A, B Interaural intensity difference (IID) as a function of azimuth for two cats at the indicated frequencies. Measurements of variation in tympanic membrane sound pressure level (SPL) were made using cochlear microphonic potentials, and sound source position was varied using a "hoop" system of the type described in Sect. 7.3. Azimuth was varied with sound source at a fixed elevation 20° above the Horsley-Clarke horizontal plane; a consequence of the hoop system is that positions of equal elevation fall on meridians meeting at the poles and not on horizontal parallels (see Fig. 5). Note that low-frequency IIDs increase monotonically with azimuth and peak at or near the poles, but higher frequency IIDs are nonmonotonic. Maximum IID varies with frequency; in the cat, it increases to a maximum at 8 – 10 kHz and decreases at higher frequencies. Variation between animals is probably due in part to nonalignment of the acoustic axis of the pinna with the elevational plane of measurement. (Calford et al. 1986) **C, D** Iso-IID contours for frontal quadrant in one cat at two frequencies as indicated. Note that IID – azimuth relationship is both nonmonotonic and dependent on frequency. (Courtesy of M. B. Calford, derived from data of Phillips et al. 1982)

higher frequencies (where the wavelength becomes progressively smaller than these dimensions) (e.g., Calford et al. 1986; Moore and Irvine 1979; humans: Shaw 1974a). Masterton, Heffner, and their colleagues (Heffner and Heffner 1982; Heffner and Masterton 1980; Masterton 1974; Masterton et al. 1969) have shown that across a large number of mammals there is a highly significant negative relationship between head size and the high-frequency limit of hearing, and have argued that this reflects selective pressure in animals with small heads for hearing at the higher frequencies at which IIDs are generated.[5] In humans there is good evidence that azimuthal localization of high-frequency pure tones (at least from 2 to 6 kHz) is based on IIDs (see Mills 1972 for review). One line of evidence is that the IIDs corresponding to MAAs at these frequencies correspond closely to the minimal IIDs discriminable in dichotic studies, viz., 0.5 – 0.8 dB (e.g., Hershkowitz and Durlach 1969; Mills 1960). Because of the marked dependence of IID on frequency, azimuthal displacement of a broad-band source results in different IIDs for different frequency components, the consequence of which is interaural spectral differences. It is not clear, however, whether the more accurate localization of broad-band signals is attributable to the detection of interaural spectral differences per se or whether it reflects the availability of IID and ITD information in different frequency bands.

The human IID and ITD thresholds cited in the preceding discussion were obtained using dichotic stimulation via sealed earphones, which allows independent specification of the stimulus at each ear. When identical stimuli are presented to the two ears through earphones, a single "fused" sound image is perceived which appears to originate within the listener's head and in the median plane. Introduction of an IID or an ITD results in the sound image being *lateralized* toward the ear receiving the earlier or the more intense stimulus, the extent of lateral movement depending on the magnitude of the interaural disparity. Colburn and his colleagues (Domnitz and Colburn 1977; Stern and Colburn 1978, 1985) have argued that the change in lateral position is in fact the primary cue on which discrimination of interaural disparities is based. A given lateral position of a fused sound image can be achieved by various combinations of time and intensity differences, reflecting what has been termed "time – intensity trading." In the extreme case of such trading, a time difference favoring one ear can be offset by an intensity difference favoring the other ear, to produce an image in the median plane. Evidence on time – intensity trading has been a major factor in the development of models of the neural processing of interaural disparity cues (e.g., Deatherage and Hirsh 1959; Domnitz and Colburn 1977) that will be considered in later sections of this review (see Sect. 8.2.2). Durlach and Colburn (1978) have provided a detailed examination of evidence and issues relating to lateralization and the discrimination of interaural disparities in their excellent review of

[5] The fact that interaural disparities are dependent on the physical dimensions of the head and pinnae also has interesting developmental implications. The values of ITD and IID associated with any given frequency at any given azimuth other than the median plane must vary in the course of development, as head and pinna size increase, and some degree of plasticity must be required to adjust to the changing correspondences between cues and sound locations. This development issue is beyond the scope of this review, but see, e.g., Knudsen (1984c) and Moore and Irvine (1981) for discussion.

binaural phenomena, and for this reason only a very general account has been given here.

Because of the technical difficulties of achieving dichotic stimulus conditions in awake animals, only limited interaural discrimination (lateralization) data are available for nonhuman subjects. The available data generally indicate higher IID and ITD thresholds in animals than in humans (Masterton et al. 1967; monkey: Don and Starr 1972; Houben and Gourevitch 1979), although Wakeford and Robinson (1974) reported IID thresholds (of 0.5 – 1.0 dB) in cats that were almost identical to those in humans at the same frequencies, and ITD thresholds that were only slightly (10 μs) higher than those of humans. These comparisons are subject to the qualifications discussed previously in the context of frequency and spatial discrimination. In monkeys, as in humans, however, there is good agreement between ITD thresholds measured dichotically and the ITDs corresponding to MAAs for low-frequency tones (Brown et al. 1978).

The preceding discussion of interaural disparities as cues to azimuth has been restricted to the horizontal interaural plane and to the frontal hemifield. In humans, the functions relating IID and ITD to azimuth are approximately symmetric about the interaural axis (e.g., Fig. 4), and this ambiguity is the basis of the front-back reversals discussed in Sect. 2.2. The ambiguity can be resolved by head movements or – for high frequency sounds of known intensity – by level changes produced by the shadowing effect of the pinna when the source is located at azimuthal positions behind the observer (e.g., Stevens and Newman 1936). More generally, however, a given IID or ITD can in fact arise from a wide range of azimuths at different elevations. For a simple model of the head as a sphere without pinnae, all sound sources located on a cone-shaped surface (the so-called cone of confusion) give rise to the same ITD (e.g., Blauert 1969/70; Mills 1972; Wallach 1939). Because of the pinna's contribution to IIDs, the surface comprising all those points giving rise to the same IID cannot be readily approximated by the simple conical model; as Figs. 6C and D illustrate, the distribution of loci giving rise to a given IID in animals with highly directional pinnae can be extremely complex. It is usually assumed that, in the absence of head movements, the confusion as to the azimuth indicated by a given IID or ITD is resolved by independent information as to elevation. If this were the case, procedures that result in disturbance or loss of elevation discrimination should also produce decrements in azimuthal acuity in situations in which the observer does not have independent knowledge that source locations are restricted to a single elevation. As Oldfield and Parker (1984b) have pointed out, their finding that azimuthal acuity was not diminished by procedures (removal of pinna cues) that substantially reduced elevation accuracy, is incompatible with this account of the resolution of the cone of confusion.

The acoustic effects of the pinna that provide the major cue for monaural elevation and azimuthal localization of high-frequency complex sounds have been analyzed in both time domain and frequency domain terms (see Blauert 1983 for review). Batteau (1967) described reflections from the pinna as producing delayed replications (echos) of the incident sound, the duration of the delays depending on the location of the source. In spectral terms, pinna resonances (together with the effects of the head and torso) produce peaks and notches at different high-frequency bands in the sound, depending on the position of the source (Blauert 1969/70, 1982; Bloom 1977a; Butler and Belendiuk 1977; Hebrank and Wright 1974b; Searle et al. 1975; Shaw and Teranishi 1968). The importance of these cues is confirmed by psychophysical evidence that artificially produced spectral notches result in changes in perceived vertical location of sounds presented through earphones (Watkins 1978) or through a fixed speaker under monaural listening conditions (Bloom 1977b). Since delay-and-add networks of the type postulated by Batteau produce spectral peaks and notches of the type described in spectral models (e.g., Watkins 1978, 1982; Wright et al. 1974), the disagreement essentially concerns the domain in which these cues are processed. As we shall see in later sections, the uncertainty as to whether a given cue is processed in the spectral or temporal domain is a recurrent question in a number of areas of central auditory processing.

The material reviewed in the preceding sections has given a broad specification of the tasks performed by the CNS in processing spectral and spatial information. As anticipated in Sect. 1, the artificiality of considering these two types of processing entirely independently has been brought out by the importance of spectral information in many aspects of localization. Before proceeding to discuss the mechanisms by which the auditory central processor achieves its tasks, it is necessary to specify the nature of the information provided to the processor by the auditory nerve. So stated, this appears to be a straightforward task. It will become clear, however, that it is by no means easy to determine which aspects of the information provided by the auditory nerve are in fact used by the CNS.

3 Auditory Nerve Input to the Central Processor

3.1 Basic Anatomy of the Auditory Nerve

In the mammalian cochlea, a single row of inner hair cells (IHCs) and a number of rows (three in the cat) of outer hair cells (OHCs) are separated by the tunnel of Corti. The IHCs are innervated by "inner radial" afferent fibers: in adult cats, each radial fiber is unbranched, terminating with a single synapse on a single IHC, and each IHC receives a synapse from 20–25 fibers (Liberman 1980; Spoendlin 1969, 1978). The OHCs receive their afferent innervation from "outer spiral" fibers, each of which synapses on ten or more OHCs, and each OHC receives collaterals from several outer spiral fibers (Spoendlin 1969, 1978, 1979). The afferent fibers in the cochlea are derived from two types of cell body in the spiral ganglion (Spoendlin 1971, 1978, 1979). The larger Type I neurons, which comprise 90% – 95% of spiral ganglion cells in the cat, have a true bipolar shape and the cell body and axon are myelinated. The smaller Type II neurons (5% – 10%) have a pseudomonopolar shape and the cell body and axon are poorly myelinated or unmyelinated. Recent tracer studies, in which small injections of horseradish peroxidase (HRP) into the auditory nerve (AN) were used to label the cell bodies and peripheral processes of small numbers of spiral ganglion cells (Kiang et al. 1982) or intracellular HRP injections were used to label individual cells (Liberman 1982b; Liberman and Oliver 1984; Robertson 1984), have confirmed Spoendlin's (1971) proposal that Type I neurons give rise to the radial fibers terminating on IHCs. Thus, each Type I neuron innervates only a single IHC and each IHC is innervated by about 20–25 Type I neurons. Spoendlin's corollary proposal that Type II cells give rise to outer spiral fibers in the cochlea was also supported by the results of Kiang et al. (1982), although the outer spiral fibers were more difficult to trace because of their small diameter. It therefore follows that each Type II neuron innervates ten or more OHCs, and each OHC receives collaterals from several Type II fibers.

In the intracellular labeling studies cited, all but one of more than 100 *physiologically characterized* neurons whose peripheral processes could be traced within the cochlea were Type I ganglion cells giving rise to inner radial fibers (Liberman 1982b; Liberman and Oliver 1984; Robertson 1984). These data therefore confirm the view that the conventional recording techniques applied to AN do not usually record the activity of thin unmyelinated fibers such as those of Type II neurons. Robertson (1984) labeled a single neuron giving rise to an outer spiral fiber; this cell was nonspiking and was unresponsive to acoustic stimulation. Other attempts to label cells with these characteristics were unsuccessful, and it is therefore impossible to say whether they are general characteristics of Type II cells. It is clear, however, that the physiological evidence on the response characteristics of AN fibers to be reviewed in this section applies only to Type I neurons.

Although there had been some doubt (Spoendlin 1979), it is now certain that Type II ganglion cells project to CN (Jones et al. 1984; Leake-Jones and Snyder

1982; chinchilla: Morest and Bohne 1983; Ruggero et al. 1982). However, little is known about the details of their pattern of termination in CN, and the account of the primary innervation of CN given in Sect. 4.1 is therefore largely restricted to innervation by Type I cells. Whether ignorance concerning the properties and terminations of Type II cells constitutes a significant restriction on the account of central auditory processing given in this review will depend on what is revealed by further studies of the function of the OHCs and the auditory responsiveness of Type II neurons. Damage restricted to OHCs produces significant changes in the frequency tuning of the Type I neurons from which AN recordings are made (e.g., Dallos and Harris 1978; Liberman and Dodds 1984; guinea pig: Robertson et al. 1980). Although these data indicate that there must be some functional coupling between OHCs and either IHCs or radial fibers, the nature of this (mechanical or electrical) coupling remains unclear (see Brownell 1982 and Liberman and Kiang 1984 for discussion). One possibility is that the OHCs are *effectors* which influence cochlear micromechanics and are under the control of the efferent olivocochlear bundle (OCB) (e.g., Brown and Nuttall 1984; Brownell 1983; Brownell et al. 1985; Mountain 1980; Siegel and Kim 1982). If this were the case, and Type I neurons alone were responsible for the flow of afferent auditory information to the CNS, lack of knowledge of the properties of Type II ganglion cells would be a restriction on our understanding of cochlear transduction mechanisms and the extent to which they are under central control, but not on understanding of the central processing of auditory information.

3.2 Basic Physiological Properties of Auditory Nerve Fibers

Mammalian AN fibers are relatively homogeneous in terms of their response properties, and it is therefore convenient to describe briefly some of their basic characteristics before considering the manner in which particular stimulus parameters are coded in the activity of the nerve. Some of the basic characteristics are illustrated in Figs. 7, 8, and 9. More detailed accounts of AN fiber anatomy and response properties will be found in a number of recent reviews (Evans 1975; Kiang 1984; Manley 1983). Rhode and Kettner (1986) have obtained recordings from AN fibers in the CN of the awake cat, and report that the basic response characteristics are the same as those observed in anesthetized animals and described in this section.

A preliminary comment on methods used for delivering and monitoring acoustic stimuli is necessary at this point. In early studies of auditory system neurophysiology, stimuli were presented in various ways (e.g., via free-field speakers or transducers coupled to hollow stereotaxic ear bars) that did not allow accurate specification of the stimulus at the tympanic membrane. In the last 10 – 15 years, most laboratories have used sealed systems and have employed the procedure of removing the pinna and transecting the meatus to allow the end of a sound delivery speculum to be positioned within 1 – 2 mm of the tympanic membrane. The sound delivery speculum usually incorporates a probe microphone assembly to allow measurement of sound pressure level (SPL; expressed in dB *re* 20 μPa unless otherwise noted) adjacent to the tympanic membrane. Dallos (1973) provides a general description of systems of this sort and of calibration procedures;

sources of inaccuracy in probe microphone measurements of tympanic membrane SPL at high frequencies have been discussed by Johnstone and Taylor (1969) and Khanna and Stinson (1985). Most of the recent neurophysiological evidence examined in later sections of this review has been obtained in experiments using such systems; only in the last few years has interest in spatial coding prompted a return to the use of free-field stimulation and the development of methods for the controlled presentation of such stimuli (discussed in Sects. 7.3 – 7.5).

1. *Spontaneous Activity.* All AN fibers exhibit spontaneous activity in the absence of controlled or measurable acoustic stimulation. The time pattern of the spontaneous discharges is irregular, and interspike interval histograms (ISIHs) are asymmetric with an exponential decay, resembling those generated by a Poisson process with deadtime (e.g., Kiang et al. 1965a; Rodieck et al. 1962; Walsh et al. 1972). The distribution of spontaneous discharge rates (SR) is bimodal, with a boundary between two subpopulations in the region of 15 – 18 spikes per second (Evans and Palmer 1980; Kiang et al. 1965a; Kim and Molnar 1979; Liberman 1978; Rhode and Smith 1985). Fibers with SR below 18 spikes per second can be further divided into two groups on the basis of threshold, those with SR < 0.5 spikes per second having higher and more heterogeneous thresholds (Kiang et al. 1976; Liberman 1978; Rhode and Smith 1985), although the size of the difference depends on the threshold criterion used (Geisler et al. 1985). Liberman (1978) therefore distinguished three groups of "low" (<0.5 spikes per second), "medium" (0.5 – 18 spikes per second), and "high" (>18 spikes per second) spontaneous activity, and these terms will be used in referring to the other properties of these fibers in subsequent sections. Correlations between spontaneous rate and radial fiber characteristics (viz., fiber diameter and the locus of synaptic contact on the IHC) have recently been demonstrated by means of intracellular HRP labeling of physiologically characterized single fibers (Liberman 1980, 1982b; Liberman and Oliver 1984).

2. *Responses as a Function of Frequency.* Individual AN fibers are highly frequency selective and this frequency selectivity is reflected in sharp frequency tuning curves (FTCs), in which the intensity required to elicit a specified discharge rate – most commonly that exceeding the spontaneous discharge rate by an arbitrary "threshold" value (see Geisler et al. 1985) – is plotted as a function of frequency (e.g., Fig. 7A). The tip of the tuning curve specifies the "characteristic frequency" (CF) of the neuron (i.e., that frequency at which the intensity required to produce the specified discharge rate is lowest). The CF of a given fiber reflects its locus of innervation in the cochlea (Kiang 1984; Liberman 1982a; guinea pig: Robertson et al. 1980). At suprathreshold SPLs, fibers respond over a wider frequency range, and the FTCs of high-CF fibers have broad low-frequency tails (e.g., Fig. 7A). The total range of frequencies and intensities over which a fiber is responsive constitutes its "response area" (e.g., Kiang et al. 1965a; squirrel monkey: Rose et al. 1971; Fig. 7B and C).
Because the nerve bundle constitutes the core of the cochlea, the AN is wrapped like a roll with the more basal fibers situated peripheral to the more apical fibers in a precise *cochleotopic* fashion (Arnesen and Osen 1978; Sando 1965). Since the

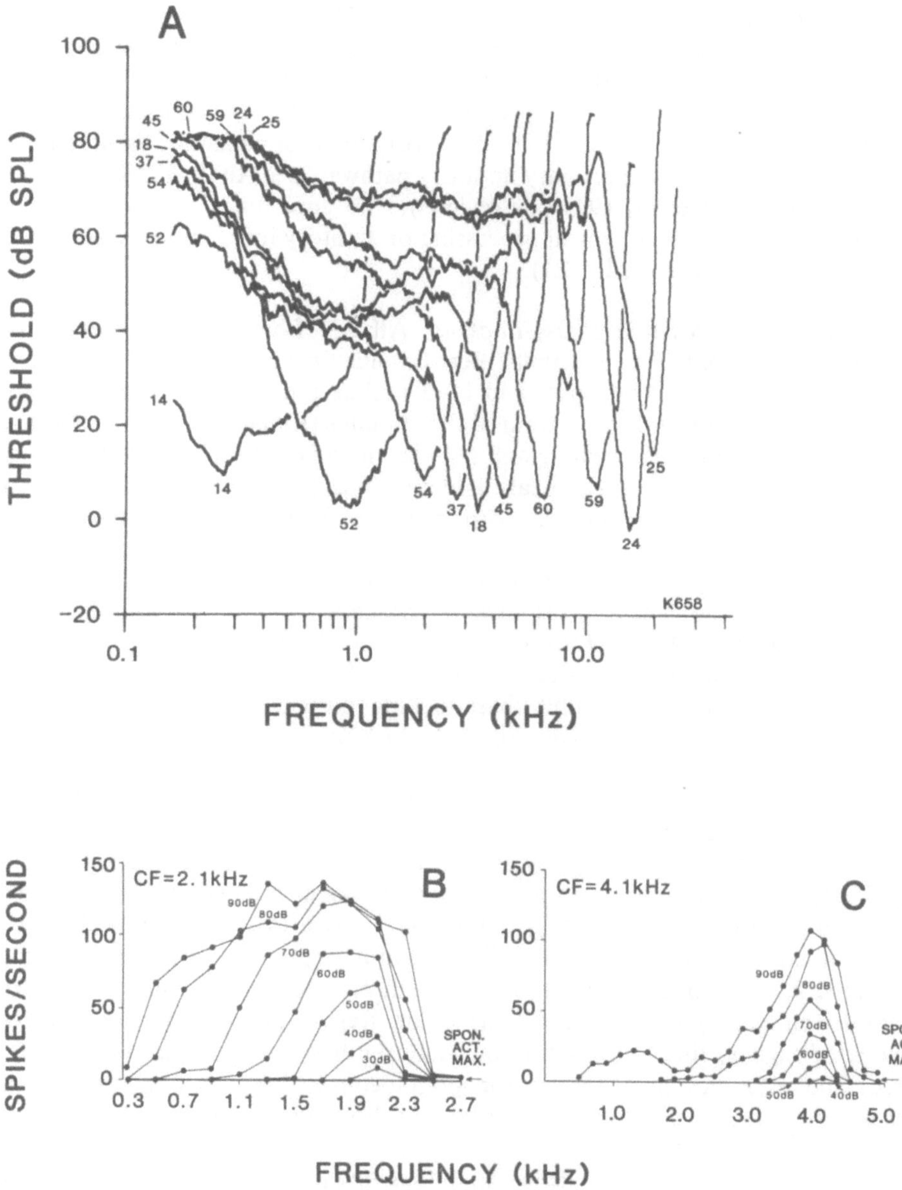

Fig. 7. A Representative frequency tuning curves for auditory nerve fibers in cat. All ten curves were obtained, using an automated procedure, from fibers with high spontaneous discharge rate in a single cat. Numbers within plot identify fibers. (Kiang 1984) **B, C** Response areas of two auditory nerve fibers in macaque as represented by isointensity contours. Each contour shows variation in discharge rate with frequency at the sound pressure level specified. For each fiber, CF is indicated in *upper left-hand corner*; *arrows* at right (marked *SPON. ACT. MAX.*) indicate maximum rate of spontaneous discharge. (Modified from Rose et al. 1971)

cochlear locus innervated by a fiber determines its CF, the nerve consequently exhibits *tonotopic* organization: microelectrode penetrations in the posterodorsal to anteroventral direction (in the cat) first encounter fibers with high CF; as the electrode is advanced, fiber CF progressively decreases and subsequently increases (Evans 1975; Kiang et al. 1965a). Tonotopic organization is preserved throughout the core of the primary auditory pathway (see Aitkin 1976; Aitkin et al. 1984b; Clopton et al. 1974 for reviews), and constitutes the basic matrix within which much of the central processing of auditory information takes place (see Sects. 4.2.2, 5.2.2, and 7.2.1).

3. *Response Patterns and Phase-Locking.* All AN fibers show sustained responses to short tone or noise bursts. Peristimulus time histograms (PSTHs) are characterized by a peak at onset and by a decline (adaptation) to a relatively stable discharge rate for the duration of the stimulus (Kiang et al. 1965a; Fig. 8A and B). In some fibers the peak is well defined and adaptation is rapid (e.g., Fig. 8A), while in others the onset peak is less distinct (e.g., Fig. 8B). Rhode and Smith (1985) have shown that the former type of PSTH is exhibited by fibers with high SR, the latter by fibers with low SR. At frequencies up to about 4 – 5 kHz, discharges occur during a restricted segment of the cycle of the stimulating sinusoid, i.e., the response is "phase-locked" to the stimulus (Fig. 8C; e.g., Kiang et al. 1965a; Johnson 1980; squirrel monkey: Rose et al. 1967). A given AN fiber exhibits phase-locking at all low frequencies within its response area; thus, high-CF fibers exhibit phase-locking in the low frequency tail region of their response areas. The extent of phase-locking and the preferred discharge phase are indicated by period histograms, which show the distribution of discharges timed relative to the stimulus period (cf. Brugge et al. 1969; see Fig. 32 for examples). A quantitative index of phase-locking is provided by a measure termed the vector strength (or synchronization index).[6] Phase-locking is also manifested in ISIHs for driven activity with peaks at integral multiples of the period of the stimulating tone (Fig. 8D, E, and F; Rose et al. 1967), reflecting the fact that under normal conditions fibers cannot respond to every cycle of the stimulus. At frequencies above those at which phase-locking occurs, the steady state driven activity of AN fibers is irregular. Regardless of frequency, the envelope of a given fiber's ISIH for continuously driven activity has exponential decay and resembles the histogram for spontaneous activity when the time scales are adjusted to take account of differences in discharge rate (Kiang 1968).

[6] The vector strength measure is derived as follows. The time of occurrence of each spike relative to a fixed point on the stimulating sinusoid is expressed as a phase angle θ. Each of the phase angles derived from a spike train is represented as a unit vector, each vector making an angle θ with the *x*-axis. Averaging of the unit vectors yields a mean vector which is characterized by a phase angle $\bar{\theta}$ (which is an estimate of the mean phase angle between the stimulus and the discharge) and a length (or vector strength). The vector strength can vary from 0 (no phase-locking, i.e., a uniform distribution of phase angles) to 1 (perfect phase locking, all spikes having the same phase angle) (Goldberg and Brown 1969; Goldberg and Brownell 1973). An identical measure called the synchronization index is obtained by computing the magnitude of the fundamental component of the Fourier transform of the period histogram and normalizing by the number of responses in the histogram (Anderson 1973; Johnson 1980).

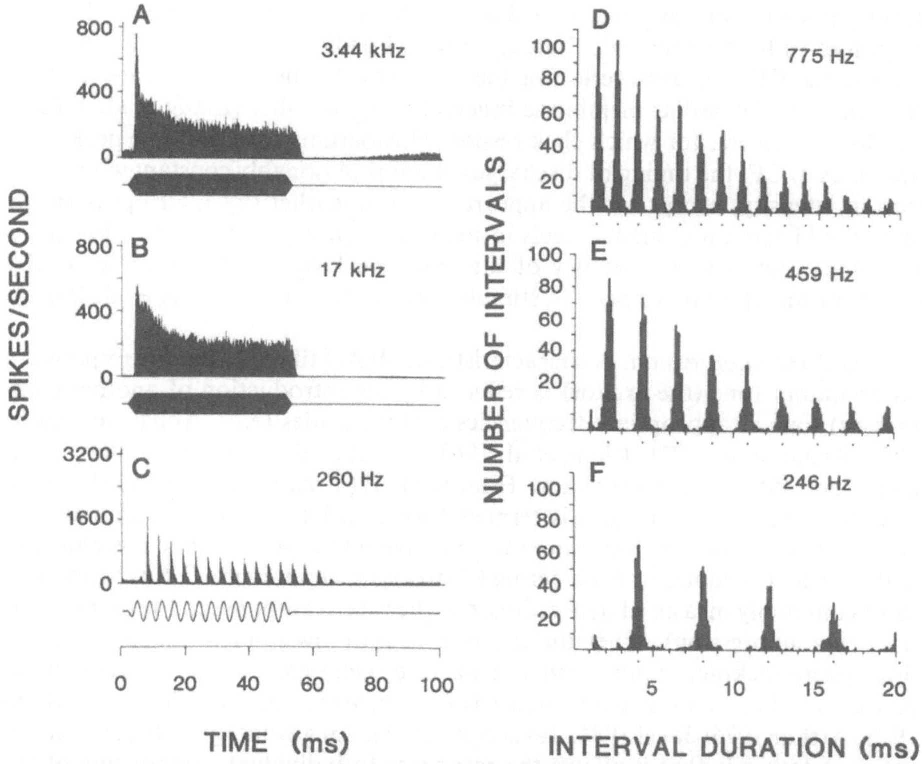

Fig. 8A–F. Peristimulus time histograms for response of three representative auditory nerve fibers to tone bursts at the CF of the fiber. Each histogram is based on responses to more than 3000 tone bursts. In **A** and **B** tone burst envelope is shown below histogram, in **C** waveform of low-frequency tone burst is shown, and phase-locked nature of discharge is apparent. Note that vertical axis in histograms is expressed as ***instantaneous spike rate*** (spikes per second), defined as follows: spike rate = number of spikes per histogram bin, divided by binwidth multiplied by the number of stimulus presentations. This definition can yield instantaneous spike rates higher than 1000 spikes per second, as in **C**. (Modified from Kiang 1984) **D – F** Representative interspike interval histograms for responses to low-frequency tones of a single auditory nerve fiber in the macaque (fiber best frequency = 600 Hz). Each histogram is based on discharges to a tone of the indicated frequency (20 s duration; 100 dB SPL). ***Dots*** below abscissa in each case mark integral multiples of the period of the stimulus frequency. (Redrawn from Rose et al. 1967)

Responses to click stimuli depend on fiber CF and on stimulus level. Click PSTHs for low-CF fibers have multiple peaks separated by intervals equal to the reciprocal of the CF at all stimulus levels (Kiang et al. 1965a; Pfeiffer and Kim 1972). Histograms for high-CF fibers usually have only a single peak at low to moderate intensities; at high levels they have multiple peaks, but the intervals are unrelated to fiber CF (Kiang et al. 1965a).

4. *Response Latency*. The response latencies of AN fiber responses to clicks (Kiang et al. 1965a) and to short tone bursts (Rhode and Smith 1985) show a similar dependence on fiber CF. For all fibers with CF above 2 – 3 kHz, the

latency is similar and short (1.3 – 1.8 ms for clicks; 1.5 – 2.5 ms for high-SR fiber responses to tone bursts with 2 ms rise time). For fibers with lower CF, latency increases as CF decreases, reflecting the time taken by the traveling wave to reach the point on the basilar membrane innervated by the fiber (Kiang et al. 1965a). For low-CF fibers, for which click response histograms have multiple peaks separated by 1/CF, the timing of discharges remains reasonably constant with variation in intensity, except for the appearance of an earlier (by 1/CF) peak in the response histogram at higher levels in some cases (Kiang et al. 1965a). For high-CF fibers, however, the latency of the initial peak of the PSTH decreases (by 0.5 ms or more) with increases in stimulus level (Kiang 1965; Kiang et al. 1965a).

5. *Two-Tone Suppression.* A characteristic of all AN fibers is that the response to an excitatory tone (the excitor) is reduced by the introduction of another (suppressor) tone at appropriate frequencies and intensities (e.g., Abbas and Sachs 1976; Arthur et al. 1971; Kiang et al. 1965a; Javel et al. 1978; Sachs 1969; Sachs and Kiang 1968; squirrel monkey: Hind et al. 1967; chinchilla: Schmiedt 1982). This phenomenon was originally termed "two-tone inhibition," but a number of characteristics indicate that it does not involve inhibitory synaptic mechanisms, and it has consequently been termed "two-tone suppression." Suppression is most commonly measured by a decrease in the rate of response to the excitor tone (i.e., *rate* suppression). When the excitor tone is of low enough frequency to produce phase-locking, suppression can also be measured by the decrease in the phase-locked (synchronized) response (i.e., *synchrony* suppression) (e.g., Abbas 1978; Arthur 1976; Javel 1981; Javel et al. 1978). An advantage of the synchronization measure is that it allows the responses to individual components of the complex to be distinguished, thus enabling suppression effects to be measured when both the nominal excitor and suppressor frequencies excite the fiber. In general, the responses of an AN fiber to a low-frequency two-tone complex, both components of which fall within the fiber's response area, will be synchronized to one or the other frequency or to both, depending on the frequency and intensity of the components (e.g., Abbas 1978; Arthur 1976; Javel 1981; Reale and Geisler 1980; squirrel monkey: Brugge et al. 1969; Hind et al. 1967).

Suppression phenomena are also seen in responses to combined noise and tonal stimuli (e.g., Costalupes et al. 1984; Geisler and Sinex 1980; Rhode et al. 1978), and apparently underlie an interesting feature of AN fiber responses to narrow-band noise bursts. Ruggero (1973) found that as the bandwidth of a narrow-band noise was symmetrically increased around CF, the responses of AN fibers (in the chinchilla) initially increased and then either reached a plateau or decreased. The initial increase was attributed to summation of the effects of different spectral components within the cell's response area, and the subsequent decrease to the fact that more energy fell within suppression regions as the bandwidth increased further.

3.3 Auditory Nerve Coding of Intensity

For almost all AN fibers, discharge rate at CF increases monotonically with stimulus intensity until high intensities are reached. For most fibers, discharge rate

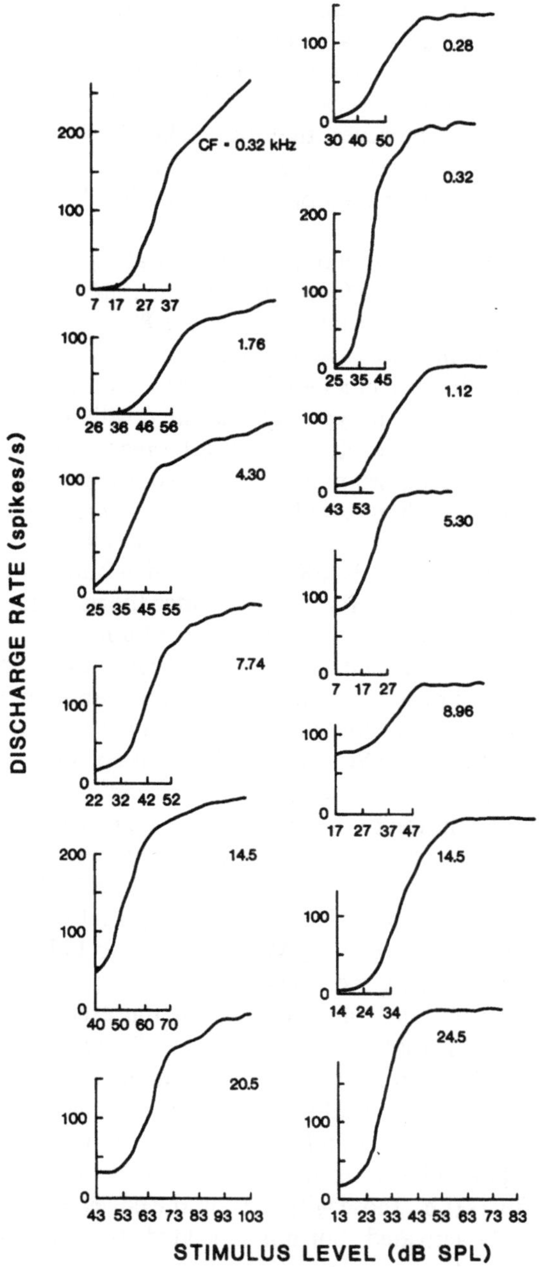

Fig. 9. Representative rate–intensity functions (i.e., plots of discharge rate against sound pressure level) for auditory nerve fibers in the cat. Stimuli were tones at fiber characteristic frequency (*CF*), which is given (in kHz) below each curve. Curves in left- and right-hand columns were chosen to illustrate sloping and flat saturation, respectively, for neurons of similar CF. (Redrawn from Sachs and Abbas 1974)

reaches a maximum (saturates) 20–50 dB above threshold (Fig. 9, right-hand column; Evans 1975; Kiang et al. 1965a; Palmer and Evans 1979; Sachs and Abbas 1974), but the functions of some fibers, chiefly those with low SR, are characterized by "sloping saturation" and extended dynamic range (Fig. 9, left-hand column; Evans and Palmer 1980; Palmer and Evans 1979; Sachs and Abbas 1974; Schalk and Sachs 1980). The proportion of fibers showing the latter characteristics is unclear and the figures given by various authors are based on different criteria and are thus not directly comparable. Although Sachs and Abbas (1974) stated that it was not possible to give general figures, they noted that 44 of 99 fibers in a typical sample showed sloping saturation. Palmer and Evans (1979) reported, however, that only 10% of a large sample of AN fibers ($N = 333$) had dynamic ranges in excess of 60 dB. The latter figure is comparable to the proportion of AN fibers (approximately 16%) having low SR (Geisler et al. 1985; Liberman 1978).

Markedly nonmonotonic behavior at high intensities has been reported by a number of authors (Javel 1981; Kiang 1984; Kiang and Moxon 1972; Liberman and Kiang 1984). At intensities above 80–90 dB, the discharge rate can decrease abruptly (in some cases to the fiber's spontaneous rate) and then return to maximum with further increases of only a few dB. These "dips" or "notches" in rate–intensity functions are typically associated with abrupt changes of approximately 180° in response phase (Kiang 1984; Liberman and Kiang 1984). Although these phenomena have important implications for the processes of AN fiber excitation, they have only recently received systematic study (Liberman and Kiang 1984).

The limited dynamic range of most AN fibers is in marked contrast to the range of approximately 100 dB over which humans can discriminate changes in stimulus level (e.g., Moore 1982). Since the variation in threshold of most fibers with a given CF (the 85% or so with medium and high SRs) is only about 20 dB (Evans 1975; Geisler et al. 1985; Kiang 1969; Liberman 1978), the limited dynamic range raises a problem with respect to the manner in which information about stimulus intensity, and consequently information about IIDs, is conveyed to the CNS[7]. As will be seen in Sect. 3.5, it also raises problems for the representation of frequency spectrum in the AN.

One solution to the dynamic range problem is that increases in intensity at higher levels are coded by the spread of activity to nonsaturated fibers of adjacent CF (e.g., Siebert 1965, 1968; Whitfield 1967; Zwicker 1970). Evidence in support of this view is provided by the finding that intensity discrimination of tones (Carlyon and Moore 1984; Moore and Raab 1974) and narrow-band noise (Viemeister 1983) at high levels is impaired by the presence of band-stop noise which disrupts information from the edges of the excitation pattern. One implication of this model is that the central coding of intensity at a particular frequency cannot be restricted to neurons with CF at or near that frequency (i.e., to a specific frequency channel), but requires processing of information in all of the fibers activated by the stimulus. Indeed, Siebert's (1965, 1968) modeling studies suggest that an ideal central processor would weight most heavily those "off-frequency" fibers (i.e., those with CFs on the edges of the excitation pattern) that are moderately activated, but not saturated. It also follows from this hypothesis that the central processing of IIDs could not be restricted to frequency-specific channels. In accordance with this conclusion, the model developed by Colburn (1973; Colburn and Durlach 1978) of binaural interaction based on AN response data

[7] Another aspect of this issue concerns dynamic range in the presence of masking noise. A brief account of some of the data related to this topic is given in Sect. 4.2.6, but see Costalupes et al. (1984), Geisler and Sinex (1980), and Palmer and Evans (1982) for discussion of relevant AN data.

achieves a good fit to human IID discrimination on the assumption of an optimal central processor utilizing average rate information from *all* AN fibers, even when the fibers are modeled as having homogeneously saturating rate–intensity functions. Note, however, that the frequency dependence of the IID–azimuth relationship requires that the resultant IID information must be combined with frequency information to yield the azimuth of the source.

Although intensity discrimination of high-level, narrow-band signals is impaired by band-stop noise, discrimination is not impossible and the decrement is not as great as that produced by a spectrally uniform masker (Moore and Raab 1974, Viemeister 1974, 1983). These observations indicate that spread of neural activity is not a *necessary* condition of intensity discrimination and that some information is in fact provided by fibers innervating the region of maximum excitation. A number of mechanisms have been proposed as enabling such coding:

1. As noted previously, fibers with low SR (<0.5 spikes per second) have thresholds that are higher than those of low- and medium-SR fibers of the same CF (Geisler et al. 1985; Kiang et al. 1976; Liberman 1978), although the magnitude of the difference depends on the threshold criterion employed (Geisler et al. 1985). It is therefore possible that intensity changes at high levels might be signaled by the recruitment of unsaturated high-threshold fibers, as proposed by Viemeister (1974, 1983). Although it is intuitively implausible that the smaller Weber fraction for tone intensity ($\Delta I/I$) at high levels (e.g., Moore 1982) could be produced by only the small proportion of fibers innervating a given cochlear locus that have high thresholds (see Evans 1981 and Moore 1982 for discussion), Viemeister (1983) has presented theoretical arguments indicating that the observed performance could in fact be accounted for by the activity of only a small number of fibers.
2. Conventional AN rate–intensity functions chiefly reflect the steady state or "adapted" component of the response. Smith and his colleagues (Smith and Brachman 1980a, b; Westerman and Smith 1984) have shown that the onset component of AN responses (in the Mongolian gerbil) has a wider dynamic range than the steady state component. The greater dynamic range of transient response components is similarly reflected in the observation by Kiang et al. (1965a) that rate–intensity functions for time-locked responses to clicks show less effect of saturation and a greater dynamic range than those for tones. It is therefore possible that onset response components could play a role in conveying intensity information to the CNS, although Carlyon and Moore (1984) have presented psychophysical evidence against a role for dynamic onset responses in intensity discrimination.
3. The "threshold" for phase-locking to low-frequency tones of spontaneously active AN fibers is lower (by 5–20 dB) than that for an increase in discharge rate (Evans 1980; Johnson 1980; squirrel monkey: Rose et al. 1967). Granted a central processor able to utilize periodicity information, such phase-locking could extend the dynamic range of AN fibers at low levels. However, the degree of phase-locking saturates at even lower levels than does mean discharge rate (Evans 1980; Johnson 1980), so that this mechanism provides no solution to the problem of intensity coding at high levels.
4. As noted previously, the latency of high-CF fiber responses decreases with increasing intensity. The obvious fact that the CNS is not provided with a timing

signal from which to measure latency means that the time of occurrence of the discharges of individual fibers could convey no information about intensity. It is possible, however, that intensity could be coded in terms of the temporal pattern of activity across a group of fibers (cf. Hind et al. 1963 for discussion of such coding in the CNS). A specific version of this suggestion is the proposal that IIDs are converted into temporal differences in the signals from the two ears as a consequence of the effects of intensity on the latency of AN responses (e.g., Jeffress 1948; Deatherage and Hirsh 1959). The conversion postulated by this "latency hypothesis" has been used to explain the psychophysical phenomenon of time–intensity trading discussed in Sect. 2.3, and is a major component of many models of binaural interaction (see Sect. 8.2.2).

This brief summary indicates that a number of different mechanisms are likely to contribute to the coding of stimulus intensity by AN fibers. Although evidence to be reviewed in later sections has provided considerable information on the ways in which IIDs are coded by central neurons, uncertainty as to the way in which intensity is represented in the activity of AN fibers is reflected in alternative models of the mechanisms underlying the central coding of IIDs (Sect. 8.2.2).

3.4 Auditory Nerve Coding of Temporal Information

The central processing of transient and ongoing ITDs requires that information as to time of occurrence at each ear be faithfully conveyed to the CNS by the AN. For low-frequency tonal stimuli, *ongoing* time information is provided by the phase-locking of AN fiber discharges. For frequencies at or near the CF of a fiber, the phase angle at which discharges occur does not vary with intensity up to the levels at which notches appear in rate–intensity functions. At frequencies above and below CF, however, the phase angle shifts (leads or lags, respectively) with increases in intensity (squirrel monkey: Anderson et al. 1971). If the central processor were to compare the times of occurrence of discharges on inputs from the two ears, changes in the relationship between these times would unequivocally signal ITDs only if the comparison were restricted to fibers with CF equal to the stimulus frequency. For fibers of CF other than the stimulus frequency, a particular time difference between discharges could be produced by either an ITD or an IID and could only be unequivocally interpreted if independent evidence as to the presence of an IID were available.

The best evidence on the neural representation of transient time information is provided by responses to clicks. For low-CF fibers, the timing of the peaks in the click PSTH is invariant with intensity, as noted in Sect. 3.2. As with low-frequency ongoing time differences, the central processor would receive unequivocal information on transient disparities if comparison were restricted to low-frequency fibers from the two ears having the same CF. For high-CF fibers, however, latency decreases with intensity and the central processor would be unable to distinguish between central time differences produced by an IID or an ITD without independent evidence on the existence of an IID.

The requirement that ITD processing at low frequencies be carried out by neurons with CF equal to the stimulus frequency has a parallel in the results of modeling studies. Colburn (1973) demonstrated that if an optimal central processor was permitted to compare temporal information from *all* AN fibers it achieved ITD discrimination performance superior to that observed in psychophysical studies. A good match to the psychophysical data was obtained when the central processor was restricted in various ways, one of which was to limit comparisons to fibers from the two ears having the same CF.

3.5 Auditory Nerve Coding of Frequency Spectrum

There is some irony in the fact that the century-long conflict between the classical "place" and "periodicity" theories of frequency coding in the AN (see Evans 1978 for review) has recently culminated in models based on a synthesis of place and temporal information. The sharp frequency tuning of AN fibers would appear to provide a basis for frequency coding according to the place principle. According to this model, the frequency of a pure-tone stimulus would be represented by increased discharge in fibers with that CF. Correspondingly, the spectrum of a complex stimulus would be reflected in the distribution of activity across the tonotopic fiber array, the amount of energy at each frequency being represented by the discharge rate of fibers with that CF. Thus, according to this model, the information provided to the central processor would be in terms of the spatial distribution of discharge rates across the fiber array, and the task of the central processor would be one of pattern recognition. Empirical evaluation of this model requires an estimate of the pattern of response across the population of AN fibers. Such an estimate can be obtained by the procedure, introduced by Pfeiffer and Kim (1975), of recording in a single animal the responses of a large sample of AN fibers to a single stimulus or a small set of stimuli. Data of this sort, showing the distribution of driven activity across the fiber array for tonal stimuli of low (1 kHz; Kim and Molnar 1979) and high (8 kHz; Evans and Palmer, reported in Evans 1981) frequency are presented in Fig. 10. The distribution for the high-frequency tone (Fig. 10A) is characterized by a clearly defined peak of activity in fibers with that CF over a reasonably wide dynamic range. Furthermore, as a consequence of the sharp high-frequency cutoffs of high-CF fibers (see Fig. 7A), the low-frequency border between active and inactive fibers remains sharp and relatively constant up to the highest level shown. For the low-frequency stimulus (Fig. 10B), however, there is a clearly defined peak in the response profile only at the lowest intensity (20 dB). At higher levels the distribution is flat and irregular, with no indication of a peak at the stimulus frequency, and the range of CFs over which fibers are activated becomes progressively broader with increasing intensity. The flattening of the distribution is the consequence of rate saturation, while the spread of activation reflects both the broad tuning of low-CF fibers and the low-frequency tails in the tuning curves of high-CF fibers (see Fig. 7A).

These data suggest that while the frequency of high-frequency pure tones may be coded in terms of the distribution of discharge rate across the tonotopic fiber array, this could only be the case for low frequencies at near-threshold levels. However, low-frequency pure tones might nevertheless be coded in place terms if the central processor were to derive critical information from changes in the positions of the edges of the active fiber array (e.g., Whitfield 1967; Zwicker 1970).

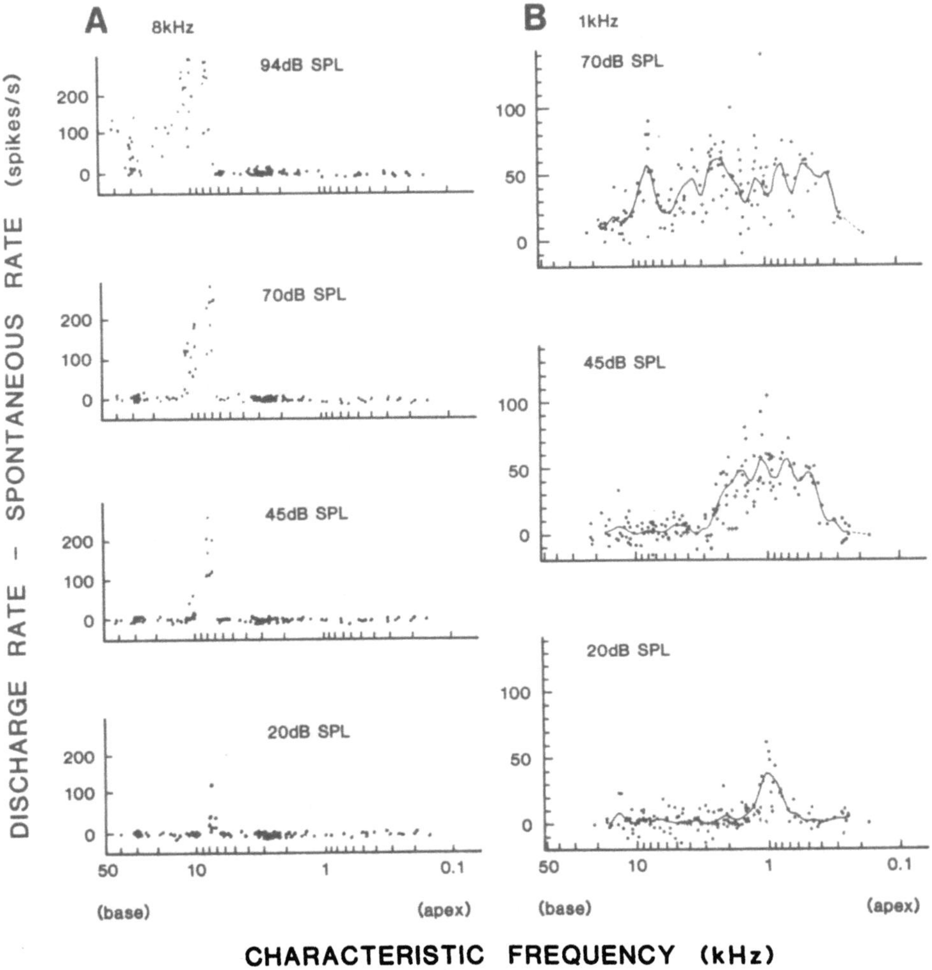

Fig. 10A,B. Distributions of evoked activity across cat auditory nerve fiber array for two frequencies: **A** 8 kHz and **B** 1 kHz, at different sound pressure levels indicated. Each data point in a given plot represents the increase in discharge rate above spontaneous activity for a single fiber plotted at its characteristic frequency; all fibers in each set of plots were recorded in a single cat. Curves drawn through data points in **B** are moving window averages that include only fibers with spontaneous rate (SR) >15 spikes per second (*dots*); data from low-SR fibers are plotted as *crosses*. (Plots in **A** redrawn from Evans 1981, based on unpublished data of Evans and Palmer; plots in **B** redrawn from Kim and Molnar 1979)

This possibility is supported by the fact that frequency discrimination of low-frequency pure tones is impaired in the presence of band-stop noise (Emmerich et al. 1983) which disrupts information at the edges of the excitation pattern. Place coding is also given more general support by Siebert's (1970) theoretical demonstration that there is adequate information in the place pattern alone (given *optimal* central processing) to account for observed frequency discrimination at low frequencies. Conversely, however, Moore (1973) has presented evidence that pure-tone frequency JNDs are in accordance with a place model at frequencies above 5 kHz, but that at lower frequencies performance exceeds that predicted by the place model.

More serious difficulties for place coding of low-frequency information are encountered when spectrally complex signals, such as speech sounds, are considered. These difficulties are illustrated in Fig. 11 by data from Sachs and Young's (1979) study of AN response profiles to synthesized steady state vowels. At levels up to 54 – 64 dB there is a clear peak in the response profile in the region of the first formant and a second peak in the region of the second and third formants. At higher levels, however, there are no peaks in the response profile corresponding to the formant peaks in the vowel spectrum. Sachs and his colleagues (Sachs and Young 1979; Sachs et al. 1980; Young and Sachs 1981) have shown that the flattening of the rate profile is due both to rate saturation and to two-tone rate suppression. Subsequent studies using consonant-vowel (CV) syllables (Miller and Sachs 1983; Sinex and Geisler 1983) have supported the conclusion that profiles of discharge rate across the tonotopic fiber array do not provide an adequate representation of the spectra of steady state vowels at higher levels, at which perceptual discrimination is good. Although this evidence argues strongly against discharge rate representation of complex low-frequency sounds with steady state spectra, Delgutte and Kiang (1984a) have pointed to a number of limitations on the available data (e.g., uncertainty as to the possible effects of anesthesia and of the efferent system) and have argued that the question cannot be considered closed. For example, Sachs and Young (1979) presented limited data indicating that formant peaks might be preserved in the discharge rate profiles of low-SR, high-threshold fibers (cf. Goldstein 1980).

In contrast to steady state vowels, Miller and Sachs (1983) and Sinex and Geisler (1983) reported that the rapidly changing spectra associated with the formant transitions in stop consonants were well represented by discharge rate profiles. The fact that time-varying consonant spectra can be represented in this fashion could reflect the fact that the consonant segments of CV syllables are of lower intensity than the vowels and/or the greater dynamic range of AN responses at stimulus onset (Miller and Sachs 1983; Sinex and Geisler 1983). Delgutte and Kiang (1984b) and Delgutte (1980) found that discharge rate distributions for responses to noise-like, voiceless fricative consonants have maxima in the frequency regions in which the stimuli have considerable energy, but do not represent formant frequency with any precision. Such a representation may be adequate, however, given evidence that phonetic distinctions between fricatives are based more on the locations of bands of energy than on formant peaks (Delgutte and Kiang 1984b).

In summary, a number of lines of evidence support the view that spectral information for high-frequency stimuli is conveyed to the CNS in terms of the distribution of discharge rate across the tonotopic fiber array. For low-frequency stimuli, however, such place coding would seem to be adequate only at near-threshold levels or for sounds with rapidly time-varying spectra. For low-frequency sounds with steady state spectra, the spatial distribution of activity in terms of discharge rate does not appear to provide adequate information to the CNS unless it is assumed that such coding is restricted to the small proportion of fibers with high thresholds or with wide dynamic range. It is, of course, precisely

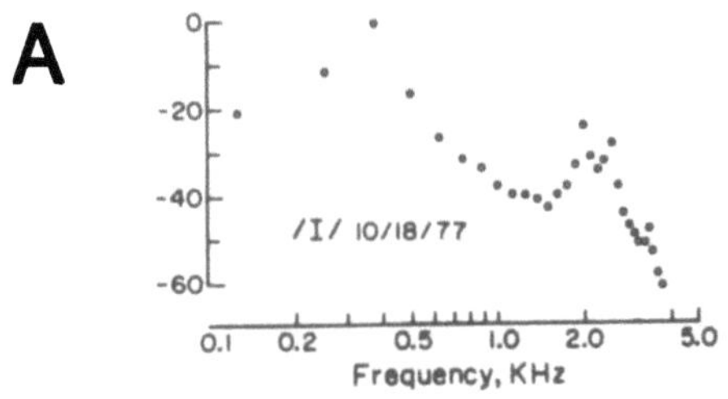

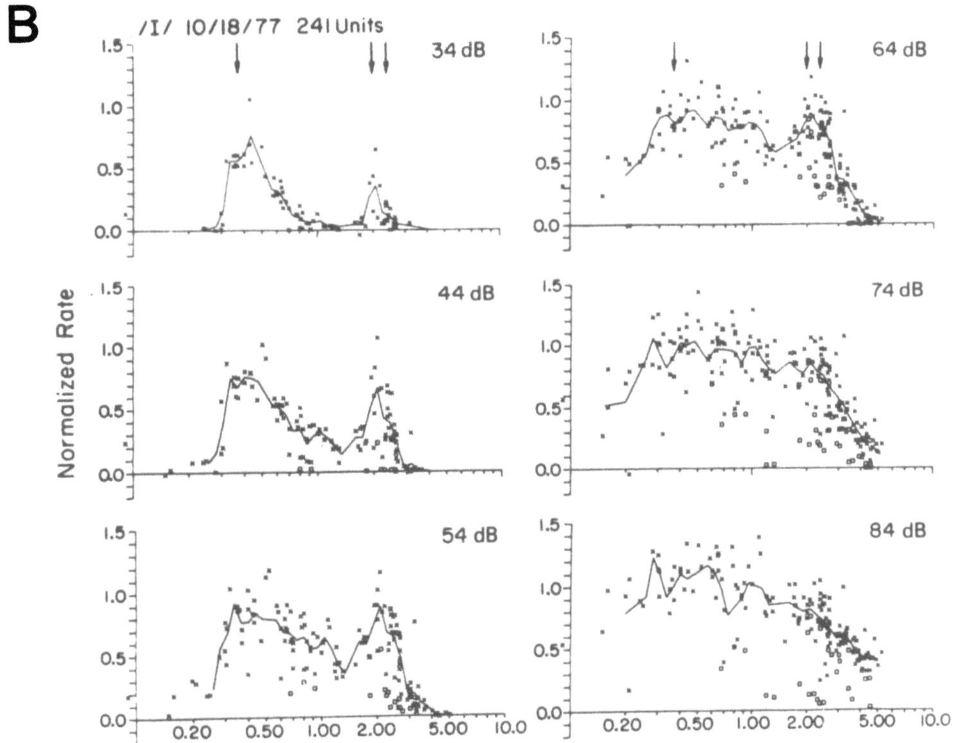

Fig. 11. A Amplitude spectrum measured near cat's tympanic membrane for synthesized vowel /I/, responses to which are shown in **B**. Each point is the amplitude of one harmonic of the steady state vowel, which is periodic with fundamental frequency 128 Hz. **B** Distribution of evoked activity across the auditory nerve fiber array in a single cat for vowel /I/ at specified intensities. Each data point in a given plot is the normalized discharge rate [(discharge rate – spontaneous rate)/(saturation rate – spontaneous rate)] for a single fiber plotted at its characteristic frequency (241 fibers). Fibers with spontaneous rate less than 1 per second are plotted with *squares*, other fibers with *crosses*. Formant frequencies of the vowel are shown by the *arrows*. Date identifies cat from which recordings were made. (Redrawn from Sachs and Young 1979)

to this aspect of frequency coding that the classical "periodicity" theory was directed.

As discussed previously (Sect. 3.2), the responses of AN fibers to pure tones up to 4 – 5 kHz exhibit precise phase-locking, such that the intervals between action potentials are integral multiples of the stimulus period. Although individual fibers usually do not respond to every cycle even at very low frequencies, precise temporal information on the period of tonal stimuli at frequencies up to 4 – 5 kHz is conveyed to the CNS by the population of phase-locked fibers, as originally postulated by Wever's (1949) "volley" theory (see Evans 1978 and Hind 1972 for reviews).

The spectra of low-frequency steady state sounds such as vowels and CV syllables are also represented in the temporal properties of AN fiber discharge (Delgutte and Kiang 1984a, c, d; Miller and Sachs 1983; Sachs et al. 1983; Sinex and Geisler 1983, 1984; Young and Sachs 1979). Sachs and his colleagues used the large-sample procedure to estimate the *temporal* response pattern across the fiber population, employing measures that assume both place and temporal coding mechanisms. Briefly, their procedure is to measure the synchronized discharge rate to individual frequency components of the signal averaged over fibers with CF close to (within 0.25 or 0.5 octaves of) that component. This restriction has the consequence that temporal features of the response are measured within the context of a place code. As illustrated in Fig. 12, the resultant plots of "average localized synchronized rate" (ALSR) against frequency provide a representation of formant peaks that, in contrast to the discharge rate profile, is not degraded by increases in intensity. Vowel representations in terms of synchronized responses are also much less affected by masking noise or by changes in preceding context than are discharge rate representations (Delgutte 1980; Delgutte and Kiang 1984c, d; Sachs et al. 1983). The fact that speech intelligibility is good at signal-to-noise ratios that result in major degradation of discharge rate representations suggests that the central processor takes advantage of the resistance to noise offered by synchronized rate representations.

A processing scheme differing from the ALSR measure was proposed by Sinex and Geisler (1983, 1984). They noted that the synchronized discharge of a fiber could be dominated by a stimulus component other than those near CF, and therefore defined the "dominant frequency" for each fiber as that frequency at which it showed the greatest synchronized response, *regardless* of the location of the frequency with respect to the fiber's CF. This procedure relaxes the strict place restriction implicit in the ALSR measure. Sinex and Geisler (1983) found that plots of dominant frequency for samples of fibers provide a good representation of formant position, a result confirmed by Delgutte (1984).

The results of these studies establish that information about the frequency spectra of simple and complex low-frequency sounds is present in the temporal properties of the discharges of AN fibers. They thus provide empirical validation of Siebert's (1970) theoretical observation that periodicity cues in the AN discharge pattern contain *in principle* much more precise information about frequency than place–rate cues. The studies differ, however, in what they take to be the important features of the temporal discharge pattern. Delgutte (1984) examined the effects of various filter characteristics within "local filtering" schemes of the type proposed by Sachs and his colleagues and compared them

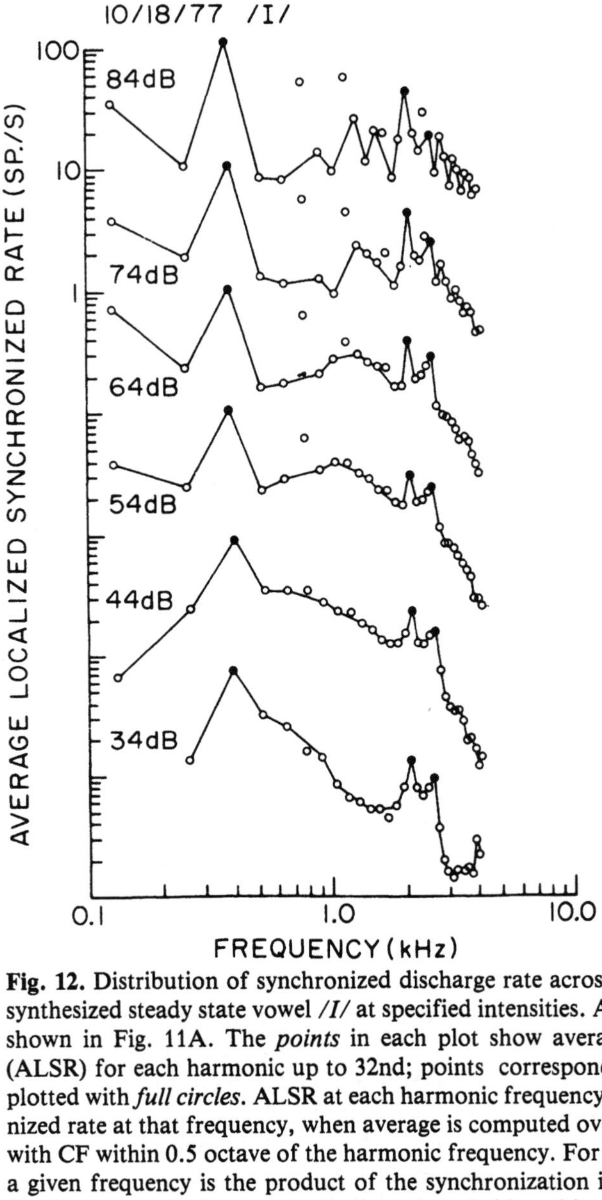

Fig. 12. Distribution of synchronized discharge rate across auditory nerve fiber array for synthesized steady state vowel /I/ at specified intensities. Amplitude spectrum for vowel is shown in Fig. 11A. The *points* in each plot show average localized synchronized rate (ALSR) for each harmonic up to 32nd; points corresponding to formant frequencies are plotted with *full circles*. ALSR at each harmonic frequency is the average value of synchronized rate at that frequency, when average is computed over those fibers in the population with CF within 0.5 octave of the harmonic frequency. For each fiber, synchronized rate at a given frequency is the product of the synchronization index at that frequency and the fiber's average discharge rate. Ordinate is scaled logarithmically, and plots are shifted vertically from one another by one order of magnitude for clarity. See Young and Sachs (1979), from which figure is reproduced, for details

with a "dominant component" coding scheme of the type proposed by Sinex and Geisler (1983). Although the effectiveness of the various schemes can be shown to differ as a function of stimulus conditions, these differences do not allow a decision to be made between them.

The availability of spectral information in the temporal discharge patterns of AN fibers does not establish that this information is in fact used by the central processor. Indeed, Siebert's (1970) theoretical studies indicated that the human JND for the frequency of pure tones was best matched by an ideal processor utilizing place information alone, and the case for the use of temporal information has been based on more complex discriminations (e.g., Goldstein 1980; Siebert 1970; Srulovicz and Goldstein 1983). A major difficulty for the use of temporal information has been to understand how the interspike intervals could be detected and processed centrally. For simple stimuli, a number of models based on some form of temporal cross-correlation have been proposed (see Langner 1983; Loeb et al. 1983; Whitfield 1970 for discussion). These models, and the possible central sites of the processing they postulate, will be considered further in Sect. 8.2.2.

Goldstein and his colleagues (e.g., Goldstein and Srulovicz 1977; Srulovicz and Goldstein 1983) have developed a psychophysical model of the representation of frequency spectrum that is in many respects complementary to the physiological evidence of Sachs and his colleagues. The two stages of processing of AN discharges postulated by this model involve the generation of an interval histogram for each fiber, which is then passed through a filter matched to the CF of the fiber. These processes clearly parallel the procedures involved in obtaining ALSR measures. According to the model, the resultant "central spectrum" is then matched with a central template to identify the spectral components of the signal. The model provides a good fit to psychophysical data on frequency discrimination of simple tones and estimation of the frequency components of complex tones (Srulovicz and Goldstein 1983), and this success has been used to argue for the importance of temporal information (Goldstein 1980; Goldstein and Srulovicz 1977; Srulovicz and Goldstein 1983). As we shall see in later sections, however, there is only limited evidence as to how the processes postulated by this and related models might be physiologically realized.

This brief review makes it clear that information about frequency spectrum is conveyed by a number of different features of the spatial and temporal pattern of AN fiber activity, and there is considerable uncertainty as to which of these features are in fact used by the central processor. Although only place–rate information seems to be available for high-frequency signals, both place and temporal information are available and seem likely to be used conjointly for low frequencies. As Siebert (1970) noted, even though discrimination of low-frequency pure tones can be accounted for by *optimal* use of place information, it is quite possible that the nervous system in fact makes *nonoptimal* use of temporal information. The question of which aspects of the activity across the AN fiber array are used by the central processor has recently assumed practical importance in the context of the development of coding strategies for electrical stimulation of AN fibers via cochlear implant auditory prostheses (see, e.g., Evans 1983; Kiang et al. 1979; Merzenich 1983).

4 Cochlear Nucleus: Anatomy and Physiology

4.1 Neuronal Architecture and Connections of Cochlear Nucleus Subdivisions

4.1.1 Preliminary Considerations

The cochlear nucleus (CN) has traditionally been divided into dorsal and ventral divisions (DCN and VCN), the latter region being further subdivided into anterior and posterior divisions (AVCN and PVCN, respectively) by the entering auditory nerve (AN) root. All AN fibers terminate in CN (Ramon y Cajal 1909; Jones and Casseday 1979a; Lorente de Nó 1933a, b; Osen 1970; Powell and Cowan 1962)[8], and each fiber bifurcates after entering the CN, sending an ascending branch to AVCN and a descending branch which traverses PVCN and terminates in DCN. The bifurcation is beautifully illustrated in Ramon y Cajal's drawing of AN fibers in the dog (Fig. 13A). The classical descriptions of the central projections of AN fibers in the cat were based on Golgi material from kittens (Lorente de Nó 1933a; Ramon y Cajal 1909). Recent studies of the central processes of individual type I ganglion cells intracellularly labeled with HRP in adult cats (Fekete et al. 1984) have confirmed many features of the traditional account, but have also revealed some departures from it. Most notably, the descending branches of a significant proportion (16%) of fibers labeled by Fekete et al. did not project into DCN. These authors, and Lorente de Nó (1981) in his most recent account of the Golgi material, have also described different groups of fibers distinguished by the morphology of the ascending branches.

As a consequence of the sequential bifurcation of AN fibers of low-to-high CF (Fekete et al. 1984; Lorente de Nó 1933a, 1979; Osen 1970; Sando 1965) and the course of the fibers (Fig. 13B and C), the innervation of each of the divisions is topographically organized, with high-to-low frequency input arrayed from dorsocaudal to rostroventral. As Fig. 13C indicates, this topography is achieved in DCN as a consequence of the fibers' turning to enter DCN in a strict low-to-high-frequency sequence (Osen 1970). This pattern of innervation is reflected in tonotopic organization of CN neurons, such that in each division microelectrode penetrations from dorsocaudal to rostroventral (i.e., on the axis indicated by the broken line in Fig. 13B) encounter cells with progressively higher CF (Evans and Nelson 1973a; Goldberg and Brownell 1973; Rose et al. 1959; Rose 1960). The frequency organization of CN is discussed in more detail in Sect. 4.2.2.

Although this tripartite division for many years provided the context for most studies of CN, it was well known from the classical Golgi studies of Ramon y Cajal (1909) and Lorente de Nó (1933b) that neurons within each division of CN were morphologically heterogeneous and received different patterns of primary

---

[8] The report by Carpenter et al. (1978) of AN projections beyond CN in the monkey seems to be attributable to transneuronal transport of the tritiated amino acids used as tracers in their studies (see Jones and Casseday 1979a for discussion), although this issue has not been unequivocally settled.

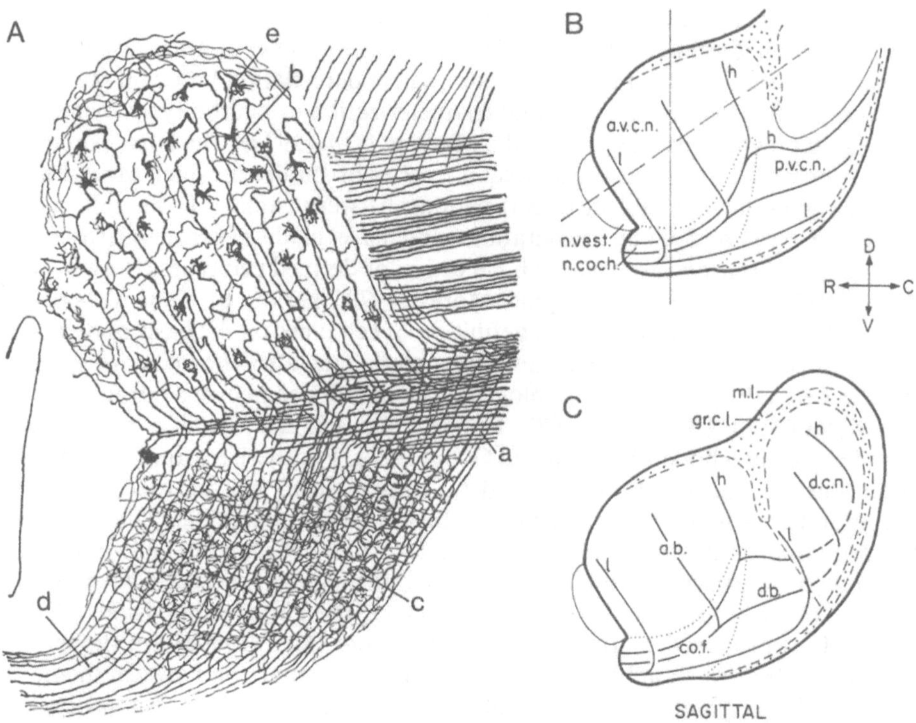

Fig. 13. A Ramon y Cajal's (1909) classical illustration of the bifurcation of auditory nerve fibers and the innervation of the ventral cochlear nucleus, from Golgi impregnation in neonatal dog. The plane of the section is approximately sagittal, with anterior up and dorsal to the left. Labeling (modified from Ramon y Cajal 1909): *a* auditory nerve fibers; *b* anteroventral cochlear nucleus, in which ascending branches terminate; *c* posteroventral cochlear nucleus, through which descending branches pass and give off terminations; *d* fibers of descending branch destined for dorsal cochlear nucleus; *e* end-bulbs of Held in anteroventral cochlear nucleus. **B, C** Schematic diagrams of the cochlear nucleus in the sagittal plane, demonstrating intranuclear course and tonotopic organization of the ascending (*a.b.*) and descending (*d.b.*) branches of the cochlear nerve fibers (*co.f.*). In **B**, ventral nucleus is shown in isolation, while in **C** it is shown in combination with dorsal nucleus. Auditory nerve branches that represent high and low frequencies are labeled *h* and *l*, respectively. Note that high-frequency fibers do not reach rostral region of anteroventral cochlear nucleus (*a.v.c.n.*) – area to left of vertical line in **B**. *Broken line* indicates oblique dorsocaudal-to-rostroventral axis of frequency organization. Other abbreviations: *d.c.n.* dorsal cochlear nucleus; *gr.c.l.* granule cell layer; *m.l.* molecular layer; *n.vest.* vestibular nerve; *n.coch.* cochlear nerve; *p.v.c.n.* posteroventral cochlear nucleus. (Osen 1970)

innervation. Lorente de Nó (1933b) distinguished 13 subdivisions on the basis of the distribution of different AN endings and of more than 50 cell types, but this parcellation had little impact on subsequent physiological studies of processing in CN. More recently, systematic study of differences in cell morphology and afferent and efferent connections have led to the identification of functionally significant subdivisions which have been related to the physiological response properties of constituent neurons.

Osen (1969a) identified nine major neuron types in Nissl preparations of cat CN and proposed a division of the nucleus into the same number of areas on the basis of the distribution of these types[9]. An alternative parcellation based on the combination of Nissl cytoarchitecture and characterization of neuron types by means of Golgi impregnations was proposed by Brawer et al. (1974). The latter parcellation has provided the basis of a subsequent series of studies in which Morest and his colleagues have established correlations between some of the cell types defined in Nissl and Golgi preparations and have made electron microscopic (EM) observations on the synaptic innervation of defined cell types (Brawer and Morest 1975; Cant 1981; Cant and Morest 1978, 1979a, b; Kane 1974a, b; Tolbert and Morest 1982a, b). The definition of cell types on the basis of detailed morphological features revealed in Golgi material has also allowed correspondences to be established with physiologically characterized neurons intracellularly labeled with HRP (Britt et al. 1983; Rhode et al. 1983a, b; Rouiller and Ryugo 1984; Smith and Rhode 1985; Wu and Oertel 1984). For these reasons, the parcellation and cell classification proposed by Brawer et al. will be used here and correspondences with Osen's scheme will be noted where possible. The parcellation proposed by Brawer et al. is too complex to be presented in full, however, and only those aspects to which functional significance can currently be attached will be treated in any detail. In Figs. 14 and 15, the major subdivisions of the parcellations proposed by Brawer et al. (1974) and Osen (1969a) are shown on approximately corresponding sections to facilitate comparison between them.

Similar cell types and parcellations of CN have been described in a variety of mammals (e.g., primates including humans: Moore and Osen 1979; Moskowitz and Liu 1972; Strominger 1978; mouse: Browner and Baruch 1982; Martin 1981; Webster and Trune 1982; Willard and Ryugo 1983: rabbit: Disterhoft et al. 1980; Perry and Webster 1981; guinea pig: Noda and Pirsig 1974; rat: Harrison and Irving 1965, 1966b; Harrison and Feldman 1970; Wouterlood and Mugnaini 1984; Wouterlood et al. 1984; American opossum: Willard and Martin 1983). Despite the similarities, there are also significant differences between species. The functional significance of these differences is unknown in all but a few cases, and the account given here will be largely restricted to the more detailed data on the cat.

These preliminary considerations raise two general points that will be made here with specific reference to CN, but which apply to a number of subsequent sections of this review. The first concerns the concept of "cell types," which has been invoked as if the use of the term and the identification of types were straightforward matters. The problems involved in the classification of neurons have received considerable attention, notably with respect to the classification of retinal ganglion cells (e.g., Hughes 1979; Rodieck and Brening 1983; Rowe and

[9] A block model of CN, based on cytoarchitectonic criteria similar to Osen's, in which CN is represented as cubes (blocks) 80 μm a side, has been constructed by Kiang et al. (1975). The purpose of this model is to provide a coordinate system allowing correlation of physiological and anatomical data from different animals and different studies (see Bourk et al. 1981; Godfrey et al. 1975a, b; Rhode et al. 1983a; Rhode and Smith 1986b for applications) rather than to propose a functionally relevant parcellation of CN, and it will therefore not be considered in detail here.

Stone 1977, 1979, 1980; Stone 1983; Tyner 1975), and a detailed discussion of these issues is beyond the scope of this review. In general terms, however, the empirical basis for the identification of cell types or classes is the covariance of a number of properties (parametric features) between cells (cf. Rodieck and Brening 1983). Ideally this covariance should be demonstrated formally, and cell types defined objectively as clusters in parametric space (as proposed by Rodieck and Brening 1983) or by similar numerical taxonomic methods. Such methods have not yet been applied to auditory brainstem neurons, although in a number of cases covariance between selected pairs of parametric features has been demonstrated quantitatively. More commonly, cell types have been characterized by listing the (qualitative or quantitative) features that are characteristic of the type, or by illustrating these features by reference to representative examples of the type. These differences in the level at which neuron types have been defined will be apparent in the following discussion of morphologically and physiologically defined cell types in CN, and in accounts of cell types in many subsequent sections of this review.

Some of the debate about the classification of retinal ganglion cells has concerned the issue of the *naming* of cell types (Hughes 1979; Rowe and Stone 1977, 1979). As Rodieck and Brening (1983) have noted, the names of cell types are *identifiers,* not *definitions*; while the use of descriptive class names involves the risk that the descriptive characteristic used might be taken to have the status of an "essential" feature, this is not a necessary consequence of the use of such terms. However, the difficulties that can arise when the names for properties are used as the names for types are well illustrated by the physiological classification of CN neurons discussed in Sect. 4.2.3.

The second general observation concerns the data obtained from studies using Golgi impregnations. Most Golgi studies employ immature animals because the best impregnations are obtained with such material (see Morest 1981). Maturational changes in the morphology of neurons and synaptic endings (see, e.g., Fig. 17; Morest 1968b; Ryugo and Fekete 1982) indicate that accounts based solely on young animals may not provide an accurate description of structures in mature animals. Although some of the major studies to be cited have employed limited adult material to check that the more extensive data from young animals provide an adequate representation of mature structures (e.g., Brawer et al. 1974), the recent intracellular labeling of individual CN neurons in mature animals has revealed complex dendritic morphology that suggests that the Golgi descriptions may not always be applicable to adults (Rhode et al. 1983a, b; Rouiller and Ryugo 1984).

4.1.2 Anteroventral Cochlear Nucleus

Brawer et al. (1974) divided AVCN on cytoarchitectonic grounds into an anterior division, comprising anterior (AA), posterior (AP), and posterodorsal (APD) parts, and a posterior division comprising dorsal (PD) and ventral (PV) parts (Fig. 14). PV contains the incoming AN fibers and their bifurcations and includes Lorente de Nó's interstitial nucleus. It is thus important to emphasize that the region of AN bifurcation does not define the division between AVCN and

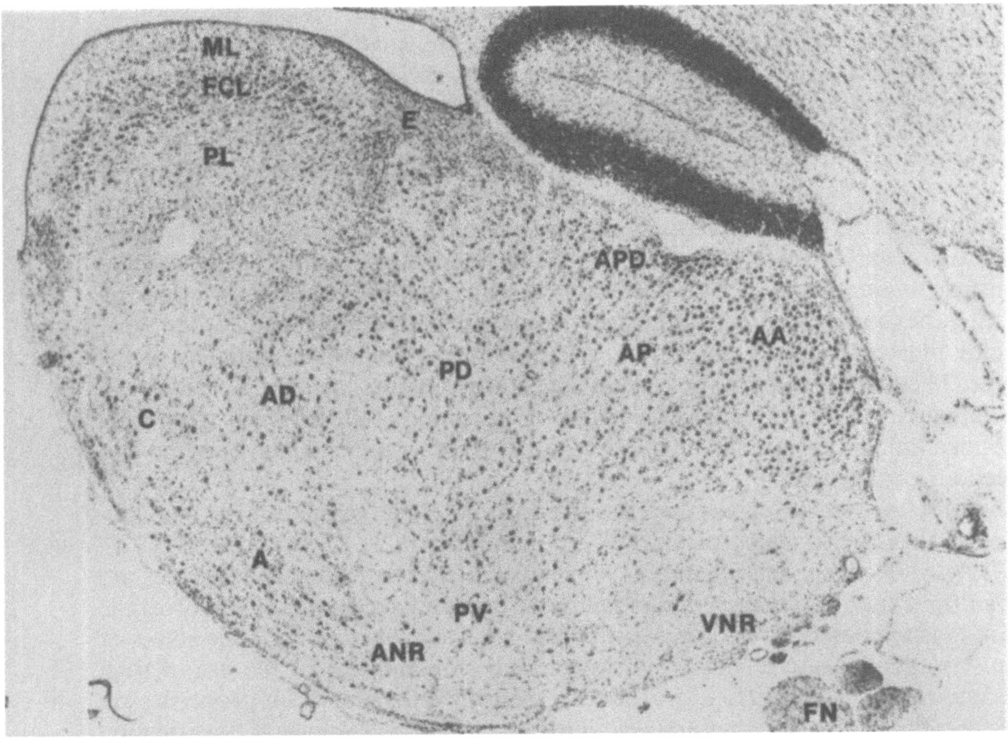

Fig. 14. Nissl-stained parasagittal section through cat cochlear nucleus showing major subdivisions of the parcellation proposed by Brawer et al. (1974). Abbreviations: *A* anterior part of posterior nucleus; *AA* anterior part of anterior division; *AD* anterodorsal part of posterior nucleus; *ANR* auditory nerve root; *AP* posterior part of anterior division; *APD* posterodorsal part of anterior division; *C* central part of posterior nucleus; *E* granular region; *FCL* fusiform cell layer of dorsal cochlear nucleus (DCN); *FN* facial nerve; *ML* molecular layer of DCN; *PD* dorsal part of posterior division; *PL* polymorphic layer of DCN; *PV* ventral part of posterior division; *VNR* vestibular nerve root. (Brawer et al. 1974)

PVCN in the cytoarchitectonically based parcellations of CN. Area PV (and the approximately equivalent "globular cell region" in Osen's parcellation discussed later in this section) includes the interstitial area and extends into adjacent parts of the traditionally defined AVCN and PVCN.

Golgi impregnations reveal that two types of large neuron, "bushy" and "stellate" cells, are intermingled with small cells throughout AVCN (Brawer et al. 1974; Cant and Morest 1979b; Tolbert and Morest 1982b). The large cell classes are not homogeneous, and both bushy and stellate cells occur in a variety of forms. Bushy cells commonly have a small number of short, stubby dendritic trunks which branch repeatedly to generate a profusion of tangled stringy appendages (Brawer et al. 1974; Cant and Morest 1979a; Fig. 16A). In immature material, most bushy cells have only a single dendrite, while others are adendritic with the soma itself covered by such appendages (Fig. 16A; Brawer et al. 1974).

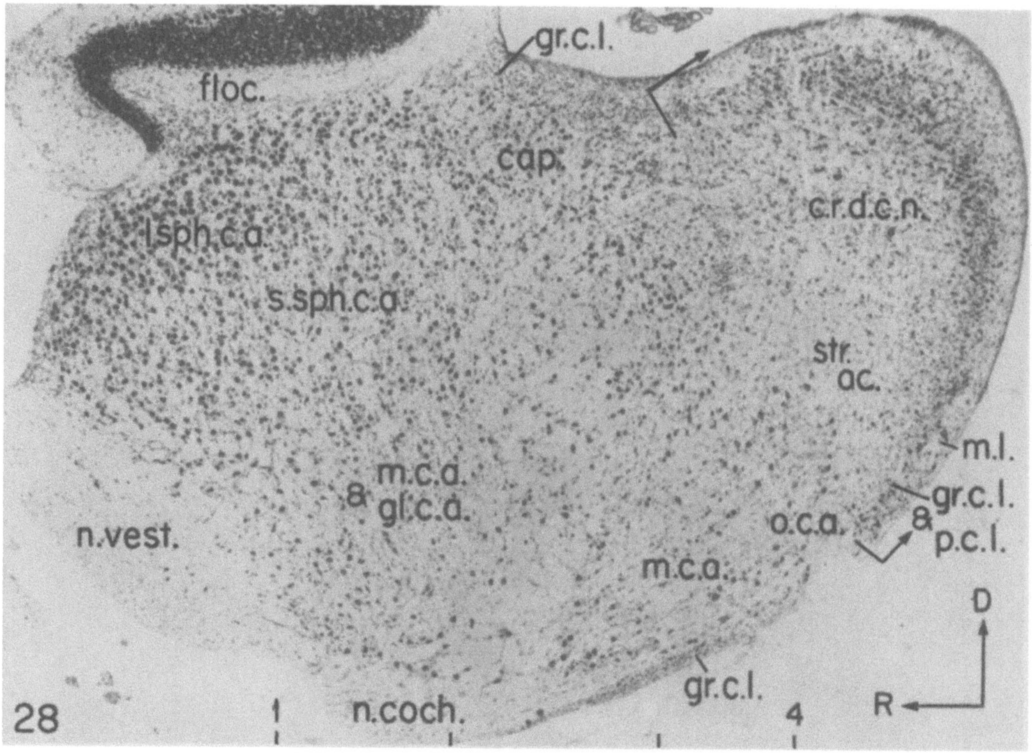

Fig. 15. Nissl-stained parasagittal section through cat cochlear nucleus at similar level to that in Fig. 14, showing Osen's (1969a) parcellation. Abbreviations: *cap* peripheral cap of small cells; *c.r.d.c.n.* central region of the dorsal cochlear nucleus; *floc.* flocculus; *gl.c.a.* globular cell area; *gr.c.l.* granular cell layer; *l.sph.c.a.* large spherical cell area; *m.l.* molecular layer; *m.c.a.* area of multipolar cells; *n.coch.* cochlear nerve; *n.vest.* vestibular nerve; *o.c.a.* octopus cell area; *p.c.l.* pyramidal cell layer; *s.sph.c.a.* small spherical cell area; *str.ac.* acoustic striae. (Osen 1969a)

In mature material, the majority of bushy cells have 1 – 3 dendrites (range 1 – 5; Cant and Morest 1979a). Stellate cells usually have a larger number of long, tapering, and less-branching dendrites, but there is considerable variation in dendritic branching patterns (Brawer et al. 1974). Although both large cell types are found throughout AVCN, the cytoarchitectonically defined subdivisions differ not only in the relative sizes and proportions of the two classes and of variants within classes, but also in the nature and complexity of the terminal axonal plexus (Brawer et al. 1974; Brawer and Morest 1975; Cant and Morest 1979a; Tolbert and Morest 1982a).

In both the anterior and posterior divisions of AVCN, correspondences between the cell types in Golgi and Nissl preparations have been established on the basis of quantitative parametric comparisons of cellular morphology and of distributional statistics (Cant and Morest 1979a; Tolbert and Morest 1982a). In the anterior division, the bushy and stellate cells correspond to the spherical and ovoid cells, respectively, identified in Nissl material (Cant and Morest 1979a).

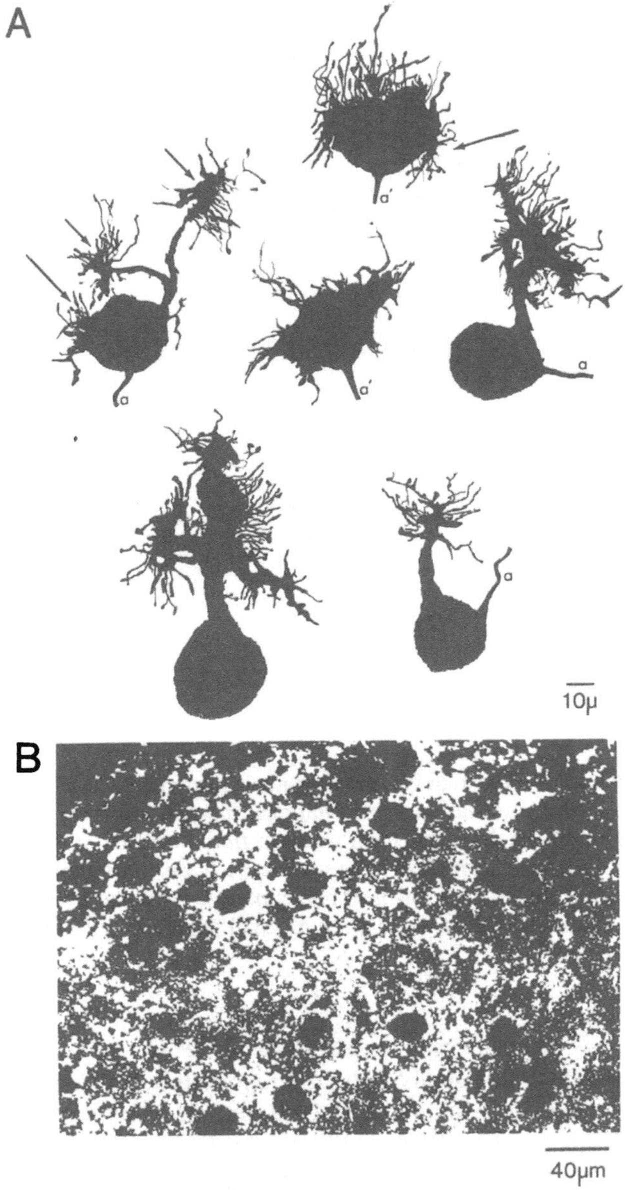

Fig. 16. A Typical sample of bushy cells from anteroventral cochlear nucleus of young kittens. The perikaryon is covered with short, spiky appendages; the stumpy dendritic trunk expands into an irregular bulb, which radiates tangled, stringy appendages. Sometimes a dendrite forms two short branches, each ending in a bulb (*short arrows*). Stringy appendages radiate from the perikaryon, often arising from elevated patches of the surface (*long arrows*). Abbreviations: *a* initial axonal segments of dendritic bushy cells; *a′* initial segments of adendritic bushy cells. Rapid Golgi impregnations from three cats, 3, 8, and 10 days old. (Brawer et al. 1974). **B** Dark-field photomicrograph of autoradiographic labeling of anteroventral cochlear nucleus after an injection of tritiated leucine into cochlea. Labeling is so dense around cell bodies of spherical/bushy cells that they are silhouetted by silver grains. (From Jones and Casseday 1979a)

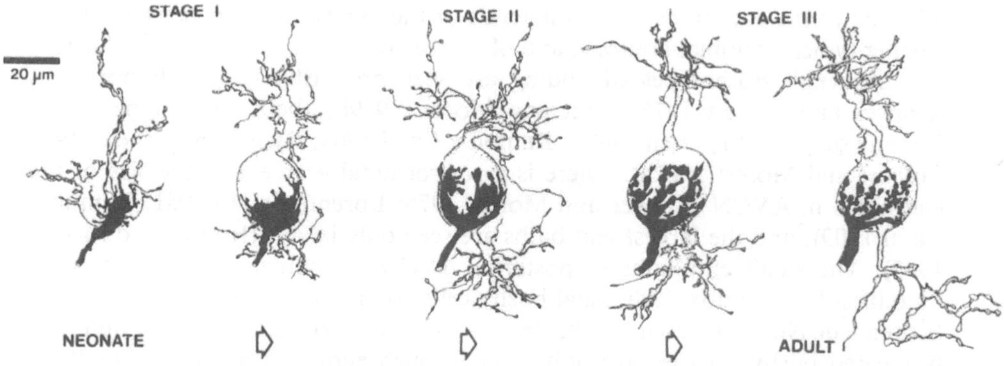

Fig. 17. Hypothetical sequence of development for a single end-bulb of Held in anteroventral cochlear nucleus of cat, abstracted from population data from animals of different ages. The four drawings on the left span 45 days of postnatal development. Note also changes in morphology of bushy cell. (Ryugo and Fekete 1982)

AA contains almost exclusively bushy cells and corresponds to Osen's (1969a, b) large spherical cell area (cf. Figs. 14 and 15)[10]. In AP, bushy and stellate cells are intermingled and occur in approximately equal numbers (Brawer et al. 1974; Cant and Morest 1979a); this subdivision can be more tentatively equated with Osen's small spherical cell area. In the posterior division, bushy and stellate or elongate cells have been identified with globular and multipolar cells, respectively (Tolbert and Morest 1982a, b; Tolbert et al. 1982). The identification of bushy cells with globular cells in this region is particularly secure because cells identified as bushy cells on the basis of diffuse HRP labeling (which can resemble Golgi impregnation) were examined by EM and found to exhibit the same ultrastructural features as globular cells (Tolbert et al. 1982). The two cell types are intermingled in both PD and PV, but PD contains a border zone comprising almost exclusively multipolar cells and PV a zone with a high concentration of globular cells (Tolbert and Morest 1982a). These observations support an approximate identification of PD and PV with Osen's multipolar and globular cell areas, respectively, as tentatively proposed by Brawer et al. (1974).

Two major types of AN terminal have been described in AVCN of the cat. The ascending branches themselves, or short thick collaterals, form large axosomatic endings called end-bulbs of Held (Figs. 13A and 17; Held 1893), the largest of which resemble the calyces of Held in the medial nucleus of the trapezoid body (MNTB; see Sect. 5.1.2 and Fig. 24). Although there are some important differences between the true calyces in MNTB and the end-bulbs in AVCN (Morest 1968b), the end-bulbs are sometimes referred to as calyces or calyceal endings. Ryugo and Fekete (1982) have described maturational changes in end-bulb morphology (Fig. 17), and have noted that many of the differences between earlier descriptions may reflect variations in the ages of the animals examined. In

[10] The large spherical/bushy cells are absent, or present only in small numbers, in some mammals with high-frequency hearing (certain bats: Feng and Vater 1985; Zook and Casseday 1982a; mouse: Webster and Trune 1982). The functional implications of this species difference are discussed in Sect. 5.3.

addition to the end-bulbs, thin collaterals of the ascending branches give rise to smaller typical terminal boutons, and often converge to form pericellular "nests" or "baskets." Both types of ending are seen on bushy/large spherical cells (Brawer and Morest 1975; Cant and Morest 1979b; Ibata and Pappas 1976; Lorente de Nó 1981; Osen 1969b; Ramon y Cajal 1909; Ryugo and Fekete 1982; Tolbert and Morest 1982a). There is a rostrocaudal gradient in the size of the end-bulbs in AVCN (Brawer and Morest 1975; Lorente de Nó 1981; Ramon y Cajal 1909), and the largest end-bulbs are seen only in AA (Brawer and Morest 1975). The small end-bulbs in posterior AVCN are sometimes referred to as "modified" end-bulbs. Individual bushy cells commonly receive 2 – 3 end-bulbs (Lorente de Nó 1981; Osen 1969b; Ramon y Cajal 1909), although some anteriorly located bushy cells receive only a single such ending (Lorente de Nó 1979, 1981). The end-bulbs invest the soma and the origins of proximal dendrites of bushy cells, and are shown by EM to contain large spherical vesicles and to make multiple, asymmetric synaptic contacts[11] (Cant and Morest 1979b; Ibata and Pappas 1976; Lenn and Reese 1966; Tolbert and Morest 1982b). The small bouton terminals, which in some cases are connected to calyceal processes by narrow stalks (Lenn and Reese 1966; Ramon y Cajal 1909), also contain large spherical vesicles and make asymmetric contacts on both the soma and dendrites of bushy cells (Cant and Morest 1979b). The density of innervation of bushy cell

[11] A comment is necessary at this point on the classification of synapses into "types" on morphological grounds and the possible correlation of these types with functionally different synapses. In aldehyde-fixed material, "asymmetric" synapses are characterized by a prominent postsynaptic density (such that the pre- and postsynaptic membranes are asymmetric), in contrast to "symmetric" synapses in which the postsynaptic density is much less pronounced (Colonnier 1968). This distinction corresponds approximately to Gray's (1959) earlier distinction between type 1 and type 2 synapses in material fixed with osmium tetroxide. Synapses in aldehyde-fixed material have also been classified into two groups on the basis of vesicle size and shape, one type containing homogeneously large, round (spherical) vesicles, whereas the other contains smaller round and ellipsoidal (flattened) vesicles (e.g., Uchizono 1965, 1967; Nakajima 1971; Walberg 1965; see Peters et al. 1976 for review). In a number of areas of cerebral cortex, these characteristics have proved to be associated: synapses making asymmetric contacts contain large round vesicles, whereas those making symmetric contacts contain smaller round or flattened vesicles (e.g., Colonnier 1968; Jones and Powell 1970; LeVay 1973; see Colonnier 1981 for review). The synapses characterized by these conjunctions of membrane specialization and vesicle characteristics are now commonly termed types 1 and 2, respectively. In a number of areas of the mammalian CNS (notably the cerebellum and hippocampus) in which the excitatory or inhibitory nature of particular synapses is known, type 1 synapses have been associated with excitation and type 2 synapses with inhibition (e.g., Eccles 1964; Gottlieb and Cowan 1972; Uchizono 1967). In accord with the former association, identified synapses with type 1 characteristics in a number of auditory brainstem nuclei – notably the end-bulbs in AVCN and the calyces in MNTB – are known to be excitatory. It must be emphasized, however, that these correlations are indirect, that they have been established only for limited areas of the CNS, and that exceptions have been noted (see Colonnier 1981; Gray 1969; Palay 1967; Peters et al. 1976 for reviews). Thus, while these cytological features can suggest physiological function, they cannot establish unequivocally whether a given class of synapse is excitatory or inhibitory.

somas by AN terminals is such that injections of tritiated leucine into the cochlea result in autoradiographs in which bushy/spherical cell bodies in AA are silhouetted by silver grains (Jones and Casseday 1979a; see Fig. 16B). Stellate cells are not contacted by end-bulbs (Brawer and Morest 1975; Cant 1981; Lorente de Nó 1981; Tolbert and Morest 1982b) and the majority receive terminal boutons on the dendrites rather than the soma. However, a second kind of stellate cell in the anterior division, the soma of which receives numerous small synaptic terminals, has been described by Cant (1981).

The end-bulb and bouton terminals of AN fibers, which contain large spherical vesicles, degenerate after cochlear ablation. A variety of other endings with smaller or flattened vesicles terminate on bushy and stellate cells and survive cochlear ablation (Cant 1981; Cant and Morest 1979b; McDonald and Rasmussen 1971; Tolbert and Morest 1982b). Noncochlear terminals with acetylcholinesterase (AChE) activity are found in the major divisions of AVCN and are most concentrated in the superficial granule cell region (McDonald and Rasmussen 1971; Osen and Roth 1969; Osen et al. 1984). Noncochlear terminals arise from multiple sources, viz., from local axon collaterals and intrinsic neurons with short axons, and from neurons in other CN divisions (e.g., Cant and Morest 1978; Lorente de Nó 1933b, 1981; Oliver 1984b), the contralateral CN (Cant and Gaston 1982), and higher brainstem centers, notably SOC (e.g., Elverland 1977; Kane 1976a). AChE-positive terminals are thought to arise from the cholinergic OCB, which gives off branches to CN on its path to the cochlea (Osen and Roth 1969; Osen et al. 1984; Rasmussen 1960, 1967). Cant and Morest (1978) described six groups of noncochlear axons projecting to AVCN in Golgi preparations, and argued that differences in the termination patterns of noncochlear axons are superimposed on the different patterns of primary innervation in the different subdivisions. However, particular terminal types on particular cell types in AVCN have not yet been associated with particular sources of noncochlear input.

Granule cells, which occur in a number of regions of CN (Brawer et al. 1974; Mugnaini et al. 1980a; Osen 1969a), form a superficial layer over the lateral free surface of AVCN. The granule cells in this layer are embedded in a dense network of mainly unmyelinated axons, dendrites, and many bouton-shaped synaptic terminals. Among these are irregularly shaped endings that resemble the endings of mossy fibers in cerebellar cortex (McDonald and Rasmussen 1971; Osen et al. 1984) and have therefore been called "mossy" endings. Some mossy endings exhibit AChE activity, as do many bouton terminals in the superficial layer (McDonald and Rasmussen 1971; Osen and Roth 1969; Osen et al. 1984). Few AN fibers terminate in the granule cell layer (Osen 1970; Rasmussen 1967), and EM observations on this layer after cochlear ablation reveal only a small number of degenerating boutons (none of which are AChE-positive), and no degenerating mossy endings (McDonald and Rasmussen 1971). In the rat, the AChE-positive fiber plexus (including those mossy endings that show AChE activity) disappears after section of the OCB (Osen et al. 1984), but equivalent observations have not been made on the cat.

4.1.3 Posteroventral Cochlear Nucleus

PVCN contains a variety of neuron types (Brawer et al. 1974; Osen 1969a) and was subdivided into eight regions by Brawer et al. on the basis of correlated Nissl and Golgi observations. Stellate cells similar to those in AVCN are widely distributed throughout almost all of these regions, as are a variety of small cells, and the superficial granule cell layer extends over the lateral free surface of PVCN. The only PVCN cell type to have been studied in detail is that to which Osen gave the name "octopus cell," on the grounds that its long, straight primary dendrites were usually gathered on one side of the cell body and gave the cell the appearance of that mollusk. The name has survived a counterclaim of ostrich-like appearance in Golgi- and EM-based reconstructions (Morest et al. 1973). Osen's octopus cell area (OCA; see Fig. 15) corresponds to the region identified by Brawer et al. as the "central part" of PVCN (C in Fig. 14), and is a region comprised almost exclusively of octopus cells (Brawer et al. 1974; Kane 1973; Osen 1969a, b).

Kane (1973, 1974b, 1977a) has provided a detailed account of octopus cell structure and innervation based on correlated Golgi and EM studies. Octopus cell somas are variable in shape and give rise to 4 – 6 thick (8 – 12 μm), straight primary dendrites which commonly bifurcate into secondary dendrites which remain reasonably thick (3 – 6 μm). The primary and secondary dendrites most frequently extend perpendicular to the descending AN branches in the region in which these fibers are most densely packed (Osen 1969a; Kane 1973). This arrangement is suggestive of a considerable degree of frequency convergence onto octopus cells (Morest et al. 1973; Osen 1969a) although, as Fekete et al. (1984) have pointed out, AN collateral ramifications in this region are perpendicular to the descending parent branches and hence parallel to the octopus cell primary and secondary dendrites. Tertiary or distal dendrites are usually perpendicular to the primary and secondary dendrites and therefore parallel to the AN parent branches.

In Golgi preparations, the descending AN branches give rise to two distinct types of collateral in OCA. Thick collaterals give rise to large boutons on the soma and basal dendrites of octopus cells, while thin collaterals give rise to small terminal boutons which characteristically aggregate into closely packed clusters on the soma and proximal dendrites (Kane 1973). Four cytologically distinct types of terminal on octopus cells have been identified in EM studies. Types 1 and 2 degenerate after cochlear ablation (Kane 1973, 1974b), and on the basis of size, spatial distribution, and associated axon diameter have been identified with the large and small boutons impregnated in Golgi material. The large type 1 terminals contain large round vesicles and have a heavily asymmetric synaptic complex; they cover about 50% of the somatic surface and about 70% of the proximal dendritic surface of octopus cells. The small type 2 terminals contain smaller flattened vesicles, make symmetric synaptic contacts, and normally cover about 15% of both the somatic and the dendritic surface (Kane 1973). This massive investment of the octopus cell soma and proximal dendrites by AN terminals is confirmed by autoradiographic evidence. Injections of tritiated leucine into the cochlea result in dense labeling in OCA (Jones and Casseday 1979a; Kane 1977b), the octopus cells in some cases being so thickly encrusted with silver grains that the soma and proximal dendrites are clearly silhouetted (Jones and Casseday 1979a). Type 3 and 4 terminals survive cochlear ablation (Kane 1974b) and appear to be derived from descending axons from subdivisions of the ipsilateral and contralateral SOC, respectively (Kane 1976b, 1977a), although OCA also receives sparse projections from IC and NLL (Kane and Conlee 1979).

The differences between type 1 and 2 terminals in vesicle characteristics and postsynaptic densities have prompted Kane (1973) to propose that single AN axons can produce morphologically (and *functionally*) heterotypic synapses upon a single cell type. Although this hypothesis provides a convenient explanation of some of the physiological characteristics of octopus cells (see Sect. 4.2.5), it should be noted that the two types of terminal cannot be traced to single collaterals in thin EM sections, and that the proposal rests on indirect evidence, viz., the identification of the type 1 and 2 terminals with the large and small boutons that *can* be traced to a single AN collateral in Golgi material. It might be possible to obtain direct evidence on this issue by combined HRP histochemistry and EM (Carson and Mesulam 1982) of OCA after intracellular injection into a single AN fiber.

4.1.4 Dorsal Cochlear Nucleus

The structure of DCN varies considerably among species. In nonprimate mammals it is stratified and has many structural similarities with the cerebellar cortex (Lorente de Nó 1933b; Mugnaini et al. 1980a, b; Osen and Mugnaini 1981; Ramon y Cajal 1909), but in primates the stratification is less clear or absent (Moore 1980; Moore and Osen 1979; Zvorykin 1964). In the cat there are three distinct neural layers beneath a layer of ependymal cells forming the floor of the lateral recess of the fourth ventricle. The following account is based chiefly on descriptions of the laminae in Nissl-stained sections by Cohen et al. (1972) and Brawer et al. (1974), and on descriptions of cell types in Golgi material by Brawer et al. (1974) and by Kane (1974a). The outermost neural layer, the molecular layer (ML; see Fig. 14) contains small cells and granule cells embedded in a fine neuropil. The fusiform cell layer (FCL; see Fig. 14) is defined by the presence of staggered rows of fusiform cell bodies (Osen's 1969a pyramidal cell layer; see Fig. 15), and also contains small cells and a great concentration of granule cells. The polymorphic or deep layer (PL; see Fig. 14) contains giant cells, small cells, and a sparse but uniform distribution of granule cells in a thick feltwork of fibers. In Golgi material, Brawer et al. (1974) distinguished three types of small neuron with different laminar distributions, and three types of giant cell. The apical dendrites of the fusiform cells branch repeatedly and extend into ML, whereas the thick basal dendrites extend into PL; occasional dendrites bifurcate into an apical and a basal branch. Fusiform cells and their processes thus span the three neuronal layers of DCN. The basal dendritic arbors of fusiform cells are markedly flattened and are oriented in parallel in a plane orthogonal to the long "strial" axis of DCN (Blackstad et al. 1984; Lorente de Nó 1981; Osen 1983). The thin axons of granule cells form the parallel fibers which course within ML and outer FCL parallel to the surface. Deeply located granule cells send their axons to ML where they bifurcate at right angles and form parallel fibers analogous to those of the cerebellar cortex (Brawer et al. 1974; Mugnaini et al. 1980a).

As noted previously, the study by Fekete et al. (1984) of HRP-labeled Type I ganglion cells revealed that the descending branches of some fibers do not reach DCN. Those fibers that do reach DCN were described as diminishing in diameter after doing so and as giving off only a few fine collaterals which ramified and terminated in PL, only occasionally extending to FCL. This account is broadly consistent with the results of anterograde degeneration (Cohen et al. 1972; Osen 1970; Powell and Cowan 1962) and autoradiographic (Jones and Casseday 1979a; Kane 1977b) studies, which have generally reported a less dense projection from the cochlea to DCN than to VCN, and a gradient in density of terminations from deep to superficial layers. Although there is some disagreement on details (see Jones and Casseday 1979a for discussion), terminations are agreed to be most dense in PL and the region of the fusiform cell basal dendrites and to be almost absent from ML. These data are in accord with the results of current source density analysis of the field potential evoked in DCN by electrical stimulation of AN, which indicate that the component corresponding to the incoming AN volley is generated predominantly in the deep layers (Manis and Brownell 1983). After noise-induced degeneration restricted to OHCs, terminal degeneration is more prominent in DCN than in VCN, suggesting that Type II ganglion cells might project preferentially to DCN (Morest and Bohne 1983).

Golgi and EM observations on normal and cochlear-lesioned animals have clarified the primary innervation of DCN neurons and have revealed some aspects of the complex circuitry of this division (Kane 1974a, b; Mugnaini et al. 1980a, b; Smith and Rhode 1985). Kane (1974a) distinguished six classes of axon in rapid Golgi preparations. The major class of AN fibers (class 1 axons) arborize within PL and the deeper FCL to form collaterals that terminate in boutons on the fusiform cells and on small cells, granule cells, and giant cells (Kane 1974a). Although Kane (1974a) described AN collaterals as terminating on fusiform cell somas and dendritic trunks, Smith and Rhode (1985) have reported that AN terminals were found on *only* the basal dendrites of a small number of HRP-labeled fusiform cells examined by EM. The axons of small cells project in turn to fusiform cell bodies and/or dendritic shafts, providing an additional, indirect primary input to fusiform cells. Kane (1974a) also reported that some "mossy" endings in FCL and ML degenerated after cochlear ablation and concluded that they were derived from AN collaterals that extended into these layers. However, a primary origin for mossy endings in DCN has been disputed by Mugnaini et al. (1980b), and Kane (1977c) subsequently reported degeneration of some mossy terminals after lesions in higher auditory centers. As in VCN, the granule cell domain in DCN is characterized by an AChE-positive fiber plexus, and positively-stained mossy fibers in the rat degenerate after OCB section (Osen et al. 1984). The mossy endings in DCN, like those in AVCN, form the centers of complex synaptic fields resembling cerebellar glomeruli, in which they are surrounded by, and make synapses on, granule cell dendrites which also receive small terminal boutons on the periphery of the glomerulus (Kane 1974a; Mugnaini et al. 1980b; Osen and Mugnaini 1981). Mugnaini et al. (1980a) reported that granule cells from *all* parts of CN send their axons into the parallel fibers and make synaptic contacts on the apical dendrites of fusiform cells. This observation is supported by Jones and Casseday's (1979b) finding that injections of tritiated leucine into AVCN in the tree shrew resulted in dense labeling in ML. As in the cat, cochlear injections in the tree shrew resulted in dense labeling only in the deep layers, indicating a laminar segregation of at least some afferents in DCN. Neither the nature of the input to fusiform cell apical dendrites from granule cells nor the function of this cerebellar-like circuitry in DCN is known. Kane (1974a) also described a second form of complex synaptic array in FCL and ML, which she termed a synaptic nest; it comprises tightly packed neuronal profiles, including a variety of synaptic endings and distinct types of dendrites, and further differs from glomeruli in not being isolated by glial envelope.

Associated with the relatively sparse primary input to DCN and its complex internal circuitry is a more massive noncochlear input than to the other divisions (e.g., Kane 1976a; Kane and Conlee 1979; Osen 1970). At least three major classes of synaptic terminal in DCN do not degenerate after cochlear ablation (Kane 1974a, b). These endings are derived from intrinsic axons and from established projections to DCN from other CN divisions (Adams 1983a; Kane and Finn 1977; Lorente de Nó 1933b, 1981; Mugnaini et al. 1980a; Oliver et al. 1983; Warr 1969; 1972), the contralateral CN (Cant and Gaston 1982), and auditory brainstem nuclei (e.g., Adams and Warr 1976; Elverland 1977; Kane 1976a, 1977c; Kane and Conlee 1979; Kane and Finn 1977; Oliver et al. 1983; Rasmussen 1967).

4.1.5 Efferent Connections: Preliminary Description

CN neurons that project centrally send their axons in three main striae: the trapezoid body (TB), the intermediate acoustic stria (stria of Held; IAS), and the dorsal acoustic stria (stria of Monakow; DAS). An important observation in recent studies utilizing intracellular HRP labeling, which allows the axons of identified neurons to be traced over long distances, is that almost all large CN neurons projecting through these striae – other than bushy cells – have axon collaterals terminating in their own and/or other subdivisions of CN (Rhode et al. 1983a, b; mouse: Wu and Oertel 1984; see also Adams 1983a; Tolbert et al. 1982 for related observations).

In broad terms, the projections of the three major CN divisions through these striae are almost mutually exclusive, although all project to the ventral nucleus of the lateral lemniscus (VNLL) and to IC (Adams 1979; Brunso-Bechtold et al. 1981; Fernandez and Karapas 1967; Glendenning and Masterton 1983; van Noort 1969; Oliver 1984b; Osen 1972; Roth et al. 1978; Warr 1966, 1969, 1972, 1982). AVCN projects via the dorsal and ventral components of TB to the main nuclei of the superior olivary complex (SOC) (Warr 1966, 1972, 1982), whereas PVCN projects through the caudal portion of TB and through IAS to the periolivary nuclei of SOC, but not to the main nuclei (van Noort 1969; Warr 1969, 1972, 1982). The fusiform and giant cells of DCN project via DAS (Adams and Warr 1976), primarily to IC (Adams 1979; Brunso-Bechtold et al. 1981; van Noort 1969; Oliver 1984b; Osen 1972; Roth et al. 1978). Within this broad framework, the subdivisions of CN defined by cell type and primary innervation have largely independent projection patterns. These differences, together with the functional differences indicated by the physiological data to be reviewed in the following sections, support the view that CN gives rise to parallel pathways to higher brainstem auditory centers that are, to some extent at least, functionally distinct. The detailed projections of the various subdivisions of CN to the nuclei and subdivisions of SOC, NLL, and IC cannot be adequately described until the parcellation of these target structures has been considered and will therefore be deferred to later sections (see Sects. 5.1.2 – 5.1.5, 6.1 and 7.1.3). Ascending projections of CN to these structures are summarized schematically in Figs. 26, 34, and 37.

4.2 Physiology of Cochlear Nucleus Neurons

4.2.1 Preliminary Observations

The diversity of morphological cell types with different patterns of primary innervation in CN suggests that the auditory input provided by the AN is transformed in different ways in different subdivisions of the nucleus. One manifestation of these transformations is that considerable diversity in temporal patterns of response and in response characteristics is observed in CN, in contrast to the relatively homogeneous response characteristics of AN fibers. The different response patterns and the basic physiological characteristics of the cells exhibiting them are discussed in the following sections. Evidence establishing correlations between morphologically and physiologically defined cell types is critically evaluated in Sect. 4.2.5.

In many of the earlier studies of response properties of CN neurons, no evidence was presented as to the locations of the cells from which recordings were made. Such data are of limited value, and these studies will be considered only where their results can be interpreted despite this deficiency. Although a comprehensive account of CN physiology is possible only for the cat, some details have been confirmed by studies in other mammals, e.g., the rat (see Møller 1983 for review), kangaroo rat (e.g., Caspary 1972; Rupert and Moushegian 1970), rabbit (e.g., Hui and Disterhoft 1980; Perry and Webster 1981), and various bats (e.g., Feng and Vater 1985; Neuweiler and Vater 1977; Vater 1982). In accordance with the restriction specified in Sect. 1, attention will be focused on the cat and on properties relevant to the coding of spectral and spatial information. More comprehensive reviews can be found elsewhere (Aitkin et al. 1984b; Evans 1975).

4.2.2 Frequency Organization

As noted previously, the microelectrode studies of Rose et al. (1959) and of later researchers (e.g, Evans and Nelson 1973a; Goldberg and Brownell 1973) confirmed a dorsal-to-ventral, high-to-low frequency organization in each of the three major divisions of CN, in accordance with the pattern of primary innervation of the nucleus (Sect. 4.1.1). Functional mapping studies using the 2-deoxyglucose (2-DG) autoradiographic technique have indicated frequency organization in CN that is generally in accord with that revealed electrophysiologically

(Nudo and Masterton 1984; Webster et al. 1978; gerbil: Ryan et al. 1982; see Sect. 7.2.1 for more detailed account of 2-DG technique and data).

The terminals of AN fibers innervating a restricted region of the basilar membrane form a layer or "plane" in each division of CN in the cat (Lorente de Nó 1979, 1981; van Noort 1969; Osen 1970; Sando 1965), suggesting that a given frequency, or a *band* of frequencies, must be represented by a "plane" (sheet or slab) of neurons. Most microelectrode studies, however, have yielded insufficient data to reveal the three-dimensional frequency organization of CN, and a map of this sort has so far been obtained only for AVCN. Bourk et al. (1981) mapped the CFs of histologically localized units from many animals onto a computerized (and updated) version of the block model of CN proposed by Kiang et al. (1975), and estimated surfaces of constant CF with the aid of computer graphics. In the anterior division of AVCN (subdivisions AA, AP, and APD), the surfaces of constant CF were approximately planar and parallel, and were tilted at about 45° from lateral to medial and from rostral to caudal. This organization agrees qualitatively with that indicated by the anatomical studies of Sando (1965) and van Noort (1969). In the posterior division (PD and PV), however, the surfaces of constant frequency progressively deviated from this simple description, converging ventrally. In PV, in particular, abrupt shifts in CF sequence and large deviations from the CF sequence predictable from the general map occurred. The complexity of the frequency organization in PV may be related to the observations that AN fibers in this region still follow a spiral course and that some neurons are innervated by pre-bifurcation collaterals of high-CF fibers that terminate separately from the ascending and descending branches of the parent axon (Bourk et al. 1981; Fekete et al. 1984).

The data of Bourk et al. also provide evidence on the frequency ranges represented in the morphologically defined subdivisions of AVCN. Osen (1970) noted that the ascending branches of high-frequency AN fibers did not reach the rostral pole of AVCN (as indicated by the vertical line in Fig. 13B) and therefore predicted that the area of large spherical cells would receive only low- and medium-frequency input. This prediction was confirmed by Goldberg and Brownell (1973) and is strongly supported by the data of Bourk et al. The distributions of CF in the three subdivisions of the anterior division of AVCN are shown in Fig. 18. In AA, which corresponds to Osen's large spherical cell area (see Sect. 4.1.2), the representation is largely restricted to frequencies below about 10 kHz. In contrast, APD contains mainly high-frequency (>10 kHz) units, and the CF distribution in AP is relatively flat from 2.5 to 30 kHz, but has virtually no representation of CFs below 1 kHz. Detailed CF distributions could not be obtained for the posterior division of AVCN because of the complexity of the frequency organization in this region, but both PD and PV appear to contain uniform CF distributions (Bourk et al. 1981). No adequate data are available on the three-dimensional frequency organization of PVCN or DCN in the cat, or on the CF distributions in particular morphological subdivisions. In the rufous horseshoe bat, however, Feng and Vater (1985) have used combined physiological and anatomical methods to describe the slab-like frequency organization of all three CN divisions and the topographically organized connections between them.

A final, more general, point to be noted here concerns the nature of the representation of the receptor surface in CN and in higher subcortical structures. The

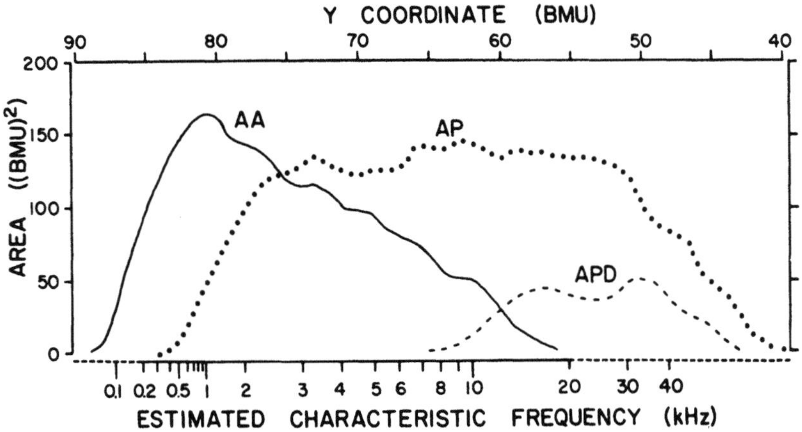

Fig. 18. Distributions of characteristic frequency (CF) in subdivisions of anterior division of anteroventral cochlear nucleus in cat (parcellation of Brawer et al. 1974). Note that the distributions are in terms of volumes – in units derived from the block model of Kiang et al. (1975), i.e., block model units or BMUs – and therefore take no account of differences in cell density in the subdivisions. The estimated CF scale at the bottom of the figure was derived from the *y*-coordinate of the block model coordinate system (*upper scale*), and is uniformly logarithmic above and below 1.7 kHz. Reproduced from Bourk et al. (1981), which should be consulted for detailed account of derivation of distributions

region of the cochlea activated by a particular frequency at low stimulus levels is essentially a point on the single line of IHCs. This point is represented in each division of CN by a sheet of neurons. This point-to-sheet representation involves a higher-order transformation than the central representations of the visual and somatosensory receptor surfaces (Kaas 1977; Merzenich et al. 1977), and raises the question of what features of the stimulus are represented across the two dimensions of each isofrequency plane. It should also be noted that despite the common use of the term "isofrequency plane" it seems likely that the representation should be thought of in terms of frequency-*band* planes or sheets. This issue will be discussed in the context of the more detailed data on midbrain frequency organization in Sect. 7.2.1.

4.2.3 Response Patterns and the Classification of Cochlear Nucleus Neurons

A number of schemes have been proposed for the classification of CN neurons in terms of response characteristics. The most influential has been that proposed by Pfeiffer and his associates (Kiang et al. 1965b; Pfeiffer 1966a) on the basis of the shape of the PSTH generated in response to short (usually 25 ms) tone bursts at the cell's CF (Fig. 19). Pfeiffer's (1966a) initial classification of response patterns comprised four classes, which he designated primarylike, chopper, onset, and pause. These terms have subsequently been used to describe the units exhibiting particular patterns, although confusion can arise from the use of the same terms to specify response patterns and unit types. The convention proposed by Godfrey et al. (1975a), according to which the terms are enclosed in quotation marks

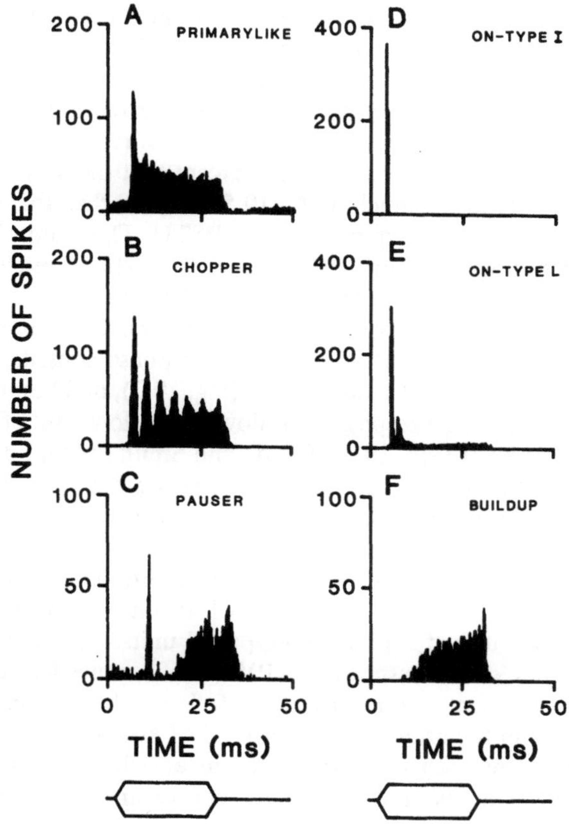

Fig. 19 A – F. Representative peristimulus time histograms illustrating major response patterns identified in the expanded version of Pfeiffer's (1966a) classification scheme for cochlear nucleus neurons. Response in each case is to tone burst of 25 ms duration with envelope as indicated beneath abscissa of lower histograms

when used to label unit types and without quotation marks when applied to response patterns, will therefore be adopted here. A further difficulty is that increased knowledge of the response characteristics of some unit types means that they have come to be defined in terms of a constellation of properties, of which response pattern is only one (cf. Godfrey et al. 1975a; Rhode and Smith 1986a, b). It is therefore sometimes important to distinguish, for example, a "primarylike" unit from a unit that has a primarylike pattern, but does not exhibit the other general characteristics of the class. These complications provide a vivid illustration of the difficulties in categorizing neurons and defining neuron types that were discussed in Sect. 4.1.1.

The primarylike pattern, as the name implies, resembles that of AN fibers, in which a peak at onset is followed by a gradual decline to a steady state level (Fig. 19A). The chopper pattern is characterized by distinct, regularly spaced PSTH peaks (reflecting regular, time-locked firing), the interval between which is not related to stimulus frequency (Fig. 19B). The pause pattern comprises an initial peak, followed by a pause and a resumption of activity for the duration of the stimulus (Fig. 19C); units exhibiting this pattern are commonly referred to as "pausers" and the pattern as pauser. Finally, the onset pattern – now more commonly designated on – is characterized by a peak or peaks at stimulus onset with little or no activity thereafter (Figs. 19D and E). The primarylike, on, chopper, and pauser response patterns have also been described in CN of unanesthetized decerebrate (Ritz and Brownell 1982) and awake restrained (Webster 1977; Rhode and Kettner 1986) cats.

Subsequent studies have necessitated extension and modification of these categories to deal with more subtle features of response patterns. Thus, some units have a primarylike response pattern except that the initial peak is followed by a pause in activity that is much briefer (0.5 – 2.0 ms) than that defining the pauser pattern and appears as a sharp notch in the PSTH. These units have been termed "primarylike with notch" (Bourk 1976; Kiang 1975; Rhode and Smith 1986a). Several classes of "on" unit have been distinguished in PVCN (Godfrey et al. 1975a) and AVCN (Bourk 1976; Rhode and Smith 1986a). Godfrey et al. (1975a) distinguished between "on-type I" units, whose response consists solely of a response (usually a single spike) to stimulus onset (Fig. 19D), and "on-type L" units, in which a clearly defined onset response is followed by a low level of activity for the duration of the tone burst (Fig. 19E). Rhode and Smith (1986a) distinguished a further group of "on" units in which the onset response has a multipeaked chopper pattern. These units also exhibit later sustained activity and were presumably included in the "on-type L" category by Godfrey et al. (1975a). However, cells of this type comprised approximately 60% of Rhode and Smith's (1986a) large sample of VCN "on" units, and have a number of distinctive properties. They will therefore be distinguished here as "on-type C" units. Ritz and Brownell (1982) described two types of on response in PVCN of unanesthetized decerebrate cats, the onset component in one case being followed by a suppression of spontaneous activity and in the other by a lower level of excitation. The equivalence of these response patterns with those described in anesthetized preparations is unclear, however, since in two units observed before and after an injection of barbiturate, on responses followed by suppression of spontaneous activity were converted into on responses with later discharge (on-type L). Three classes of "chopper" unit have been distinguished on the basis of the duration of the chopping pattern (sustained or transient) or of a longer interval between peaks (Bourk 1976; Rhode and Smith 1986a; Romand 1979). All of the unit types thus far described show secure phase-locking at frequencies below approximately 1 kHz, and units with CF in this frequency range are therefore difficult to classify on the basis of PSTH pattern alone (Bourk 1976; Godfrey et al. 1975a; Kiang et al. 1973; Rhode and Smith 1986a). One solution to this problem is to assign all low-CF phase-locked units to a special "phase-locked" category (Bourk 1976). However, Rhode and Smith (1986a) have reported that low-CF units of the various types can be distinguished on the basis of their ISIHs and synchronization coefficient values, and have proposed that a low-frequency subcategory of each type be distinguished. Finally, Rose et al. (1959) had reported that some DCN units had a long latency and a response that increased (or built up) in the course of the stimulus; these neurons have subsequently been termed "buildup" units (Fig. 19F). Neurons exhibiting these various response patterns are not homogeneously distributed throughout CN, and in some cases distributional statistics suggest a correlation between response pattern and morphological cell type. The distributional and other evidence bearing on these correlations is considered in Sect. 4.2.5.

The basis of classification in Pfeiffer's scheme is restricted to a single feature of the cell's response area, viz., the *excitatory* response pattern exhibited by the cell at its CF. A good deal of early evidence (e.g., Gerstein et al. 1968; Greenwood and Maruyama 1965; Moushegian et al. 1962; Rose et al. 1959) had indicated that many CN neurons, particularly those in DCN, exhibited suppression of spontaneous activity at frequencies other than CF, or at CF itself at some levels. Evans and Nelson (1973a, b) emphasized this characteristic in proposing a classification based on the extent to which inhibitory[12] effects are produced by single tones in the cell's response area. They distinguished five types ranging from cells exhibiting sustained excitation only (type 1) to those exhibiting a wholly inhibitory response at threshold (type 5). Their studies also indicated major differences between VCN and DCN, both in the proportions of the different cell types and in the effects on responses of barbiturate anesthesia. The major response types in VCN of pentobarbital-anesthetized cats were type 1 and type 2 (the latter defined by an excitatory response followed by "off inhibition" at tone offset), and there was only a slight increase in the proportion of cells showing inhibitory responses in unanesthetized or chloralose-anesthetized animals. Other studies have also reported response patterns in VCN of unanesthetized cats that are similar to those in anesthetized animals (Rhode and Kettner 1986; Ritz and Brownell 1982; Webster 1977). Evans and Nelson (1973a, b) described the majority of cells in DCN of barbiturate-anesthetized animals as type 2 and type 3 (defined by the presence of "sideband inhibition" at frequencies above and/or below the excitatory frequency band). In unanesthetized or chloralose-anesthetized preparations, however, inhibitory effects were far more marked: the majority of cells (approximately 50%) were of type 5, the remainder being mainly type 3 or type 4 (predominantly inhibitory; see Fig. 20). The predominance of inhibitory input and the effects of barbiturate anesthesia in DCN have been confirmed by others (van Gisbergen et al. 1975a; Rhode and Kettner 1986; Voigt and Young 1980; Young and Brownell 1976; Young and Voigt 1982). However, Young and Brownell (1976) and Rhode and Kettner (1986) reported that they encountered very few type 5 cells in DCN of unanesthetized cats, and Rhode and Smith (1986b) reported a significant proportion of type 4 cells in anesthetized cats. These data therefore suggest that the reduction in inhibition in DCN produced by anesthesia, although undoubtedly marked, might be less profound than was suggested by Evans and Nelson's data (see Rhode and Kettner 1986 and Rhode and Smith 1986b for discussion).

A difficulty with Evans and Nelson's classification scheme is that detection of inhibitory responses to pure tones in extracellular recordings depends on the presence of spontaneous activity. The distinction between type 2 and 3 cells (viz., the presence of inhibitory sidebands in the response areas of the type 3 cells) might therefore merely reflect the latter's spontaneous activity and not be a basic difference between cell types. For this reason, Young and Brownell (1976) proposed a

[12] When presentation of a test stimulus results in suppression of spontaneous activity or of the excitatory response to another stimulus, the response to the test stimulus is commonly described as inhibitory, although the processes underlying the effect of the stimulus have not in fact been established as involving synaptic inhibition. This terminology is inexact, but it is so well established in the auditory system literature that systematically to avoid it would be impracticable. The term "inhibition" will therefore be used in this and subsequent sections to refer to stimulus-evoked suppression of activity (whether spontaneous or driven) without implying anything about the nature of the underlying mechanisms.

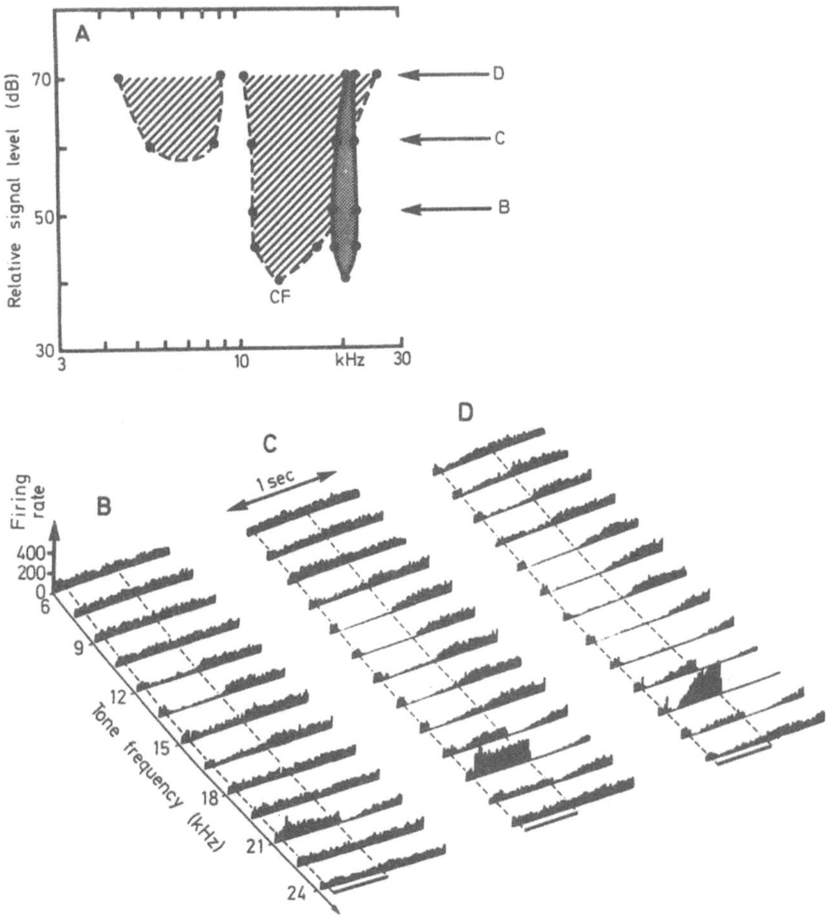

Fig. 20 A – D. Frequency response **A** and temporal pattern of response as a function of frequency and stimulus level **B**, **C**, and **D** of a cell in dorsal cochlear nucleus of cat (chloralose anesthesia) classified as type 4 (predominantly inhibitory) by Evans and Nelson (1973 a, b) **A** frequency threshold curves of excitatory region (*full line* and *stippled area*) and inhibitory region (*broken lines* and *hatched areas*). Threshold is expressed in terms of relative signal level to the transducer, and values given represent dB sound pressure level (SPL) at the unit's characteristic frequency (CF); SPLs at other frequencies were within 5 dB of these values, **B – D** three-dimensional arrays of peristimulus time histograms at frequencies 6 – 24 kHz and at levels 10, 20, and 30 dB above CF threshold (**B**, **C**, and **D**, respectively – see *arrows* indicating levels in **A**). Tone burst duration (0.4 s) is indicated by *bar* and *broken lines*. Ordinate is firing rate in spikes per second, and each histogram represents average of ten tone presentations. Note extensive bandwidth of "on" and "off" inhibition, particularly at 30 dB above threshold (in **D**). (Evans 1975, based on earlier figures of Evans and Nelson 1973a)

modification of Evans and Nelson's system, in which cells that have a V-shaped excitatory region and are excited by CF tones at all suprathreshold levels are designated type II/III, regardless of whether or not spontaneous activity allows the detection of inhibitory sidebands. This proposal is supported by two observations. First, intracellular recordings from some cells without spontaneous activity reveal hyperpolarization in response to frequencies adjacent to the excitatory region (Starr and Britt 1970). Second, some type II/III cells without spontaneous activity are unresponsive to broad-band stimuli, suggesting the presence of strong surround inhibition (Young and Brownell 1976). Young and Voigt (1982) have subsequently emphasized differences between type II and type III cells, but these differences do not include the presence or absence of sideband inhibition. The second class of DCN units distinguished by Young and Brownell (1976) and designated type IV are excited by CF tones near threshold, but are always inhibited by CF stimulation at higher levels and by stimulation at other frequencies, and therefore correspond approximately to Evans and Nelson's type 4. Young and Brownell studied a small number of type IV cells in decerebrate cats before and after administration of barbiturate anesthesia. After pentobarbital administration, all of these cells showed pauser or buildup patterns at CF at all levels, suggesting that type IV cells can be tentatively identified with the unit class or classes defined by these response patterns in anesthetized animals. However, units with type IV response areas and an onset response in the excitatory region have also been described in DCN of barbiturate-anesthetized cats (Rhode and Smith 1986b). Type II/III cells had response patterns within their excitatory area that could be classified as either primarylike, pauser, or what Goldberg and Brownell (1973) termed steady. Because the DCN cell types identified by Young and Brownell can thus be tentatively correlated with those defined in terms of excitatory response patterns in anesthetized animals, and because subsequent studies have established correlations of their cell types with morphologically defined cell types having distinct projection patterns (Young 1980), their terminology will be used here.

It seems clear that some form of combination of the two classes of classification scheme that have been described is necessary to give an adequate account of the responses of CN neurons to pure tones. The extended version of Pfeiffer's scheme appears to provide an adequate basis for description of a cell's response pattern within excitatory regions of its response area, although variation in the response patterns of some cells across their response area, and the occurrence of borderline cases, ultimately limit the value of the system (see Sect. 4.2.5). Description of the cell's excitatory responses must be combined with an account of the presence and distribution of inhibition within its response area. Van Gisbergen et al. (1975a) proposed a composite classification scheme of this sort, but it used a restricted set of response pattern categories and has not achieved general acceptance.

4.2.4 Spatial Distribution and Basic Response Characteristics of Physiologically Defined Cell Types

The diverse response patterns exhibited by CN neurons are a manifestation of the transformations imposed on AN input by the different patterns of synaptic innervation of neurons of different morphology. It is therefore not surprising that the cell types defined in terms of response patterns also differ in their responses to basic parameters of the acoustic stimulus and in their spatial distribution. These characteristics of the major response types will be considered in this section as a prelude to examination of attempts to correlate physiologically and morphologically defined cell types.

"Primarylike" units are encountered most frequently in AVCN, although cells with a primarylike response pattern are also encountered in DCN (Goldberg and Brownell 1973; Rhode et al. 1983b; Rhode and Smith 1986b; Young and Brownell 1976) and occasionally in PVCN (Godfrey et al. 1975a; Ritz and Brownell 1982). "Primarylike" cells are particularly heavily concentrated in AA (Osen's large spherical cell area; Bourk 1976; van Gisbergen et al. 1975a; Goldberg and Brownell 1973), whereas "primarylike with notch" cells are more common in posterior AVCN, particularly PV (Bourk 1976; Godfrey et al. 1975a; Kiang 1975). Many "primarylike" cells in AVCN have response characteristics that are indistinguishable from those of AN fibers. Thus, they have narrow FTCs without inhibitory sidebands (Bourk 1976; van Gisbergen et al. 1975a; Goldberg and Brownell 1973), and have the same response pattern throughout their response area (Britt and Starr 1976a). Rate–intensity functions are monotonic and sigmoid, saturating at levels 20–30 dB above threshold (Bourk 1976; Goldberg and Brownell 1973). Like AN fibers, the discharge of "primarylike" cells is irregular and ISIHs for spontaneous and continuously driven activity are, in the majority of cases, asymmetric with exponential or faster-than-exponential decay[13] (Bourk 1976; van Gisbergen et al. 1975a; Goldberg and Brownell 1973). Most "primarylike" cells exhibit phase-locking that is as precise (as indicated by the synchronization measures discussed in Sect. 3.2) as that in AN fibers, and extends to frequencies in excess of 3 kHz (Bourk 1976, 1977; van Gisbergen et al. 1975b; Goldberg and Brownell 1973; Rose et al. 1974). The faithfulness with which these "primarylike" neurons preserve the phase information provided by AN fibers has allowed their use as a means of investigating cochlear nonlinearities (e.g., Greenwood et al. 1976; Rose et al. 1974; Smoorenburg et al. 1976). Not all AVCN neurons are sensitive to phase, however, and sensitive and insensitive cells can be intermingled along the same electrode track (Rose et al. 1974). Some cells with primarylike responses have also been reported to be non-phase-locking or to phase-lock only weakly (Britt 1976). However, Rose et al. (1974) did not report the response patterns of their cells, and Britt (1976) did not specify the histological locations of units. It is therefore unclear whether any significant proportion of "primarylike" units in AVCN fail to exhibit secure phase-locking. "Primarylike with notch" units have response characteristics very similar to those of "primarylike" units and AN fibers, except that their spontaneous discharge rate is generally much higher (Rhode and Smith 1986a).

A striking feature of the extracellularly recorded activity of many cells in AVCN with a primarylike response pattern is that each action potential is preceded (by 0.4–0.6 ms) by a positive-going "prepotential" (Fig. 21A and B; Goldberg and

[13] Bourk (1976) reprocessed the AN data of Walsh et al. (1972) and showed that the proportions of AN fibers whose ISIHs showed exponential and faster-than-exponential decay were the same as the proportions he obtained for "primarylike" units in AVCN.

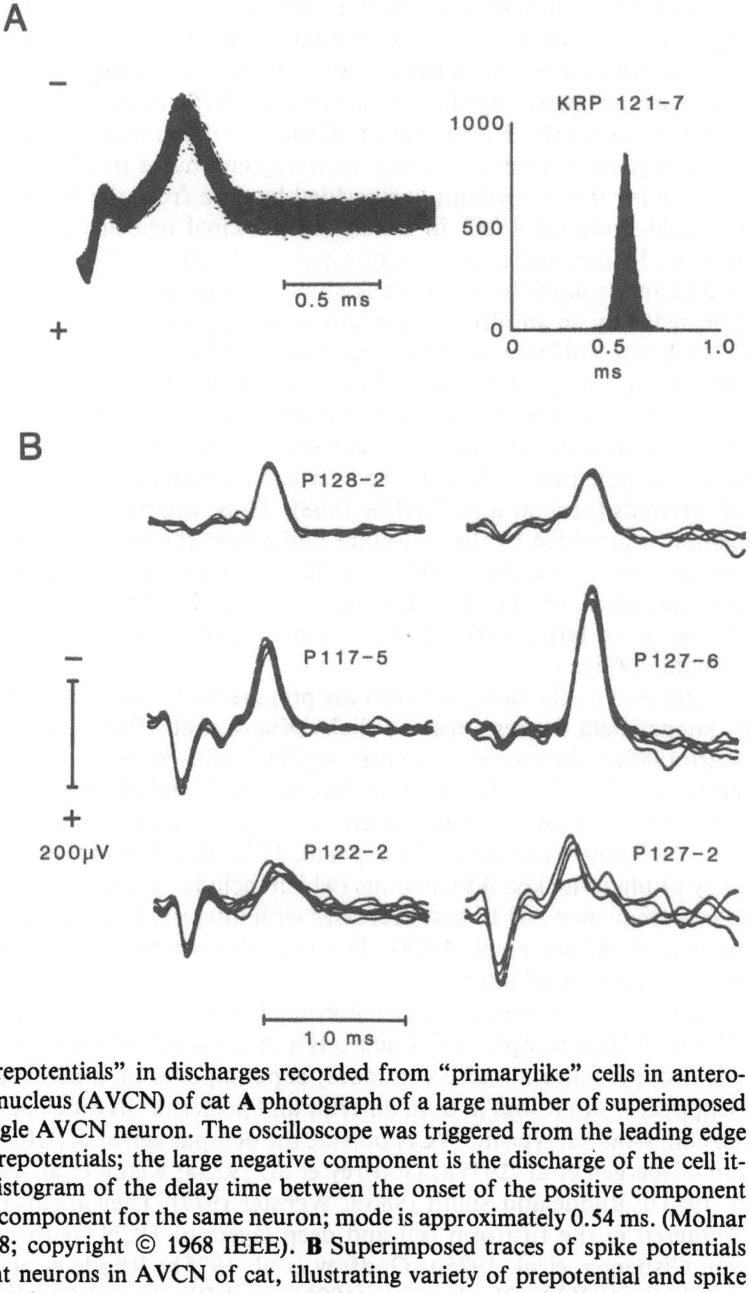

Fig. 21 A, B. "Prepotentials" in discharges recorded from "primarylike" cells in anteroventral cochlear nucleus (AVCN) of cat **A** photograph of a large number of superimposed traces from a single AVCN neuron. The oscilloscope was triggered from the leading edge of the positive prepotentials; the large negative component is the discharge of the cell itself. On right: histogram of the delay time between the onset of the positive component and the negative component for the same neuron; mode is approximately 0.54 ms. (Molnar and Pfeiffer 1968; copyright © 1968 IEEE). **B** Superimposed traces of spike potentials from six different neurons in AVCN of cat, illustrating variety of prepotential and spike waveforms obtained. In each case, traces were synchronized to an arbitrary point on the waveform. (Pfeiffer 1966b; copyright 1966 by the AAAS)

Brownell 1973; Molnar and Pfeiffer 1968; Pfeiffer 1966b). Whereas "primarylike" units have unequivocal prepotentials of the type illustrated in Fig. 21, "primarylike with notch" units have smaller and less secure prepotentials that can only be discerned in averaged responses (Bourk 1976). A number of lines of circumstantial evidence (Pfeiffer 1966b) indicate that the prepotential reflects discharge of a presynaptic terminal in cells receiving end-bulbs of Held. The most direct evidence for this association is provided by data from the medial nucleus of the trapezoid body (MNTB), in which the principal neurons receive true calyceal endings. In this nucleus, the action potential, but not the prepotential, can be elicited antidromically and is blocked when an orthodromically elicited response is preceded by an antidromic response at an appropriate interval (Li and Guinan 1971). Bourk (1976) reported having made similar observations in AVCN. Examination of trains of action potentials in those CN neurons with prepotentials large enough to be distinguished in single traces (e.g., Fig. 21B) indicates that there is perfect association between the occurrence of the prepotential and the postsynaptic action potential, unless the neuron has discharged within the preceding few milliseconds (Molnar and Pfeiffer 1968). Thus, extremely secure synaptic transmission is provided by the end-bulbs and allows the faithful preservation of AN fiber properties. Further evidence of this precision is provided by intracellular recordings, which reveal that the response to sound of "primarylike" units consists of a series of large, very rapid (1–2 ms) depolarizations (Rhode et al. 1983b; Romand 1978).

As indicated by the delay between the prepotentials and postsynaptic spikes, and by latency data for responses to clicks (Kiang et al. 1973) and tones (Rhode and Smith 1986a), the latencies of "primarylike" units exceed those of AN fibers with the same CF by 0.5–1.0 ms. For "primarylike" and all other CN units, latency decreases as intensity is increased, usually reaching asymptotic values some 20–30 dB above threshold (Kitzes et al. 1978; Rhode and Smith 1986a). The latency of phase-locked AVCN units (which include "primarylike" units), like that of low-frequency AN fibers, decreases with intensity in steps equal to the stimulus period (Kitzes et al. 1978). For non-phase-locked units the decrease is a smooth function of intensity.

"Chopper" units were the most prevalent neuron type in Rhode and Smith's (1986a, b) large sample of CN neurons and are widely distributed in the nucleus. The chopper pattern is also commonly superimposed on other response patterns. "Choppers" are found in both anterior and posterior divisions of AVCN and are the dominant cell type in PVCN outside the octopus cell area (OCA) (Bourk 1976; van Gisbergen et al. 1975a; Godfrey et al. 1975a; Kiang et al. 1965b; Rhode et al. 1983b; Rhode and Smith 1986a; Webster 1977). They are also commonly encountered in the fusiform cell and deep layers of DCN, particularly the latter (van Gisbergen et al. 1975a; Godfrey et al. 1975b; Rhode et al. 1983a; Rhode and Smith 1986b). Rhode et al. (1983a) and Rhode and Smith (1986b) found that many fusiform cells in DCN showed a sustained chopper pattern after microelectrode impalement, apparently as a consequence of membrane depolarization. Cells with a chopper response pattern in DCN are generally characterized by wider modes in the PSTHs, and their other response properties resemble those of "pauser" and "buildup" units rather than those of VCN "choppers" (Rhode and Smith 1986b). The following account applies only to VCN "choppers."

"Choppers" in VCN exhibit sharp frequency tuning, but – unlike "primarylike" units – commonly have inhibitory sidebands (van Gisbergen et al. 1975a; Godfrey et al. 1975a; Rhode and Smith 1986a). At frequencies below about 800 Hz, the phase-locking of "chopper" units is comparable to that of AN fibers and "primarylike" units, but at higher frequencies it is much poorer (as reflected in lower vector strength values; Bourk 1976, 1977; Rhode and Smith 1986a). Most "choppers" have a limited dynamic range of 20–30 dB, like that of AN fibers (Bourk 1976; Godfrey et al. 1975a; Rhode et al. 1983b; Rhode and Smith 1986a), although Bourk (1976) reported that for many AVCN "choppers" the changes in discharge rate over this limited dynamic range were much greater (i.e., the rate–intensity functions had steeper slopes) than those of AN fibers or other CN unit categories. The response latencies of "chopper" units in both AVCN and PVCN to clicks (Bourk 1976; Godfrey et al. 1975a; Kiang et al. 1973) and to tones (Rhode and Smith 1986a) are longer and vary over a wider range than those of all other VCN cell types.

Both the spontaneous and driven activity of "chopper" units is more regular than that of "primarylike" units, and ISIHs are either asymmetric with slower-than-exponential decay or are relatively symmetric (van Gisbergen et al. 1975c; Pfeiffer 1966a; Rhode et al. 1983b; Webster 1977). The symmetric ISIHs reflect the more regular firing of those units in which the chopper pattern is sustained for the duration of the stimulus. Molnar and Pfeiffer (1968) proposed a Markov process model that accounts for regular discharge in some CN neurons in terms of temporal integration of a number of irregularly firing inputs, each of which produces a depolarization which is small relative to that required to discharge the neuron. Subsequently, van Gisbergen et al. (1975c) obtained evidence that temporal integration was an important mechanism in generating the regular discharge of "chopper" cells, and suggested that this mechanism might be responsible for the deterioration of phase-locking in "choppers." Intracellular recordings from "chopper" cells show sustained depolarization to tone-burst stimuli (Romand 1978, 1979), which is compatible with temporal integration of a large number of small inputs. However, recent intracellular recordings in slice preparations of AVCN of the mouse have revealed one class of cells that respond to intracellularly injected current pulses with regularly firing action potentials (Oertel 1983; Wu and Oertel 1984). These observations suggest that the regular discharge of "choppers" (and of other regularly firing cells in CN) can arise from the intrinsic membrane properties of the cells.

"On" units are concentrated in the OCA, but they are also encountered elsewhere in PVCN, in AVCN, and in the deep layers of DCN (Bourk 1976; Godfrey et al. 1975a, b; Pfeiffer 1966a; Rhode and Smith 1986a, b). The properties of DCN "on" units differ from those of "on" units in VCN, and the following account is restricted to the latter. As noted previously, "on" units can respond synchronously to individual cycles of low-frequency tone bursts, i.e., exhibit phase-locking. They consequently have sustained discharge patterns at low frequencies, a fact that contributes to the difficulties in classifying the response patterns of low-CF units (Bourk 1976; Godfrey et al. 1975a; Kiang et al. 1973; Rhode and Smith 1986a). Phase-locking in "on" units does not extend to such high frequencies as it does in "primarylike" units, but at low frequencies (up to approximately 1 kHz) is equally precise (Bourk 1976, 1977; Rhode and Smith 1986a). Indeed,

Rhode and Smith (1986a) reported that the synchronization coefficients of low-frequency "on" units and of "on-type I" units whose response areas included frequencies below 1 kHz were often higher than those of AN fibers, reflecting the fact that these units discharged a single spike per stimulus cycle at a very precise phase. Some "on" units respond to *every* cycle at frequencies up to 1 kHz, and their ISIHs have a single narrow peak at the stimulus period. They thus achieve maximum discharge rates considerably higher than those of AN fibers or "primarylike" units. Both secure phase-locking and entrainment have also been reported in "on" units in unanesthetized cats (Rhode and Kettner 1986). Møller (1969a) reported similar entrainment for "on" units in CN of the rat, which responded in a synchronized fashion to repetitive transient stimuli (clicks or short tone bursts) up to rates as high as 700 stimuli per second. He found that individual units differed in their "repetition rate selectivity," the maximum rate at which each stimulus elicited a response, over a range of 200 – 800 clicks per second. Godfrey et al. (1975a) observed similar rate selectivity in PVCN "on" units in the cat. Some "on" units thus preserve periodicity information both for pure tones and for repetitive transient stimuli and, as Møller (1969a, 1970) suggested, could contribute to the mechanisms underlying discrimination of periodicity.

Although some "on" cells have sharp frequency tuning (Bourk 1976; Rhode et al. 1983b; Rhode and Smith 1986a), many PVCN "on" units have considerably broader FTCs than AN fibers, but have a sharp tip at CF (Godfrey et al. 1975a). Rhode and Smith's (1986a) data indicate that broad frequency tuning is characteristic only of "on-type C" units (i.e., those with a multipeaked chopper pattern), which comprised the largest subgroup (60%) of their sample of "on" units. "On-type I" units (those responding with a single spike to tone onset) have a very limited dynamic range of 10 – 20 dB (Godfrey et al. 1975a; Rhode and Smith 1986a), but Rhode and Smith (1986a) reported that "on-type C" units commonly have monotonic rate – intensity functions with dynamic range as large as 90 dB. At higher levels the increase in response occurs in the sustained activity following the multipeaked onset response. Godfrey et al. (1975a) reported that some of their "on-type L" units in PVCN had much larger dynamic ranges that AN fibers, and noted that in all such units for which the response pattern was examined ($N = 15$) it assumed a chopper pattern at high levels. These units presumably correspond to those in Rhode and Smith's (1986a) "on-type C" category. The broad frequency tuning and extended dynamic range of "on-type C" units are suggestive of convergence from AN fibers of different CF and dynamic range. Rhode and Kettner (1986) have reported that "on-type C" units in unanesthetized cats also show these characteristics.

Response latencies of "on" units exceed those of AN fibers by only 0.4 – 0.7 ms and show a similar dependence on frequency as do those of AN fibers (Rhode and Smith 1986a). A distinctive feature of "on" units in VCN is the precise timing of the onset spike. Rhode and Smith (1986a) reported that the mean standard deviations of first-spike latency for "on-type C" and "on-type I" units were 0.09 and 0.12 ms, respectively, values much lower than that for AN fibers (0.73 ms) or those of other CN unit types. This precise timing is also seen in awake cats (Rhode and Kettner 1986), and is reflected in the high synchronization coefficients of "on" units.

"On" units generally have no spontaneous activity in anesthetized preparations (van Gisbergen et al. 1975a; Kiang et al. 1965b; Pfeiffer 1966a), so that extracellular recordings can give no evidence of inhibitory input. In unanesthetized decerebrate cats, however, Ritz and Brownell (1982) found that most "on" units exhibited spontaneous activity, and one response type was characterized by sustained suppression of spontaneous activity following the onset response. Intracellular recordings from VCN "on" units in anesthetized preparations in most cases reveal sustained depolarization that is maximal at tone onset, with no evidence of hyperpolarization during the course of the tone (Britt and Starr 1976a; Rhode et al. 1983b; Romand 1978). If the transformation from sustained AN responses to the onset pattern is produced by synaptic inhibition, the inhibition must be generated at sites remote from the recording electrodes in these studies. Alternatively, it is possible that the on pattern is a consequence of particular membrane properties. Ritz and Brownell (1982) have suggested that a depolarization block might be the mechanism responsible for the suppression of spontaneous activity after the onset response in some PVCN cells in decerebrate unanesthetized cats.

"Buildup" and "pauser" units, although occasionally encountered in PVCN (Godfrey et al. 1975a), are largely restricted to DCN, and together with "choppers" comprise the major cell classes in this division (van Gisbergen et al. 1975a; Godfrey et al. 1975b; Goldberg and Brownell 1973; Kitzes et al. 1978; Pfeiffer 1966a; Rhode et al. 1983a; Rhode and Kettner 1986; Rhode and Smith 1986b). Many other DCN units have PSTHs that are intermediate between these three forms, notably "pauser/buildup" units, in which the sustained component following the pause has the buildup pattern (Godfrey et al. 1975a; Rhode and Smith 1986b). As noted previously, a variety of other response patterns (primarylike and various sorts of on responses) are also encountered in DCN, particularly in DL (Godfrey et al. 1975b; Rhode et al. 1983a; Rhode and Smith 1986b). For a number of reasons, the classification of DCN units in terms of response characteristics presents considerable difficulties: a given cell's excitatory response can vary at different points in its response area (Godfrey et al. 1975a; Goldberg and Brownell 1973; Rhode et al. 1983a; Rhode and Smith 1986b); the response areas of many cells are dominated by inhibition (Evans and Nelson 1973a; van Gisbergen et al. 1975a; Rhode and Smith 1986b; Young and Brownell 1976; see, e.g., Fig. 20); and barbiturate anesthesia has marked effects on responses (Evans and Nelson 1973a; Rhode and Kettner 1986; Young and Brownell 1976). These characteristics presumably reflect the more complex internal circuitry of DCN and the fact that it receives massive input from noncochlear sources (see Sect. 4.1.4). The latter feature is also reflected in the fact that spontaneous activity in DCN is relatively unaffected by cochlear destruction, whereas that in VCN is almost totally eliminated (Koerber et al. 1966). Because of these complexities, the response characteristics of DCN neurons generally will be discussed here, rather than those of "buildup" and "pauser" units as particular classes.

Most DCN neurons are influenced by a wide range of frequencies. Some neurons have broad excitatory FTCs (van Gisbergen et al. 1975a; Goldberg and Brownell 1973), but in those with narrow excitatory FTCs the total response area, including both excitatory and inhibitory regions, tends to be large (see, e.g., Fig. 20; Evans and Nelson 1973a; Goldberg and Brownell 1973; Rhode and Smith 1986b; Young and Brownell 1976). Latencies of "pauser" and "buildup" units to clicks (Godfrey et al. 1975b) and tones (Rhode and Smith 1986b) are longer than those of VCN cells and have a wider range of values; click latencies are up to 6 ms longer than the latencies of AN fibers of comparable CF (Godfrey et al. 1975b; Kiang et al. 1973). In both anesthetized and unanesthetized preparations, many DCN units have nonmonotonic rate–intensity functions (Goldberg and Brownell 1973; Greenwood and Maruyama 1965; Rhode and Kettner 1986; Rhode and Smith 1986b; Rose et al. 1959; Young and Brownell 1976). In unanesthetized animals, nonmonotonicity in type IV cells reflects the fact that the cell's responses at CF are excitatory near threshold, but inhibitory at higher levels (Rhode and Kettner 1986; Young and Brownell 1976). Rate–intensity functions for type IV cells consequently peak within 20 dB of threshold and then decline to zero.

The phase-locking of DCN neurons is much poorer than that in VCN, and many "pauser" and "buildup" units have synchronization coefficients near zero at all frequencies (van Gisbergen et al. 1975b; Goldberg and Brownell 1976; Lavine 1971; Rhode et al. 1983a; Rhode and Smith 1986b). Rhode et al. (1983a) also observed poor phase-locking in DCN units with primarylike response patterns. The spontaneous and continuously driven activity of most DCN units tends to be more regular than that of VCN cells other than "choppers," and many units have relatively symmetric ISIHs (van Gisbergen et al. 1975c; Goldberg and Brownell 1973; Pfeiffer 1966a; Pfeiffer and Kiang 1965; Rhode and Smith 1986b). This discharge regularity presumably underlies the appearance of the chopper pattern in at least some part of the response area of many DCN neurons (e.g., Rhode et al. 1983a). As in PVCN "choppers," it could reflect temporal integration of multiple inputs (van Gisbergen et al. 1975c) and/or the membrane properties of the cells (Oertel 1983; Wu and Oertel 1984).

Young and Brownell (1976) reported that most type II/III and type IV cells were influenced by stimulation of the contralateral ear at levels that precluded interaural crosstalk[14]. Mast (1970, 1973) also reported that some units in DCN of the chinchilla could be excited or inhibited by contralateral stimulation, while Klinke et al. (1969) and Pfalz (1962; guinea pig and cat) had earlier demonstrated inhibition of CN activity by contralateral acoustic stimulation in studies in which recording sites were not histologically localized to CN divisions. In Pfalz's (1962) study, contralateral inhibition of spontaneous activity was examined after destruction of the ipsilateral cochlea; since spontaneous activity survives deafferentation only in DCN (Koerber et al. 1966), it seems likely that his observations

[14] Determination of the influences exerted on a binaural neuron by input from each ear requires that stimuli be delivered independently to each ear through sealed systems of the type described in Sect. 3.2. By convention in such studies, the ear to which stimulation is presented is defined as "contralateral" or "ipsilateral" relative to the recording site in the brain. A critical issue in the use of such systems for studies of binaural input concerns the amount of interaural crosstalk, i.e., the extent to which sound presented to one ear is conducted (via air- and/or bone-conduction pathways) to the contralateral cochlea. The amount of interaural attenuation provided by dichotic systems depends on a number of factors, chief among which are the degree of coupling between the acoustic drivers and the head-holding frame, and whether or not the bullae are open (Caird et al. 1980; Gibson 1982). These recent studies indicate that under optimal conditions (driver-frame isolation and closed bullae) interaural attenuation is greater than 50 dB at all frequencies. Under nonoptimal conditions, however, interaural attenuation can be as low as 25 dB at some frequencies.

were made on DCN neurons (cf. comment by Evans in Pfalz 1973). These observations are in accord with evidence that electrical stimulation of various brainstem auditory structures can influence spontaneous and evoked discharges in all three major divisions of CN (e.g., Comis 1970; Comis and Whitfield 1968; Starr and Wernick 1968). Although some of the inhibitory effects observed in these studies are undoubtedly mediated by OCB effects in the cochlea, the occurrence of such effects on spontaneous activity in deafferented CN, and certain pharmacological observations (e.g., Comis 1970), establish that there are also direct effects on CN neurons. The occurrence of these effects is not surprising, given the evidence for descending projections to CN – particularly DCN – from higher brainstem structures that receive binaural input. The fact that units were not physiologically characterized in any of these studies other than that of Young and Brownell (1976) precludes more detailed consideration of these effects, and their significance for brainstem processing remains unclear.

4.2.5 Correlations Between Morphologically and Physiologically Defined Cell Types

Attempts to relate physiologically defined cell types in CN to morphologically defined cell types have largely depended on distributional characteristics and on other forms of evidence which, however compelling, are basically indirect (e.g., Godfrey et al. 1975a; Goldberg and Brownell 1973; Kiang et al. 1973). More recently, the intracellular labeling of single, physiologically characterized neurons has provided direct evidence on this question (Britt et al. 1983; Rhode et al. 1983a, b; Rouiller and Ryugo 1984; Wu and Oertel 1984). The technical difficulty of the intracellular labeling and recording procedures has the consequences that the data base is as yet relatively small and that the physiological characterization of some labeled cells is limited to PSTH pattern rather than the range of parametric features that have come to define at least some cell types. It nevertheless seems likely that the accumulation of further data of this sort within the next few years will allow correlations to be established with greater certainty. The data already available, however, indicate that the extent to which correlations can be established between physiological and morphological types is ultimately limited by the lack of absolute distinctions between the supposed "types" in either domain.

"Primarylike" and "primarylike with notch" units are encountered most frequently in those areas of AVCN in which bushy (spherical and globular) cells predominate (Bourk 1976; Godfrey et al. 1975a; Goldberg and Brownell 1973) and are the only CN units with prepotentials. Given the evidence for the origin of the prepotentials in calyceal terminals and the restriction of these endings to bushy cells, these observations provide strong support for the identification of these two unit classes with bushy cells. "Primarylike" units are encountered most frequently in anterior AVCN, where bushy cells correspond to large spherical cells and receive a small number of large end-bulbs. "Primarylike with notch" units are encountered most frequently in posterior AVCN, where bushy cells correspond to globular cells. The fact that "primarylike with notch" units tend to have small prepotentials (Bourk 1976) is in accord with the fact that bushy/globular cells in posterior AVCN receive multiple small "modified" end-bulbs. Rhode and

Smith (1986a) have suggested that the high spontaneous rates of "primarylike with notch" units could reflect the convergence of a number of end-bulbs each of which is individually capable of driving the cell. The identification of "primarylike with notch" units with bushy/globular cells is also supported by the observations that these units have the shortest antidromic latencies to trapezoid body (TB) stimulation (Bourk 1976) and that the largest (and therefore fastest-conducting) TB axons are derived from globular/bushy cells (Osen 1969a; Tolbert et al. 1982).

Rhode et al. (1983b) were able to label intracellularly three AVCN units which they defined as "primarylike", although the identification was tentative in the third case because the cell had low CF (700 Hz) and a phase-locked response. The two unequivocally "primarylike" units were bushy cells, whereas the third unit was a small cell. AVCN cells with primarylike PSTHs labeled by Britt et al. (1983) and by Rouiller and Ryugo (1984) were also identified as bushy cells. Intracellularly labeled bushy cells in slices of mouse AVCN respond to intracellularly injected current pulses with a single, brief action potential and repolarize rapidly after the end of the pulse (Wu and Oertel 1984). These characteristics correspond closely with the intracellular response characteristics of "primarylike" units in cat AVCN (Rhode et al. 1983b). Although additional intracellular labeling data are required, it seems certain that AVCN "primarylike" units, which are defined by preservation not only of the response pattern, but also of the other properties of AN fibers, are bushy cells receiving end-bulbs. However, the fact that the primarylike response pattern is seen in DCN indicates that this characteristic can be associated with cells other than bushy cells; indeed, Britt et al. (1983) have reported labeling a DCN granule cell with primarylike PSTH. Furthermore, it is not clear whether all bushy cells have primarylike PSTHs. Rouiller and Ryugo (1984) classified the PSTHs of two labeled bushy cells as on-type L and that of another as a chopper. In each of these cases, however, they noted that some uncertainty attached to the response classification: there are difficulties in distinguishing between some on-type L and primarylike with notch patterns (cf. Rhode and Smith 1986a), and the pattern classified as chopper had a number of unusual characteristics that suggested the possibility of pathology. These results illustrate the difficulties already alluded to in connection with the definition of cell types.

"Chopper" units in VCN are encountered in areas in which stellate cells are widely distributed, suggesting an association between these classes (Bourk 1976; Godfrey et al. 1975a; Kiang 1975). The fact that the primary innervation of stellate cells is via many small terminal boutons (Sect. 4.1.2) is also consistent with models in which the regular discharge of "choppers" is produced by temporal integration of a large number of individually subthreshold inputs (van Gisbergen et al. 1975c; Molnar and Pfeiffer 1968). Although these indirect lines of evidence are rather tenuous, they have received strong support from intracellular labeling studies. All of eleven labeled cells in PVCN classified as "choppers" by Rhode et al. (1983b) proved to be stellate cells, and an unequivocal "chopper" labeled by Rouiller and Ryugo (1984) was also a stellate cell. Furthermore, labeled stellate cells in slices of mouse AVCN respond to intracellularly injected current pulses with regularly firing action potentials. These observations establish that "chopper" units in VCN are stellate cells, although the chopper response pattern can be exhibited by nonstellate cells in DCN (Rhode et al. 1983a), and stellate cells can have responses other than chopper (see next paragraph).

"On" units are concentrated in OCA (Godfrey et al. 1975a), and IAS fibers – which are chiefly the axons of octopus cells (Osen 1969b; Warr 1972, 1982) – also exhibit on responses (Adams 1976; Kiang et al. 1973). Because of the manner in which the dendrites of octopus cells intercept the ascending AN fibers, it had been predicted that these cells would have broad frequency tuning (Morest et al. 1973; Osen 1969a; Sect. 4.1.3), and one subgroup of "on" cells ("on-type C") do in fact have the broadest FTCs encountered in CN (Rhode and Smith 1986a). The extended dynamic range of "on-type C" units is also suggestive of highly convergent input. In the study by Rhode et al. (1983b), however, only one of three intracellularly labeled "on" units in PVCN proved to be an octopus cell while the other two were stellate cells. Rouiller and Ryugo (1984) tentatively (because of incomplete filling of the dendrites) identified an "on-type I" unit in PVCN as an octopus cell, but also suggested, as noted previously, that two bushy cells in AVCN had on-type L responses. Finally, Britt et al. (1983) reported that a small stellate cell in PVCN and a horizontal giant cell in DCN had on responses. These observations, together with the existence of a number of subgroups of "on" units and the occurrence of on responses outside OCA, indicate that while most (perhaps all) octopus cells may be "on" units, stellate cells in PVCN and other cell types in other divisions can also exhibit this response pattern.

Kane (1973) suggested that derivation of the on response of octopus cells from the sustained input provided by AN fibers might involve inhibition from the small terminals on the octopus cell somas that contain pleomorphic vesicles (see Sect. 4.1.3). Since these endings are said to arise from thin collaterals of AN branches that also give rise to the large (presumably excitatory) endings on octopus cells (Kane 1973, 1974b), this hypothesis requires either that different transmitters (or sets of transmitters) be released at different terminals arising from the same parent axon, or that the same transmitter (or transmitters) have different effects at different postsynaptic sites on the same neuron (see Segal 1983 for a recent discussion of these issues). There is no *direct* evidence, however, for the inhibitory nature of the type 2 terminals on octopus cells, and intracellular recordings from PVCN "on" units only rarely give evidence of hyperpolarization during the stimulus (Britt and Starr 1976a; Rhode et al. 1983b; Romand 1978). It is at least equally plausible that on responses might be attributable to inhibitory input on the distal branches of octopus cell dendrites (Godfrey et al. 1975a), or to intrinsic membrane properties of octopus cells (cf. Ritz and Brownell 1982; Romand 1978). Extension of intracellular recordings in the slice preparation (Oertel 1983; Wu and Oertel 1984) to the OCA should help to clarify the mechanisms underlying on responses in these neurons.

"Pauser" and "buildup" units are commonly encountered in FCL, suggesting an association with fusiform cells (Godfrey et al. 1975b). This hypothesis is supported by the fact that pauser and buildup response patterns are commonly seen in DAS, which contains the axons of fusiform cells and of giant cells in PL (Adams and Warr 1976; Osen 1972). Rhode et al. (1983a) found that the majority of intracellularly labeled fusiform cells were either "pauser" or "buildup" units, although others had varied response properties. Two fusiform cells had primarylike response patterns, but their phase-locking was much weaker than that of "primarylike" units in AVCN. The observations that not all fusiform cells are "pauser" or "buildup" units and that these response patterns are also found in PL indicate that the correlation with fusiform cells is partial. Young (1980) found that most type IV cells could be antidromically activated from DAS, indicating that these responses are recorded from the DCN projection cells, viz., fusiform cells and giant cells (Adams 1979; Adams and Warr 1976; Osen 1972; Roth et al. 1978). These observations therefore support Young and Brownell's (1976) tentative identification of type IV cells with "pauser/buildup" units. In contrast, very few type II/III cells, particularly those without spontaneous activity, could be activated from DAS. These and other observations suggest that type II/III cells are small interneurons in DCN. In an elegant series of experiments, Voigt and Young (1980) obtained cross-correlograms of the simultaneously recorded activity of pairs of neurons comprising a type II/III unit and a type IV unit. In many of these pairs there was an inhibitory trough in the cross-correlogram that corresponded to a reduced probability of discharge in the type IV cell following a discharge in the type II/III cell. These data suggest that type II/III cells are inhibitory interneurons that are presynaptic to type IV cells.

4.2.6 Spectral Coding and Responses to Complex Stimuli

Technical difficulties have prevented recording from large numbers of CN neurons in order to estimate population response patterns in the manner employed in the AN. It is apparent, however, that the properties of AVCN "primarylike" units are such that the patterns of AN activity in which spectral information is coded – whether in terms of average rate or synchronized rate – would be faithfully preserved across the "primarylike" population. The function of the "primarylike" units would then be one of relaying the information rather than of processing, and the extraction of important spectral features would be deferred to higher centers. A model of this sort was proposed by Loeb et al. (1983) and is discussed in Sect. 8.1. On the other hand, the properties of other CN cell types indicate significant transformation of AN fiber input and therefore suggest the possibility of significant processing and extraction of critical features at this level. Thus, Greenwood and Maruyama (1965) suggested that the sideband inhibition exhibited by many cells in DCN and PVCN might serve to delimit the population excited by a pure tone and to segregate populations excited by different frequency components of a multicomponent stimulus. As discussed further in Sect. 4.3, the complex organization and response properties of DCN, particularly, have suggested to some that it might be involved in the processing of dynamic and complex acoustic stimuli (e.g., Evans 1974; Evans and Nelson 1973a, b). Investigation of the responses of CN neurons to complex stimuli, predominantly noise and amplitude- or frequency-modulated (AM or FM) tones, has illustrated some of the ways in which such processing might occur. Unfortunately, many of these studies have not identified the loci from which recordings were made or the response pattern and other characteristics of the units. These omissions, and the limited nature of the data base, prohibit any systematic account of the processing of complex stimuli in CN. Because of their basically illustrative nature, these data will be only briefly considered here. Responses to AM and FM stimulation have been reviewed by Evans (1975) and, in the context of signal analysis techniques, by Brugge and Geisler (1978). Møller (1983) presents an integrated account of much of his earlier work on the responses to complex stimuli of CN neurons in the rat.

Van Gisbergen et al. (1975b) showed that the responses of many CN neurons to stationary and AM noise could be predicted from their pure-tone responses. The correlation between predicted and actual responses was better for units in VCN than DCN. Other accounts of responses to noise have been restricted to PVCN and DCN. In anesthetized cats, the responses of neurons in these divisions to increases in the bandwidth of noise centered at CF is related to the nature of their rate–intensity functions. For neurons whose rate–intensity functions for CF tones or narrow-band noise are nonmonotonic, increases in bandwidth generally lead to a progressively increasing suppression and, in some cases, total disappearance of the response (Greenwood and Goldberg 1970; Greenwood and Maruyama 1965; Goldberg and Greenwood 1966). For neurons with monotonic rate–intensity functions, increasing bandwidth commonly results in an increase in response over some range, beyond which further increases in bandwidth result in suppression that varies from slight to very marked. The latter behavior is also exhibited by AN fibers, in which the decline in response presumably reflects the phenomenon of "two-tone" suppression (chinchilla: Ruggero 1973; see discussion in Sect. 3.2). The pure suppression exhibited by nonmonotonic units and suppression below spontaneous activity in some nonmonotonic units cannot be accounted for in terms of AN characteristics. These phenomena presumably reflect the fact that more energy falls into inhibitory regions of the cell's response area as the bandwidth is increased, as proposed by Greenwood and Goldberg (1970).

Greenwood and Goldberg (1970) examined the responses of PVCN and DCN units to combinations of a CF tone and a band of noise centered at a frequency above or below CF, and described complex interactions as a function of the relative intensity of the two frequency components. Related observations were made by Palmer and King (1982), who found that the range over which some DCN units could respond to increases in the intensity of a CF tone was increased in the presence of band-stop noise. They attributed this effect to the fact that the noise activated inhibitory sidebands and thus "biased" the cell's discharge to below saturation level, allowing it to be driven more strongly with an increase in tone intensity. Under the conditions of their experiments, such enhancement of dynamic range was a rare occurrence in VCN units or AN fibers. However, Geisler and Sinex (1980) and Costalupes et al. (1984), using different procedures, have shown that increased dynamic range in the presence of background noise is a common feature of AN fibers and reflects "two-tone" suppression. It is therefore unclear to what extent increased dynamic range in the presence of noise reflects inhibitory sidebands or is unique to DCN.

Given the marked effects of anesthesia on inhibitory phenomena in DCN, it is perhaps not surprising that responses to noise in unanesthetized animals differ from those reported by Greenwood and Goldberg (1970). Young and Brownell (1976) found that type IV cells (which have nonmonotonic rate–intensity functions for tones) responded *more* strongly to broad-band noise than to pure tones. Conversely, type II/III cells, which have monotonic rate-intensity functions for tones, responsed weakly or not at all to noise. These characteristics fit well with the proposal that type II/III cells provide the inhibitory input to type IV cells, since if the former are unresponsive to broad-band stimuli, the type IV cells will receive little inhibitory input in association with such stimuli. The differences between the results obtained in anesthetized and unanesthetized preparations again stress the importance of anesthetic effects in DCN: to the extent that the responses to complex stimuli of neurons in DCN (or any other nucleus) are dependent on inhibition, their responses are likely to be dramatically modified by anesthesia. Evidence on the interactions between type II/III and type IV cells also point to one of the factors that make it impossible to predict the responses of some DCN cells to complex stimuli on the basis of their pure-tone response areas.

The responses of AN fibers and of "primarylike" units to FM tones in which the frequency sweep follows a triangular or trapezoidal waveform are symmetric in that their response at any given frequency within the sweep is approximately equal, regardless of the direction of frequency change (Britt and Starr 1976b; Sinex and Geisler 1981). "Mirror symmetric" responses of this sort preserve information about the direction of frequency change in the temporal pattern of response, but indicate that direction information has not been extracted at that level (Erulkar et al. 1968a). In contrast, many other CN neurons respond differently to ascending and descending sweeps (i.e., their responses show various forms of asymmetry), in extreme cases having no response at all to a sweep in one direction (Britt and Starr 1976b; Erulkar et al. 1968a; Fernald and Gerstein 1972; Rhode and Smith 1986a, b). Such units can be said to have extracted information about, or detected, the direction of frequency change. Britt and Starr (1976b) reported that most of their "on" units had asymmetric responses. Given the partial correlation between "on" units and octopus cells, these data are in agreement with Szentagothai and Arbib's (1975) prediction that octopus cells would show directional FM preference, a prediction based on the orientation of octopus cell dendrites relative to the tonotopic AN fiber array and on Rall's (1964) spatiotemporal dendritic simulation model. However, Rhode and Smith (1986a) reported that only "on-type I" units, which comprised a very small proportion of their sample of "on" units in VCN (12 of 182), had asymmetric responses. Britt and Starr (1976b) reported that the responses to FM tones of "pauser" and "buildup" units were diverse, being symmetric, asymmetric, or unidirectional. A cell's response to FM stimuli depends in part on the relationship of the center frequency and modulation depth of the signal to the excitatory and inhibitory regions of the cell's response area (e.g., Britt and Starr 1976b; Suga 1968, 1971; Vater 1982). On this basis alone, the complex response areas of some DCN "pauser" and "buildup" neurons would be expected to result in nonsymmetric responses to FM sweeps.

Only limited data are available on the sensitivity of CN neurons in the cat to the modulation frequency of periodic FM stimuli (Britt and Starr 1976b; Erulkar et al. 1968a, b), but this variable has been studied in greater detail in the horseshoe bat (Vater 1982) and in the rat (Møller 1971, 1972a, b, 1974c, 1977). In bats whose echolocation signal comprises constant-frequency and FM components, periodic frequency and amplitude modulation are produced in the echo of the constant-frequency component by the wing beats of insect prey. Vater (1982) reported that the discharge of "primarylike" units in CN of the horseshoe bat was synchronized to the modulation frequency of periodic (sinusoidal) FM signals at frequencies up to 700 Hz. At low modulation frequencies, the histogram envelopes were nearly sinusoidal and reproduced the time course of the frequency change, but at higher modulation frequencies, discharges were restricted to a narrower frequency range in each modulation cycle. This finding is in agreement with Møller's (1969b, 1971, 1974c) finding, in CN of the rat, that at higher modulation rates the response is restricted to a narrower range of frequencies around CF. The locations and PSTH patterns of the units from which Møller's data were obtained were unfortunately not specified, except in one case in which they were described as having sustained responses to tone bursts (Møller 1969b). In contrast to the response of "primarylike" units, none of the "buildup" units examined by Vater (1982) in CN of the bat were synchronized to the modulation frequency and "pausers" were synchronized only at low frequencies. "On" units were heterogeneous, some exhibiting synchronized discharges to modulation frequencies as high as 800 Hz whereas others showed synchrony only at modulation frequencies below about 400 Hz.

The responses of many high-frequency CN neurons are synchronized to the modulation envelope of AM stimuli with carrier frequency at or near CF, but neurons differ in the highest modulation frequency at which they continue to show synchrony (Fernald and Gerstein 1972; Glattke 1969; Hirsch and Morton Gibson 1976; bat: Vater 1982; rat: Møller 1972a, b, 1974b, 1977). As with FM stimuli, the variation in discharge rate for some neurons reproduces the modulation envelope in an apparently linear fashion, but for other neurons the discharge rate variation shows evidence of nonlinearity (Fernald and Gerstein 1972; Hirsch and Morton Gibson 1976; bat: Vater 1982; rat: Møller 1972a, 1974b). The PSTH types of the units studied with AM stimuli were reported only in Vater's (1982) study of the horseshoe bat: as with FM stimuli, "primarylike" units were synchronized to modulation frequencies up to 700 – 800 Hz, "buildup" units exhibited no synchronization; "on" units were heterogeneous. Møller (1972b) reported that some units in CN of the rat were "tuned" to a particular modulation frequency, in that a given stimulus modulation depth produced greatest discharge variation at that modulation frequency. Some of these units showed discharge rate modulation to AM tones over a wider intensity range than their dynamic range for steady tones (Møller 1974b, 1977).

The only studies of CN units using speech sounds appear to be those of Moore and Cashin (1974, 1976) on the responses to vowels of low-frequency units of unspecified location in the guinea pig CN. They presented qualitative data indicating that information about formant frequencies is present in the temporal pattern of discharge of some units. However, although they also obtained data on PSTHs to pure tones, they did not establish correlations between unit types and the manner in which vowel properties were coded. Their data cannot readily be related either to the evidence on AN representation of vowel sounds or to the body of evidence on cat CN units.

An interesting approach to the study of responses to complex stimuli has been the study by Aertsen and his colleagues (Aertsen et al. 1979; Smolders et al. 1979) of responses of cat CN neurons to an ensemble of sounds occurring in the natural environment of the animal (what they call the "acoustic biotope"). Preliminary data from a small number of units have established the value of this approach, but do not permit of correlations with the main body of evidence on CN units.

This brief, and highly selective, review makes it clear that the responses of some neurons in CN of the cat, particularly in PVCN and DCN, are sensitive to particular features of spectrally complex, time-varying sounds. The asymmetric responses to FM stimuli and greater responsiveness to noise than to tones of some DCN units provide some support for the notion that they might serve to detect or extract information about the features of such stimuli. Conversely, the properties of "primarylike" units indicate that they preserve information about modulation rate and waveform. As stated earlier, however, the information provided by these studies is as yet too limited to allow any sort of developed account of even the earliest stages of the central processing of complex stimuli.

4.3 Hypotheses Concerning Parallel Processing in Cochlear Nucleus

As we have seen, the information conveyed to the CN by AN fibers is transformed in different ways by morphologically different cell types receiving different patterns of synaptic innervation and having different patterns of projection to higher brainstem centers. These observations suggest that different cell classes might be engaged in the parallel processing of different aspects of the information conveyed by the acoustic stimulus and might therefore be the source of functionally distinct, parallel pathways to higher levels of the auditory system. Unfortunately, our knowledge of the *functional* significance of the transformations carried out by different cell types remains so limited that, except in the simplest cases, we can only speculate about the nature of the processing carried out by those cells. Furthermore, as we shall see, even when a function can reasonably be ascribed to a given cell group, it is by no means clear either that this is the *only* function performed by those cells or that other cell groups do not also contribute to the defined function.

In describing the three major striae, Poljak (1926, p. 467) proposed on anatomical grounds that they were "the starting point of three different, but not completely isolated, central systems." In elaboration of Poljak's proposal, and by analogy with the notion of there being "two visual systems," responsible for form and place processing (e.g., Held 1968; Held et al. 1967; Schneider 1967, 1969), Evans (1974, 1975) and Evans and Nelson (1973a, b) proposed a distinction between "ventral" (comprising TB and IAS) and "dorsal" (i.e., DAS) auditory pathways. Because of the preservation of temporal information in the ventral pathway, it was postulated to be involved in sound localization (place processing), while the dorsal pathway from DCN was identified as the first stage in a pattern (form) processing system.

The only cell type in CN to which a function can be ascribed with any degree of confidence is the AVCN "primarylike" units whose faithful preservation of AN fiber discharge characteristics suggests that they serve to relay information to other neurons rather than to process it in any way. The major projections of "primarylike" units are via TB to the major nuclei of the SOC. As we shall see in subsequent discussion of SOC (Sects. 5.2.2 – 5.2.4), the basic processing of interaural time and intensity difference cues for spatial location takes place in these nuclei and the pathways from AVCN to SOC are specialized to convey precise time and intensity information. These data therefore support the suggestion of Evans (1974, 1975) and Evans and Nelson (1973a, b), in that at least one part of the "ventral pathway" is critically involved in spatial coding. Note, however, that place–rate and place–periodicity information as to the spectral properties of the stimulus is also faithfully preserved and transmitted to the SOC by AVCN "primarylike" units. As noted previously, some contemporary models (e.g., Loeb et al. 1983; Srulovicz and Goldstein 1983) suggest that spectral information is extracted centrally from information coded in this fashion. On this view, the component of the "ventral" pathway originating in "primarylike" units would be as important for spectral (i.e., form) as for spatial processing.

It should also be noted that the preservation of AN properties by "primarylike" units reflects the secure excitatory drive provided by the end-bulbs. Bushy cells also receive a variety of bouton terminals from other sources (some of which contain flattened vesicles and might therefore be inhibitory) that provide a possible means of modulating the end-bulb input. Although such modulation is not apparent in the available physiological data, these data are derived almost exclusively from anesthetized preparations. It is possible that significant departures from a simple "relay" function will be revealed when detailed information on the response properties of "primarylike" units in awake (and behaving) animals becomes available.

The proposed role of the dorsal pathway in the processing of dynamic and complex acoustic stimuli derives some support from the fact that many DCN neurons have complex response characteristics suggestive of feature extraction (Sect. 4.2.6) and project via DAS mainly to NLL and IC, bypassing the SOC nuclei involved in binaural integration. However, DCN neurons receive a sparser primary afferent input and denser descending projections from higher brainstem centers than other CN neurons, and it is consequently unclear to what extent the response characteristics of DCN neurons reflect intrinsic processing as opposed to processing at higher centers. As pointed out already, there are also reasons to suppose that AVCN "primarylike" units are involved in representing the spectrum of complex signals. Even if DCN and its projections should prove to play a critical role in the processing of spectrally complex stimuli, the fact that spectral transformations provide major cues for sound location suggests that only a partial distinction between place and pattern systems is possible.

These considerations suggest that the functions that can be ascribed to the pathways originating in AVCN "primarylike" cells and in the projection cells of DCN are neither as discrete nor as clearly defined as required by the concept of parallel processing. Furthermore, these proposals deal with only two of the major cell types in CN. No functions have so far been ascribed to "choppers," the most commonly encountered unit type in CN, and it is not even clear whether their regular discharge is of functional significance or simply an epiphenomenal consequence of some other specialization. Ascription of functions to "on" units is also extremely tentative. Octopus cells project via IAS, mainly to the periolivary cell groups from which descending auditory pathways to the cochlea and CN arise (see Sect. 5.1.5), and it has therefore been suggested that "on" units are involved in the control of these centrifugal systems. The ability of many "on" units to entrain to low-frequency tones and to repetitive transient stimuli suggests that they might be involved in the coding of periodicity (e.g., Møller 1969a). It has also been suggested that the very precise onset timing of "on" units might serve to provide information about onset time disparities to binaural centers (Rhode and Smith 1986a). As will be seen in later sections, however, "on" units do not appear to provide significant input to the major SOC nuclei in which interaural comparisons are made. Finally, Kiang and his colleagues (Godfrey et al. 1975a; Kiang 1975) have suggested that "on" units might have an arousal or alerting function, signaling a change in the acoustic environment. This suggestion finds some support in the fact that the majority of acoustically responsive cells in the reticular formation of chloralose-anesthetized cats exhibit on responses (Irvine and Jackson 1983). The diverse and speculative nature of these suggestions reflects the fact that we simply do not have an adequate data base to allow anything other than speculation about the function of many of the major cell groups in CN.

In summary, although there is some support for the notion of parallel processing in, and projections from, defined cell groups in CN, we can so far offer only a limited account of the nature of the processing involved and there is considerable uncertainty as to the extent to which the pathways can be considered *functionally* distinct. These issues are considered in more detail in the light of information concerning the higher brainstem centers that are the targets of CN efferents in later sections (viz., Sects. 5.4, 7.7, 8.1, and 8.2.1).

5 Superior Olivary Complex: Anatomy and Physiology

5.1 Neuronal Architecture and Connections of Superior Olivary Complex Subdivisions

5.1.1 General Description and Comparative Differences

In all mammals, the superior olivary complex (SOC) comprises a number of closely grouped nuclei in the lateral part of the ventral pons (see Fig. 1). The three most distinctive nuclei in the cat are commonly designated the lateral superior olivary nucleus (LSO), the medial superior olivary nucleus (MSO), and the medial nucleus of the trapezoid body (MNTB; see Figs. 1 and 23). These major nuclei are clearly defined in the drawing of the afferent axonal plexus in SOC by Ramon y Cajal (1909) (Fig. 22A), by whom they were termed the superior olive, the accessory superior olive, and the nucleus of the trapezoid body, respectively. Surrounding the major nuclei are a number of smaller and less clearly defined aggregations of cells which have been given a variety of names. Ventral to MNTB and LSO, respectively, are situated what Ramon y Cajal (1909) termed the internal and external preolivary nuclei (Fig. 22A). They are now more commonly known either as the medial and lateral preolivary nuclei, or as the ventral and lateral nuclei of the trapezoid body (VNTB and LNTB; Fig. 1; Morest 1968a, 1973). The latter terms will be used here as they more accurately describe the location of the nuclei and have the further advantage of avoiding confusion with the terms used to describe the final group of SOC neurons. These consist of scattered groups of neurons at the margins of the SOC (corresponding in part to Ramon y Cajal's retro-olivary region), which together comprise the *periolivary* region. This region has conventionally been divided into a number of nuclei identified on the basis of their position relative to the major nuclei of the complex. The most extensively studied of these is situated on the dorsomedial border of the complex and is consequently termed the dorsomedial periolivary nucleus (DMPO; see Fig. 26). The approximate locations of the dorsal periolivary cell groups is indicated in Fig. 1, and a number of individual periolivary "nuclei" are identified in Fig. 26. As discussed in more detail in Sect. 5.1.5, the absence of clear nuclear boundaries in the periolivary region has recently prompted proposals that the designation of specific periolivary "nuclei" be abandoned in favor of the identification of "zones" or of cell groups characterized by neuronal morphology and connections.

Comparative studies of the mammalian SOC have revealed marked differences between species in the relative sizes of the major nuclei (Harrison and Feldman 1970; Harrison and Irving 1966c; Irving and Harrison 1967; Moore and Moore 1971). A striking feature of these differences is the consistently parallel development of MNTB and LSO, and three major groups of mammals can be distinguished on the basis of the development of these two nuclei relative to that of MSO. In one group, which includes the mouse, hedgehog, dolphin, and some

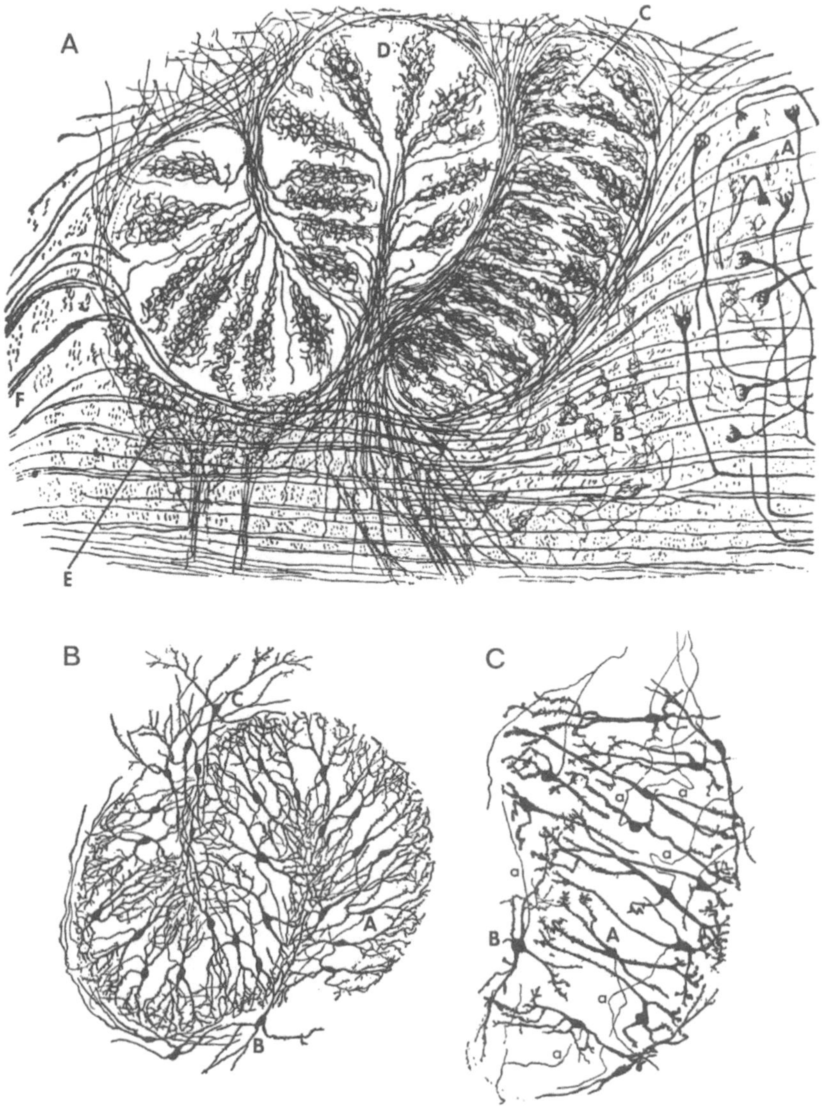

Fig. 22 A – C. Ramon y Cajal's classical illustrations of **A** the afferent axonal plexus, and of cell types in **B** lateral and **C** medial superior olivary nuclei, based on Golgi material from young kittens. Labeling in figures has been modified from Ramon y Cajal (1909), and identification of regions uses modern terminology as discussed in text. In **A**: *A* medial nucleus of trapezoid body (note calyces of Held); *B* periolivary cell region; *C* medial superior olivary nucleus; *D* lateral superior olivary nucleus; *E* lateral nucleus of trapezoid body; *F* fibers of trapezoid body. In **B**: *A* central fusiform cells; *B*, *C*: marginal cells. In **C**: *A* central cells; *B* marginal cells; *a* axons of central cells

echolocating bats[15], MNTB and LSO are relatively large whereas MSO is small or absent (Fig. 23; Harrison and Howe 1974a; Irving and Harrison 1967; Moore and Moore 1971). At the other extreme, MSO is large and MNTB/LSO is small or absent in a group of primates including humans (Fig. 23; Harrison and Howe 1974a; Moore and Moore 1971). The cat, dog, chinchilla, and guinea pig are members of the largest group of mammals, in which all three nuclei are well developed (Fig. 23; Irving and Harrison 1967; Harrison and Howe 1974a; Moore and Moore 1971).

Two hypotheses have been put forward to account for this variation between species. Harrison and Irving (1966c) and Irving and Harrison (1967) reported a near-linear relationship across a number of species between cell counts in MSO and in the abducens nucleus, and therefore proposed that MSO was involved in the control of acoustically elicited visual orienting reflexes in predominantly visual species. Conversely, Masterton and his associates (Masterton 1974; Masterton and Diamond 1967; Masterton et al. 1975) and Zvorykin (1964) have pointed out that the ratio of the sizes of the major nuclei is related to the frequency range of hearing of the species. Thus, those species in which LSO/MNTB is large and MSO vestigial (viz., certain echolocating bats, the hedgehog, and dolphin) have the highest upper frequency limits of hearing among mammals. Conversely, those species in which MSO is large and LSO/MNTB vestigial have the lowest high-frequency limits of hearing. They have therefore proposed that the relative sizes of LSO/MTNB and MSO reflect the relative dependence of the species on high-frequency and low-frequency cues (IID and IPD respectively) for auditory azimuthal location. That this correlation is incomplete is indicated by recent evidence that MSO is well developed in some high-frequency, echolocating bats (Schweitzer 1981; Zook and Casseday 1982a). However, as indicated in later sections, physiological evidence on the sensitivity to interaural disparities of LSO and MSO neurons supports a slightly modified version of this hypothesis (see Sects. 5.2.3, 5.2.4, and 5.4).

---

[15] In the mouse and other rodents, a large and conspicuous nucleus called the superior paraolivary nucleus, which is absent in the cat, is situated dorsomedial to the small MSO (Harrison and Feldman 1970; Ollo and Schwartz 1979; Osen et al. 1984; Willard and Ryugo 1983). This nucleus has been identified as MSO by some authors (see Ollo and Schwartz 1979 for discussion), and it is likely that confusion about nomenclature is responsible for some of the disagreements as to the relative size of MSO in certain species (see discussion by Harrison and Feldman 1970).

There appears to be considerable variation in the relative size of MSO between species of echolocating bats. MSO was reported by Harrison and Irving (1966b) and Irving and Harrison (1967) to be small or absent in a number of species (viz., *Myotis lucifugus, Phyllostomus hastatus,* and *Carollia perspicillata*) and similar observations were made on *Artibeus jamaicensis* by Zook and Casseday (1980). However, recent evidence has indicated that MSO is well developed in some other species, viz., *Rhinolophus ferrumequinum* (Schweitzer 1981), *Pteronotus parnellii* (Zook and Casseday 1980, 1982a), and *Molossus ater* (Harnischfeger et al. 1985). Bats of both types are found within a single family (*Phyllostomus, Carollia,* and *Artibeus* are all members of the family *Phyllostomidae*), and Zook and Casseday (1982a) have suggested that the basic difference might be between groups of bats employing different kinds of echolocation signal. Systematic studies applying standardized structural and connectional criteria for identification of nuclei to members of different groups are required to resolve this issue.

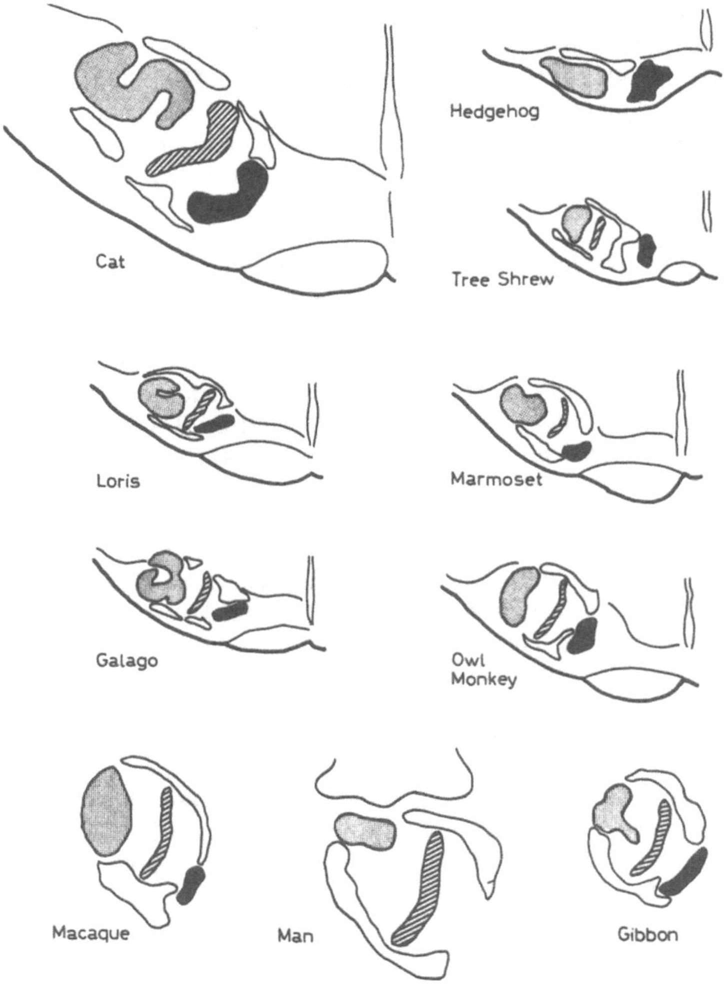

Fig. 23. Comparison of the nuclear morphology of the superior olivary complex (SOC) in the cat, hedgehog, and a series of primates. The drawings were made from transverse sections through the greatest extent of the SOC, all projected at the same magnification. In each drawing, the *stippled area* represents the lateral superior olivary nucleus, the *solid black area* represents the medial nucleus of the trapezoid body, the *cross-hatched area* represents the medial superior olivary nucleus, and the *open areas* represent the periolivary cell groups. (Moore and Moore 1971)

Despite the variation in the relative sizes of the major nuclei, their cellular morphology and connections are similar to those in the cat across a broad range of mammals (e.g., comparative: Harrison and Howe 1974a; cat, rat, rabbit, and opossum: Morest 1968b; dog: Goldberg and Brown 1968; rabbit: Borg 1973a; kangaroo rat: Browner and Webster 1975; mouse: Ollo and Schwartz 1979; Willard and Ryugo 1983; echolocating bats: Schweitzer 1981; Zook and Casseday 1982a, b). In the following accounts of SOC anatomy and physiology, the emphasis will again be on the cat. However, the variations already discussed and the fact that some of the most important physiological data have been obtained in species other than the cat (notably the dog; Goldberg and Brown 1968, 1969) require that rather more attention be devoted to other species than has been the case in previous sections.

5.1.2 Medial Nucleus of the Trapezoid Body

MNTB is distinguished by the fact that it contains the calyces of Held, probably the largest synaptic terminals in the mammalian brain (Morest 1973). The detailed cellular morphology and synaptic innervation of MNTB neurons in the cat have been described by Morest and his associates on the basis of Golgi and EM studies (Jean-Baptiste and Morest 1975; Morest 1968a, b, 1973). Morest (1968a) described three types of neuron in MNTB which he termed *principal, elongate,* and *stellate* neurons. The principal neurons are characterized by compact, tufted dendrites and receive the large axosomatic calyces of Held, which were described by the classical anatomists (see Fig. 22A; Held 1893; Ramon y Cajal 1909; see Morest 1968a, b for historical reviews). Each calyx derives from a large-diameter, heavily myelinated axon in TB, and is a cup-shaped expansion of the axon which extends over the surface of the principal cell soma in a number of thick, petal-like and finger-like processes (Fig. 24; Jean-Baptiste and Morest 1975; Morest 1968a). Each calyx ends on only a single principal neuron and each principal neuron receives only a single calyx. Similar descriptions have been given of principal cells and calyces in MNTB of the mouse (Ollo and Schwartz 1979; Willard and Ryugo 1983).

EM studies reveal that calyces in a number of species contain large spherical vesicles and that the large area of apposition between the calyx and the principal cell soma contains many discrete asymmetric synaptic contacts (Jean-Baptiste and Morest 1975; Morest 1973; cat, rat, and chinchilla: Lenn and Reese 1966; bat: Nakajima 1971). Collateral processes from the body of the calyx arborize in the neighboring neuropil and terminate as small boutons on principal neurons other than that embraced by the parent calyx (Fig. 24; Jean-Baptiste and Morest 1975; Morest 1968a). The ultrastructural characteristics of these calycine appendages resemble those of the calyces and they are therefore readily distinguishable from other small endings (Jean-Baptiste and Morest 1975). In a number of species, boutons other than calycine appendages are distinguished by small vesicles, which may be round, flattened, or pleomorphic, depending on the fixative employed, and by synaptic contacts with less-pronounced postsynaptic densities (Jean-Baptiste and Morest 1975; cat, rat, and chinchilla: Lenn and Reese 1966; bat: Nakajima 1971). As indicated in the following discussion, many axon terminals in MNTB form boutons and the origins of the small-vesicle endings are not known.

A variety of axonal endings in MNTB in addition to the calyces and calycine appendages were described by Morest (1968a) in Golgi material from (mainly immature) cats. Collaterals of various kinds arise from the calyciferous axon proximal to the calyx (Fig. 24); many of these collaterals ramify in MNTB, but the most common type terminates in relation to the dendrites of neurons in DMPO. These calycine collaterals may be less prevalent and extensive in adults than in

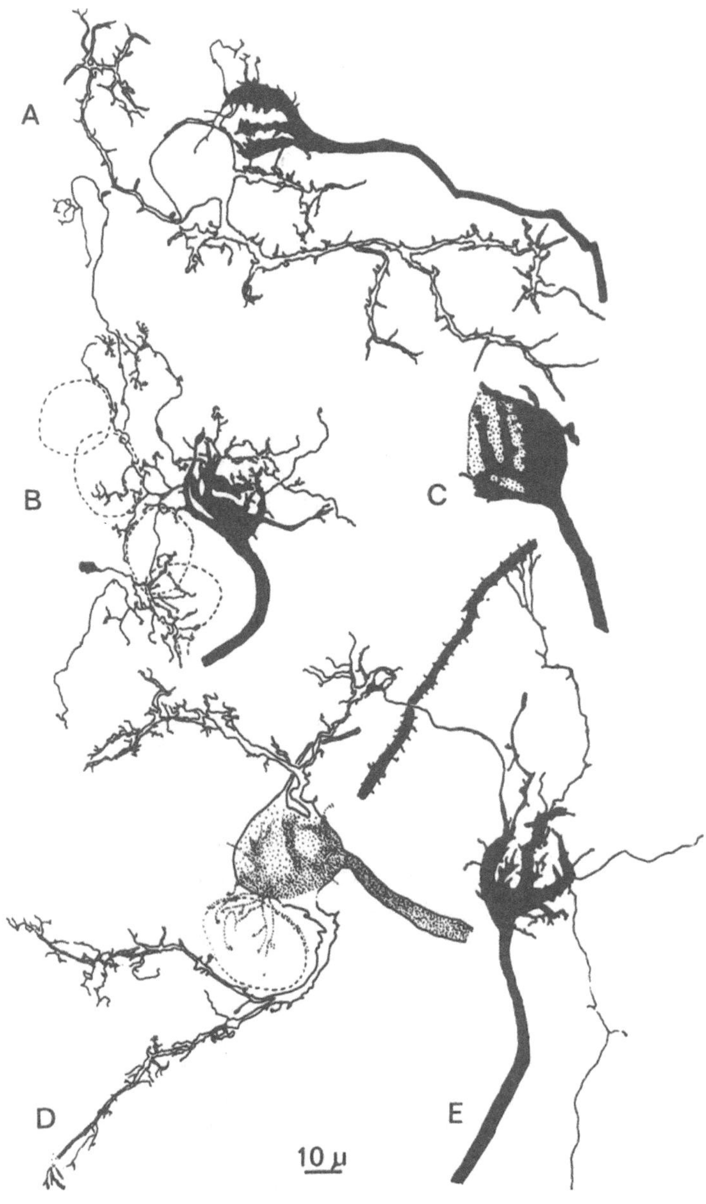

Fig. 24A – D. Calyces of Held and calycine collaterals in Golgi-stained transverse sections of medial nucleus of trapezoid body of 14-day-old cats. **A** Calycine collaterals (*in black*), ending in relation to the soma and dendrite of a neighboring principal neuron (*in outline*), **B** an extensive terminal network of calycine collaterals, associated with the profiles of osmium-stained perikarya (*dashed lines*), **C** calyx of the classical description, clasping the *stippled* profile of an osmium-stained perikaryon, **D** simultaneous impregnations of a principal neuron (*sparse stippling*) and a calyx (*dense stippling*), the collaterals of which (*dotted lines*) closely associate with the osmium-stained profile of an adjacent perikaryon (*dashed line*), **E** calyx with collaterals, one of which branches in relation to the dendrite (*in black*) of an elongate neuron, the other, in relation to a dendrite of a principal neuron. (Morest 1968a, p. 298)

kittens (Tolbert et al. 1982). Collaterals of axons proceeding to MSO form *peridendritic plexuses,* which associate primarily with the dendrites of principal cells. Both the tufted dendrites of the principal cells and the peridendritic plexuses are flattened in the horizontal plane, forming laminae. Pericellular nests, which envelop the cell bodies and dendrites of stellate neurons, are formed by fine TB axons. Finally, other axonal endings derive from collaterals of TB axons, and some of these end in relation to the elongate neurons. Both stellate and elongate neurons have long dendrites which can span the entire extent of the nucleus and hence encounter the collaterals of many horizontally coursing axons.

In the cat and other species, lesions of the globular cell region of CN (corresponding approximately to PV in the parcellation proposed by Brawer et al. 1974) result in degeneration of the large-diameter TB axons and loss of calyces in the contralateral MNTB (van Noort 1969; Warr 1972, 1982; rat: Harrison and Irving 1966c). As noted in Sect. 4.1.2, there is strong evidence for the identification of bushy cells with globular cells in posterior AVCN. The lesion evidence, in conjunction with the fact that bushy/globular cells in the cat (Osen 1969a; Tolbert et al. 1982) and rat (Harrison and Warr 1962) have thick axons, strongly suggests that bushy/globular cells in PV give rise to calyces in the contralateral MNTB. Further support for this hypothesis is provided by the finding that HRP injections limited to MNTB result in granular labeling only of bushy cells in the contralateral posterior AVCN (Tolbert et al. 1982; see also Glendenning et al. 1985). Because calyces are not the only terminal in MNTB, these data fall short of *unequivocally* establishing that PV bushy cells give rise to calyces, although the circumstantial evidence that this is the case is overwhelmingly strong. The fact that each large-diameter axon forms a single calyx that terminates on a single principal cell indicates that a one-to-one relationship is established between individual bushy cells in PV and individual principal neurons in the contralateral MNTB.

In Golgi material, the axons of principal cells give off a number of collaterals, some of which arborize in MNTB while others project to DMPO (Morest 1968a). The main axons leave the nucleus laterally, but their destination cannot be established in this material. Anterograde degeneration studies in the cat and other species (van Noort 1969; Rasmussen 1967; rabbit: Borg 1973a; kangaroo rat: Browner and Webster 1975) indicate that the major projections of MNTB are to ipsilateral LSO and DMPO, the latter projection concurring with Morest's description of this nucleus as a major target of principal cell axon collaterals. The projection from MNTB to LSO is confirmed by tracer studies in the cat. Injections of various anterograde tracers into MNTB label a topographically organized projection to LSO, which is dense to all parts of the nucleus except its lateral limb (Elverland 1978; Glendenning et al. 1985). Glendenning et al. (1985) also reported that injections of HRP restricted to LSO label many neurons in ipsilateral MNTB: most, but not all, of the labeled cells in MNTB were principal cells, but the labeled nonprincipal cells were not identified. It is consequently not clear whether stellate or elongate cells participate in this projection. Some stellate cells, at least, contribute axons to the OCB (Luk et al. 1974; Warr 1975), but nothing appears to be known about the projections of elongate cells. MNTB also has ascending projections to NLL and to IC (Glendenning et al. 1981, 1985; Henkel and Spangler 1983); these projections will be discussed in more detail when the target structures are considered (see Sects. 6.1 and 7.1.3).

The specificity of the bushy cell – principal cell projection and the secure excitatory transmission suggested by the nature of the synaptic contacts between the calyx and the principal cell soma suggest that MNTB serves primarily as a relay between posterior AVCN and LSO (cf. Goldberg and Brown 1968). As Morest et al. (1973) have noted, however, the terminals on the principal cells made by the peridendritic plexus and by other axons provide a possible mechanism for modulation or transformation of the calyceal input. The precise origins of these inputs to the principal cells have not been established, and the nature of the information conveyed over them is consequently unknown.

5.1.3 Medial Superior Olivary Nucleus

The crescent-shaped MSO seen in transverse sections of the cat's brainstem (see Figs. 1, 22, and 23) is the cross section of a rostrocaudally elongated plate of neurons angled at about 45° from dorsomedial to ventrolateral. The standard account of the cellular morphology of the nucleus is derived from the classical account of Ramon y Cajal (1909) and the later study of Stotler (1953), both based on transverse sections. The center of the plate is occupied by irregular rows of 4 – 5 fusiform cells, which at each of their poles give rise to 2 – 3 dendrites which are oriented perpendicular to the plane of the nucleus and extend to its medial and lateral boundaries (see Fig. 22C). The horizontal layering of the dendrites of these cells (which are commonly termed "bipolar") is matched by that of the terminal axonal plexus (see Fig. 22A; Morest 1973; Ramon y Cajal 1909; Scheibel and Scheibel 1974; Schwartz 1984). At the extremes of the perpendicular dendrites there is a small population of *marginal* cells whose major dendritic arborization is in the plane of the nucleus, but which can also have perpendicular dendrites oriented toward the central cell band (see Fig. 22; Ramon y Cajal 1909; Schwartz 1977).

More recent Golgi and EM studies in which sagittal and horizontal sections were examined have questioned the exclusively "bipolar" nature of the central cells and have clarified the three-dimensional structure of MSO (Morest 1973; Scheibel and Scheibel 1974; Schwartz 1977). Schwartz (1977) reported that bipolar cells were the most commonly impregnated cell type in her material, but that many central cells were *multipolar* whereas others were predominantly rostrocaudally oriented. Most multipolar cells had some dendrites that spanned the width of the nucleus and some of these cells could appear bipolar in transverse sections. Kiss and Majorossy (1983) also reported considerable numbers of multipolar cells in the cellular core of MSO. Both studies described cells at the periphery of the central band that had perpendicular dendrites restricted to one-half of the width of the nucleus and were thus distinct from the marginal cells defined by dendritic arborization in the plane of the nucleus. Scheibel and Scheibel (1974) described marked longitudinal extension of the dendritic domains of MSO (and LSO) neurons (Fig. 25) and, although they presented no quantitative data, implied that all central cells must be considered multipolar rather than bipolar. They also described the terminal axonal plexus as forming horizontally compressed, longitudinally extended sheets (Fig. 25). These data, and the earlier report of Morest (1973), indicate that MSO is organized as a series of horizontal

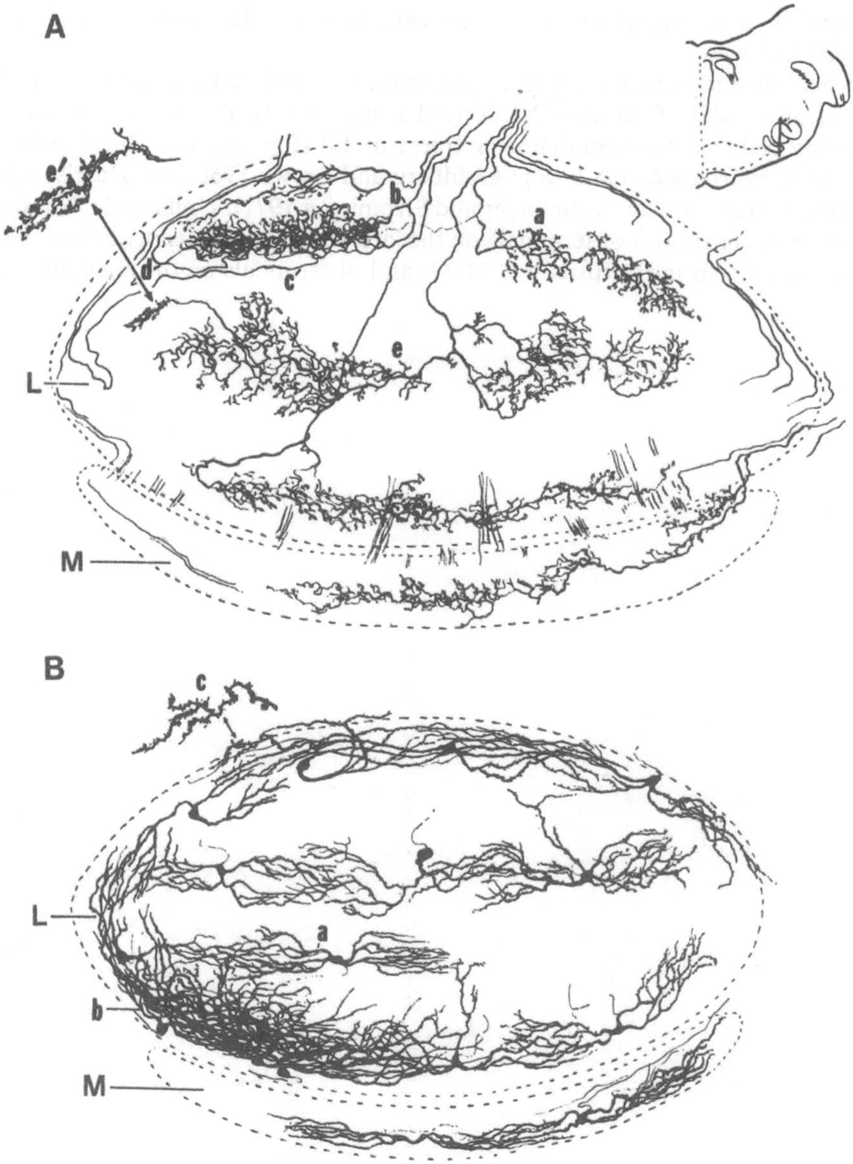

Fig. 25 A, B. Drawings from sagittal sections of Golgi-impregnated superior olivary complex in 10-day-old cat, showing **A** terminal patterns of afferent axons and **B** dendritic domains of neurons in lateral (*L* in each) and medial (*M* in each) superior olivary nucleus. *Line* through superior olivary complex in *insert* shows plane of section. In **A**: individual terminal arbors extend for approximately half (*a*) and the full extent (*e*) of the nucleus. Fibers (*b, c,* and *d*) mutually generate a heavily overlapped neuropil sheet which is believed to represent a more usual picture than the simple isolated arbors at *a* and *e*. The enlarged terminal at *e'* shows some details of a terminating axon system. In **B**: neuron at *a* shows typical dendritic domain of olivary cells, while more heavily stained part of section at *b* shows density of dendritic neuropil and degree of overlap. At *c*, terminal dendrite segment is enlarged to show dendrite spines and excrescences. (Scheibel and Scheibel 1974)

laminae or sheets, of which the classical transverse illustrations provide a cross-sectional view.

Ablation studies in a number of species have established a massive bilateral projection to MSO from AVCN, derived from the rostral area of large spherical/bushy cells via medium-diameter fibers in TB (Fig. 26; van Noort 1969; Osen 1969b; Warr 1966, 1982; dog: Goldberg and Brown 1968; rat: Harrison and Irving 1966a; monkey: Strominger and Strominger 1971). Unilateral CN lesions result in degeneration of terminals on the laterally oriented dendrites and the soma of central cells in the ipsilateral MSO, and of terminals on the medially oriented

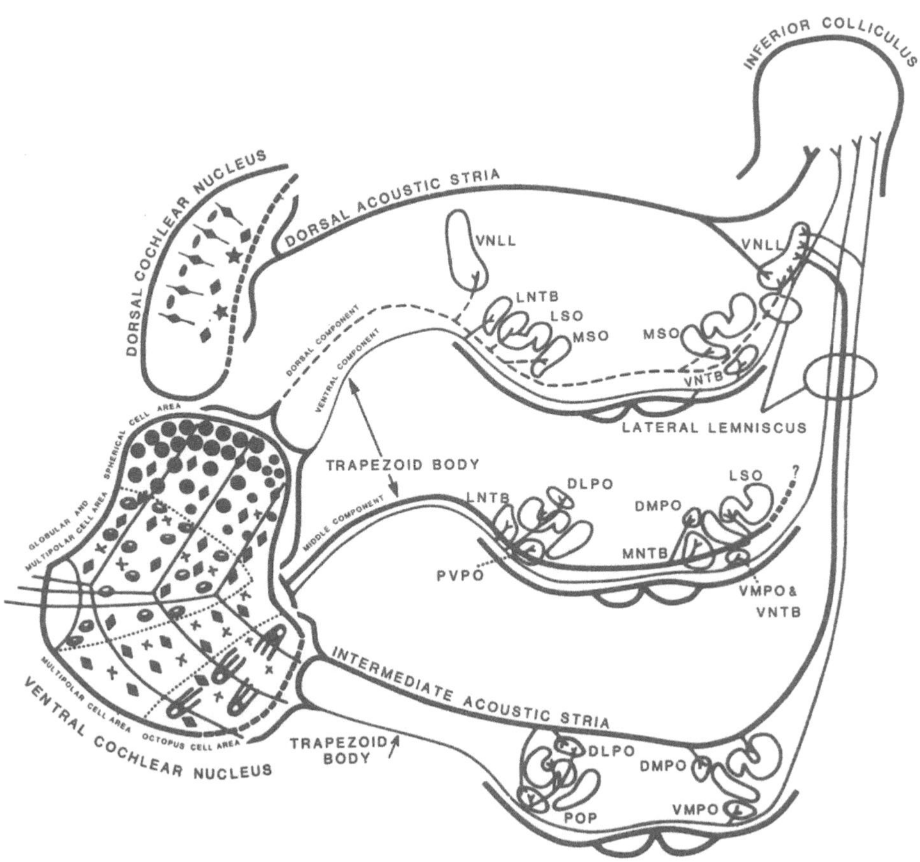

Fig. 26. Schematic summary of projections from major subdivisions of cat cochlear nucleus (CN), which are identified by labels along margins of nucleus. Periolivary cell groups, distinguished by their position relative to the major SOC nuclei, are here designated as discrete nuclei. Abbreviations: *DLPO* dorsolateral periolivary nucleus; *DMPO* dorsomedial periolivary nucleus; *LNTB* lateral nucleus of trapezoid body; *LSO* lateral superior olivary nucleus; *MNTB* medial nucleus of the trapezoid body; *MSO* medial superior olivary nucleus; *POP* posterior periolivary nucleus; *PVPO* posteroventral periolivary nucleus; *VMPO* ventromedial periolivary nucleus; *VNLL* ventral nucleus of the lateral lemniscus; *VNTB* ventral nucleus of the trapezoid body. (Modified slightly from Warr 1982)

dendrites and the soma of principal cells in the contralateral MSO (Lindsey 1975; van Noort 1969; Schwartz 1984; Stotler 1953; Warr 1982; rabbit: Borg 1973a; dog: Goldberg and Brown 1968; rat: Harrison and Irving 1966a; chinchilla: Perkins 1973; monkey: Strominger and Strominger 1971). In some cases a small number of "crossed" terminals are seen on the proximal portion of dendrites oriented away from the side of the lesion (Stotler 1953; Warr 1966; dog: Goldberg and Brown 1968; chinchilla: Perkins 1973). AVCN on each side of the brain therefore projects to each MSO, and central cells in each MSO receive input from each CN. The segregation of input from ipsilateral and contralateral AVCN to the laterally and medially oriented dendrites of the central cells suggests that these neurons are ideally suited to act as comparators of the input from the two ears, a suggestion that is supported by physiological data (see Sect. 5.2.3). In contrast to the central cells, marginal cells receive CN afferents predominantly from only one side, those on the lateral and medial borders from ipsilateral and contralateral CN, respectively (Lindsey 1975).

As Fig. 26 illustrates, much of the innervation of SOC nuclei is provided by collaterals of axons proceeding either to higher centers or to other SOC divisions (Morest 1968a; van Noort 1969; Ramon y Cajal 1909; Warr 1966, 1982), and Schwartz's (1984) Golgi impregnations indicate that the input to MSO from ipsilateral AVCN is derived from collaterals of TB axons that continue across the midline. Warr (1966) reported that the projections to ipsilateral and contralateral MSO from rostral AVCN were derived from different fiber components of the TB; although this observation appears to make it unlikely that single bushy cells in AA send collaterals to MSO on both sides, this question can only be unequivocally settled by double-retrograde labeling experiments.

A number of classes of axon additional to the principal CN input have been described in Golgi impregnations of MSO. Schwartz (1984) identified three classes of axon as extrinsic, some of which in kittens appeared to be derived from DMPO, and one group as intrinsic. The sources of extrinsic projections will be hard to establish in adults because of the difficulty of making deposits of retrograde tracers restricted to MSO.

Most EM studies have distinguished a number of types of terminal on MSO neurons, although the distinctions have been based on different criteria and the number of identified classes has varied (Clark 1969; Kiss and Majorossy 1983; Lindsey 1975; Schwartz 1972, 1980, 1984; chinchilla: Perkins 1973). There is general agreement that the most common class consists of large, often elongated endings that contain large spherical or oval vesicles and make multiple asymmetric synaptic contacts. These terminals degenerate after destruction of CN (Kiss and Majorossy 1983; Lindsey 1975; Schwartz 1984; chinchilla: Perkins 1973), and are consequently identified as the terminals of the principal CN input. They have been described as having many ultrastructural features in common with MNTB calyces (Schwartz 1972), and comprise the majority (85% – 90%) of all terminals on the central cells, the somas and dendrites of which are densely covered with an almost uninterrupted layer of terminals (Schwartz 1980; chinchilla: Perkins 1973). The remaining terminals are smaller and contain smaller vesicles; they have been divided into varying numbers of subgroups by various investigators on the basis of either vesicle size or shape (Clark 1969; Kiss and Majorossy 1983; Lindsey 1975; Schwartz 1980; chinchilla: Perkins 1973). As in other structures, the characteristics of the small vesicles are dependent on fixation procedures (Schwartz 1984; cf. Nakajima 1971). Although there is general agreement that the smaller endings do not degenerate after CN ablations, the differences between studies in the classification of these endings, and in accounts of their distribution and origins, do not permit of an integrated account of their characteristics. Schwartz's (1980) report of a much higher proportion of terminals with small vesicles on marginal than on central cell somas (22.6% against 3.5%) is of interest, however, in respect to the binaural input to some MSO cells (Sect. 5.2.2).

Degeneration (van Noort 1969; Stotler 1953; kangaroo rat: Browner and Webster 1975) and tracer studies (Adams 1979; Aitkin and Schuck 1985; Brunso-Bechtold et al. 1981; Elverland 1978; Glendenning and Masterton 1983; Henkel and Spangler 1983; Roth et al. 1978; bat: Zook and Casseday 1982b) in a number of species indicate that the ascending projections of MSO neurons are almost entirely restricted to the ipsilateral IC and NLL, with only a very small projection to contralateral IC. The possibility of a significant species difference is, however, suggested by the recent report of Willard and Martin (1984) that almost all MSO neurons projecting to IC in the American opossum send collaterals to IC on both sides. There is some disagreement concerning the cell types giving rise to ascending projections in the cat. Adams (1979) and Henkel and Spangler (1983) reported that both bipolar (fusiform) and multipolar central cells, and some marginal cells, were labeled after HRP injections into IC. Aitkin and Schuck (1985) also reported labeling of both bipolar and multipolar central cells after such injections, but Kiss and Majorossy (1983) reported that no multipolar cells were labeled. The detailed organization of MSO's ascending projections will be considered in association with accounts of the target structures in Sects. 6.1 and 7.1.3 (see Figs. 34 and 37).

Although there have been some reports of descending projections from MSO, the data are open to question. Kane and Finn (1977) reported that some central MSO cells were labeled after HRP injections into CN. However, this finding is not supported by other HRP studies (Adams 1983b; Warr 1975; guinea pig: Strutz and Spatz 1980), and Adams (1983b) has suggested that it might be an artifactual consequence of unusually high concentrations of chromogen (diaminobenzidine) in their incubation medium. Borg (1973b) reported terminal degeneration in the facial and trigeminal motor nuclei in the rabbit after medial SOC lesions, and argued that MSO neurons were involved in the pathway for acoustically elicited reflex activity of the middle ear muscles. This conclusion must be seen as tentative, however, as the data do not exclude the possibility that the relevant neurons were located in periolivary cell groups adjacent to MSO. Although the middle ear muscles have important regulatory and antimasking effects (e.g., Borg and Zakrisson 1975; Irvine 1976; Møller 1974a), they are outside the scope of this review.

5.1.4 Lateral Superior Olivary Nucleus

In transverse sections, the LSO appears as a plate of cells arranged in an S-shaped formation, with the medial hilus (concavity) directed ventrolaterally and the lateral hilus directed dorsomedially (see Figs. 1 and 22). Our knowledge of the cellular morphology of LSO rests largely on the classical Golgi studies of Ramon y Cajal (1909) and on Ollo and Schwartz's (1979) observations on the mouse. Transverse sections of LSO show a central band of apparently fusiform cells with polar dendrites oriented approximately perpendicular to the axis of curvature of the nucleus, and marginal cells at the borders of the nucleus (see Fig. 22 B). As in MSO, sagittal sections reveal that the central cells are predominantly multipolar, with dendritic plexuses forming rostrocaudally oriented sheets (Scheibel and Scheibel 1974; mouse: Ollo and Schwartz 1979; see Fig. 25 B). The axonal plexus forms similarly oriented sheets that mimic the organization of the dendritic arbors (see Figs. 22 A and 25 A, Ramon y Cajal 1909; Scheibel and Scheibel 1974). Thus LSO, like MSO, is organized as a series of rostrocaudally extended laminae. Scheibel and Scheibel's (1974) account was largely restricted to organization of the neuropil in LSO, however, and there has been no detailed modern study of cell types in cat LSO, although Cant (1984) confirmed the presence of fusiform cells with disk-shaped dendritic fields in Golgi impregnations of adult cats. Ollo and Schwartz (1979) described a population of spindle-shaped cells in addition to multipolar and marginal cells in LSO of the mouse.

The major sources of afferents to LSO are ipsilateral AVCN and MNTB; in the recent studies of Glendenning et al. (1985), over 95% of cells labeled after HRP injections restricted to LSO were located in these two nuclei. The projections from MNTB were described in Sect. 5.1.2 and are the second link in a pathway from contralateral AVCN that is apparently specialized for rapid, secure transmission. An important observation by Glendenning et al. (1985) is that this projection is much less strong to the lateral limb of LSO than to the rest of the nucleus. The projection to LSO from ipsilateral AVCN is derived from the spherical/bushy cell region (Glendenning et al. 1985; Osen 1969 b; Warr 1966), and the HRP data of Glendenning et al. indicate that most of the fibers projecting to LSO are collaterals of TB axons that continue across the midline. Glendenning et al. (1985) did not identify the particular cell classes in AVCN that were labeled after their HRP injections, and the precise source of the projection is unclear. Osen (1969 b) suggested that it is derived from small spherical/bushy cells, in contrast to the projection to MSO from large spherical/bushy cells. Although the major projection to LSO is probably from small spherical/bushy cells, it is not clear to what extent stellate cells – which are intermingled with bushy cells in the region corresponding to Osen's small spherical cell area (viz., AP; see Sect. 4.1.2) – might contribute to the projection. Furthermore, both Warr (1982) and Harrison and Irving (1966 a; in the rat) reported degeneration in both ipsilateral LSO and MSO after lesions restricted to the large-cell rostral pole of AVCN (region III in the rat, which corresponds approximately to area AA and Osen's large spherical cell area; see Sect. 4.1.2). Whether the same bushy cells send collaterals to both nuclei (as suggested by Warr 1966, 1982), or whether there is partial segregation (associated with size or some other characteristic) of the large bushy cells projecting to MSO and LSO, are issues that await investigation with double-retrograde tracers.

In the cat and a number of other species, it has generally been reported that terminal degeneration in LSO after unilateral AVCN lesions is restricted to the nucleus ipsilateral to the lesion (van Noort 1969; Osen 1969 b; Stotler 1953; Warr

1966; kangaroo rat: Browner and Webster 1975; rat: Harrison and Irving 1966a; monkey: Strominger and Strominger 1971). However, Warr (1972, 1982) and Glendenning et al. (1985) have reported a small area of preterminal degeneration in the ventromedial tip of contralateral LSO in the cat after lesions in the globular cell area, and Goldberg and Brown (1968) reported a sparse projection throughout contralateral LSO in the dog. The fact that the projection described by Warr (1972, 1982) and Glendenning et al. (1985) has not been reported in other studies of the cat suggests that sparse or restricted contralateral projections to LSO might have gone undetected in other species.

There has been little attention to the structure and distribution of synaptic terminals in LSO. Stotler (1953) and Harrison and Warr (1962) described the afferent fibers in the cat and rat, respectively, as terminating by means of boutons on the cell bodies and dendrites. Cant (1984) has recently presented what appears to be the only report of synaptic fine structure in the nucleus. She described three types of terminal on the principal (fusiform or multipolar) cells in the cat. Terminals containing small, flattened, or cylindrical vesicles and making slightly asymmetric contacts were found predominantly on the soma and proximal dendrites. Terminals with large round vesicles and making strongly asymmetric contacts were found predominantly on distal dendrites and only occasionally on the soma. A third class of terminal with small round vesicles was encountered only rarely. The origins of these terminals were not determined experimentally, although physiological evidence that will be considered in Sect. 5.2.2 led Cant to suggest that the terminals containing flattened vesicles and located on the soma were derived from the ipsilateral MNTB, and those with large round vesicles contacting the distal dendrites were from ipsilateral AVCN. The former suggestion is supported by the finding of Glendenning et al. (1985) that terminal labeling in LSO after injections of anterograde tracers into MNTB is mainly perisomal.

In a number of species, LSO has been shown to project *bilaterally* to DNLL and IC (Adams 1979; Brunso-Bechtold et al. 1981; Elverland 1978; Glendenning et al. 1981; Glendenning and Masterton 1983; van Noort 1969; Roth et al. 1978; Stotler 1953; kangaroo rat: Browner and Webster 1975; American opossum: Willard and Martin 1983, 1984; bat: Zook and Casseday 1982b). The detailed organization of these projections will be discussed in later sections (see Sects. 6.1 and 7.1.3 and Figs. 34 and 37). The topographic organization of the projecting neurons in the cat has recently been described in studies using a range of tract-tracing techniques (Glendenning and Masterton 1983). Cells projecting ipsilaterally are concentrated in the lateral limb of LSO, while those projecting contralaterally are concentrated in the medial limb (see Fig. 38). A small proportion of LSO neurons ($<5\%$) are shown by double-retrograde tracing techniques to project bilaterally via axon collaterals. In the American opossum, a higher proportion of central LSO neurons project bilaterally to IC, but no topographic organization of projection neurons has been described (Willard and Martin 1984). Whether marginal cells contribute to the ascending projections in the cat is not known.

Following injections of HRP into the cochlea in the cat, labeling of some cells on the margins of LSO and of a dense mass of periolivary cells in the dorsal hilus is seen bilaterally (Fig. 27; Adams 1983b; Warr 1975; guinea pig: Strutz and Spatz 1980). These cells, which are thus identified as sending their axons into the crossed or uncrossed OCB, are also AChE-positive (Osen and Roth 1969; Osen et

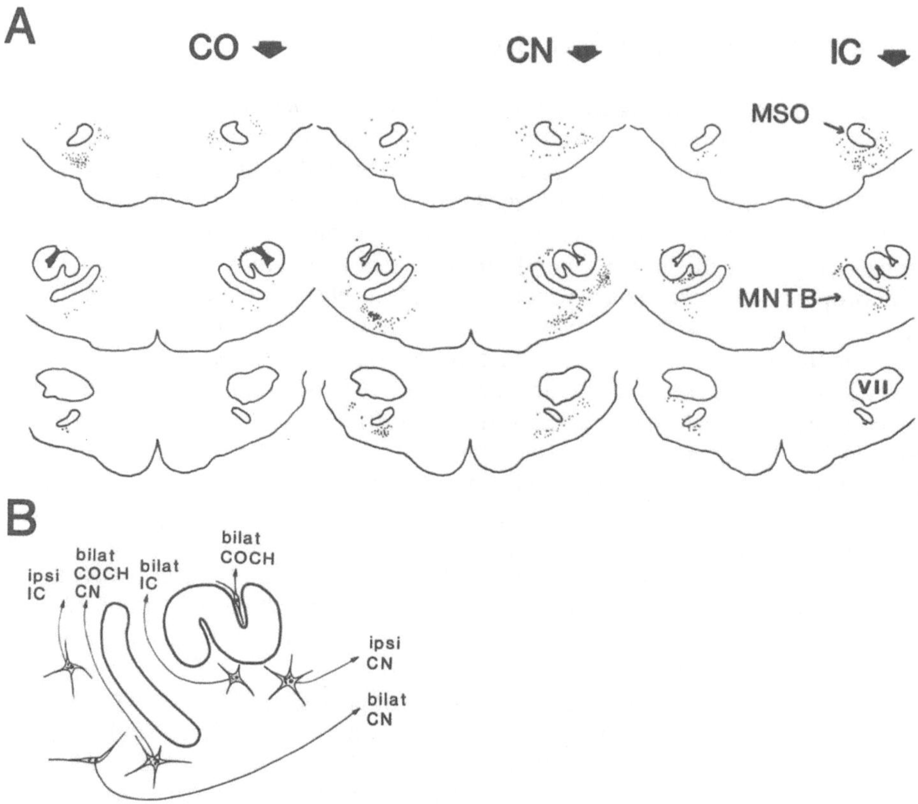

Fig. 27. A Frontal sections through three levels of the superior olivary complex of the cat, showing positions of labeled cells (*dots*) following HRP injections of the cochlea (*CO*), cochlear nucleus (*CN*), and inferior colliculus (*IC*). Rostralmost sections are at top. In all cases, the structure receiving the injection was on the right, as indicated by the *arrow*. Note that dense mass of labeled cells in dorsal hilus of LSO after cochlear injection (*left-hand column*) appears as a *solid black area*. **B** Summary diagram showing six periolivary cell classes distinguished by Adams (1983b), with their projections, either ipsilateral (*ipsi*) or bilateral (*bilat*), to various target nuclei and to the cochlea (*COCH*) indicated. Note that indication of bilateral projection of a given class does not necessarily mean that individual cells have bilateral projections. (Adams 1983b)

al. 1984; Warr 1975). Although some marginal cells thus give rise to descending projections, no principal or other cells in the interior of LSO of the cat are either retrogradely labeled after cochlear injections or exhibit AChE activity. There are major species differences in this respect, however, since cochlear injections of retrograde tracers in rodents label large numbers of cells throughout ipsilateral LSO (guinea pig: Stopp 1983; Strutz 1981; Strutz and Bielenberg 1984; rat: White and Warr 1983), and small (i.e., nonprincipal) cells within LSO are AChE-positive (rat, mouse, and chinchilla: Osen et al. 1984).

5.1.5 Other Superior Olivary Cell Groups

The cell groups comprising the lateral and ventral nuclei of the trapezoid body (LNTB and VNTB) and the various periolivary "nuclei" are poorly defined, and details of neuronal architecture and afferent organization are available only for one of the periolivary cell groups (viz., DMPO; Morest 1968a, 1973). Because of their poor definition, the validity of the conventional definition of these cell groups as specific nuclei has recently been questioned. Thus, Osen et al. (1984) have proposed that periolivary neurons should simply be identified by their location in various "zones" identified in terms of position relative to the major nuclei. The theme that has emerged most strongly from recent research on CN, and that was stressed in Sect. 4, is that morphologically different cell types with distinctive physiological characteristics and projections can be present within the boundaries of conventionally defined "nuclei". Adams (1983b) has emphasized this point in arguing that the study of periolivary neurons (broadly defined) should proceed by defining the morphology of cells with particular projection patterns, independent of their location in arbitrarily defined nuclei. These points are well taken, and the term "periolivary" will be used here to refer generally to neurons conventionally assigned to particular periolivary "nuclei". It will be necessary in referring to particular research findings, however, to use the nuclear designations of the original reports.

There is general agreement that, in a number of species, LNTB receives ascending projections only from the ipsilateral CN (Fernandez and Karapas 1967; van Noort 1969; Powell and Cowan 1962; Stotler 1953; Warr 1966, 1972, 1982; rabbit: Borg 1973a; dog: Goldberg and Brown 1968; rat: Harrison and Irving 1966a). The projections are derived principally from multipolar (stellate) cells in rostral PVCN (Warr 1972, 1982), although there is a small projection to the anterior part of LNTB from rostral AVCN (Warr 1966, 1982; rat: Harrison and Irving 1966a). The terminals on LNTB cells comprise boutons and small modified calyces resembling those in the intermediate portion of VCN (Stotler 1953; rat: Harrison and Irving 1966a). Stotler (1953) also described only ipsilateral projections to VNTB, but subsequent studies in a number of species have indicated that the major projection to VNTB is in fact from the contralateral rostral PVCN (Fernandez and Karapas 1967; Glendenning et al. 1985; van Noort 1969; Warr 1972, 1982; rabbit: Borg 1973a; dog: Goldberg and Brown 1968). Some of these studies have also indicated a weaker ipsilateral input to VNTB (Fernandez and Karapas 1967; dog: Goldberg and Brown 1968). Unpublished observations cited by Goldberg and Brown (1968) indicate that in the cat the contralateral projection is concentrated in the medial part of the nucleus, while the ipsilateral fibers terminate more laterally.

The projections to the other periolivary cell groups from the ipsilateral and contralateral CN show a similar pattern of segregation as those to LNTB and VNTB. In general, as illustrated schematically in Fig. 26, those periolivary cell groups located lateral to the long axis of MSO receive projections predominantly (but not exclusively) from the ipsilateral CN, while those located medial to this axis receive projections predominantly from the contralateral CN (Warr 1969, 1972, 1982). These projections are derived via overlapping projections from the OCA and caudal parts of the multipolar cell area of PVCN (via IAS and TB; Fernandez and Karapas 1967; Warr 1969, 1982; rabbit: Borg 1973a), and from the rostral multipolar cell area of PVCN and adjacent globular cell area (via TB; Warr 1972, 1982). A projection from globular cells to at least some contralateral periolivary cell groups is supported by Morest's (1968a, 1973) description of precalycine collaterals terminating in DMPO. The DAS has also been reported to give off collaterals to the dorsal periolivary region (Fernandez and Karapas 1967). Finally, the periolivary region (broadly defined to include LNTB and VNTB) also receives descending projections from IC and from NLL (Andersen et al. 1980a; Kane and Conlee 1979; Kudo 1981; Whitley and Henkel 1984).

Stotler (1953) reported that both LNTB and VNTB project via the lateral lemniscus, and weak ascending projections to NLL and IC from these nuclei and from other periolivary cell groups have been confirmed by subsequent tracer studies in a number of species (Adams 1979, 1983b; Brunso-Bechtold et al. 1981; Glendenning et al. 1981; Roth et al. 1978; rat: Beyerl 1978; gerbil: Nordeen et al. 1983; bat: Zook and Casseday 1982b). However, these regions are also the origin of major descending projections to the cochlea and – via the crossed and uncrossed OCBs – to CN (Adams 1983b; Adams and Warr 1976; Elverland 1977; Glendenning and Masterton 1983; Guinan et al. 1983, 1984; Kane 1976b; Kane and Finn 1977; Luk et al. 1974; van Noort 1969; Osen and Roth 1969; Osen et al. 1984; Rasmussen 1967; Warr 1975; guinea pig: Strutz and Spatz 1980). Adams's (1983b) summary representation of the locations of periolivary cells labeled following injections of HRP into the cochlea, CN, or IC is reproduced in Fig. 27. For the reasons indicated earlier, the labeled neurons have not been assigned to particular nuclei. Although there is considerable overlap in the distributions in some regions (e.g., in the region ventral to MSO in the most rostral sections), there is also a significant degree of segregation of neurons projecting to different targets. Thus, there is a major concentration of cells projecting to CN in the area ventral to MSO and LSO which corresponds to VNTB and, less certainly, LNTB. In areas in which morphologically similar cell types projected to IC and to the cochlea, injections of different retrograde tracers did not reveal any cells sending axons to both structures (Adams 1983b).

Recent anterograde tracer studies have led to a revision of the traditional distinction between crossed and uncrossed olivocochlear systems (e.g., Rasmussen 1960). Olivocochlear neurons located lateral to MSO project bilaterally (but more strongly ipsilaterally) to the region beneath the IHCs, whereas medially located olivocochlear neurons project bilaterally (but more strongly contralaterally) to the region beneath the OHCs (Guinan et al. 1983, 1984; Warr and Guinan 1979). A similar distinction between medial and lateral systems has been described in the rat by White and Warr (1983), but as noted previously, the cells of origin in the rat are mainly located in LSO and in VNTB rather than in periolivary regions as in the cat (see also Osen et al. 1984).

This brief account of the connections of the neuronal groups surrounding the major SOC nuclei points to a gross distinction between the major and surrounding nuclei. Neurons in MSO and LSO receive almost all of their input either directly or indirectly (via MNTB) from various cell types in AVCN, and have almost exclusively ascending projections to NLL and IC. Conversely, neurons in the surrounding (broadly defined periolivary) nuclei receive their input predominantly from PVCN and have predominantly descending projections to CN and to the cochlea. These differences suggest a functional distinction between these broadly defined projection systems that will be examined in more detail after consideration of the physiological evidence on processing in SOC. However, the major involvement of periolivary neurons in descending auditory systems whose function and mode of operation are little understood precludes any detailed account of their role in auditory processing. The function of the OCB remains elusive (see Dallos 1981; Klinke and Galley 1974 for reviews), but almost certainly involves modulation or control of mechanical and/or electrical events in the cochlea (e.g., Mountain 1980; Siegel and Kim 1982). The function (or functions) of descending fibers to CN, as of descending projections at other levels of the auditory system, are even less understood.

5.2 Physiology of Superior Olivary Complex Neurons

5.2.1 Patterns of Binaural Input and Interaction: Classification of Binaural Neurons

The SOC receives projections from CN on both sides of the brain, and the majority of neurons in MSO and LSO receive binaural input. The classification of binaural neurons presents difficulties that have much in common with those encountered in the classification of CN neurons in terms of response patterns. As is the case with CN, it is hoped that a classification system will emerge that is based on covariance between sets of physiological and morphological features of neurons. In the absence of the detailed knowledge necessary for the formulation of such a classification scheme, a variety of descriptive schemes based on the nature of the input from the two ears and of the interactions between these inputs have been proposed. The simplest and most influential has been that employed by Goldberg and Brown (1968, 1969), and the scheme to be developed here is an elaboration of their original classification. Although the initial impetus for such a classification scheme arose from the need to describe SOC neurons, any scheme should be sufficiently general to incorporate patterns of binaural input and interaction at other levels of the auditory pathway. For this reason, it will be necessary to refer to data from higher levels in elaborating the scheme to be used in this review.

Unless otherwise noted, all of the physiological data to be reviewed in this and following sections have been obtained using dichotic stimulation through sealed systems of the type described in Sect. 3. As noted in Sect. 4.2.4 (see footnote on p. 68), the amount of interaural crosstalk in such systems depends on apparatus and procedural factors. Very few investigators have reported crosstalk values for their particular stimulus arrangements, and the extent to which crosstalk might have contributed to particular results is consequently unclear.

Goldberg and Brown (1968, 1969) distinguished two major classes of binaural neuron in MSO and VNTB of the dog. The neurons in one class exhibited a predominantly excitatory response to stimulation of either ear, and were therefore designated excitatory–excitatory (EE). The second class comprised neurons that exhibited a predominantly excitatory response to stimulation of the contralateral ear and a predominantly inhibitory response to stimulation of the ipsilateral ear; these neurons were therefore designated excitatory–inhibitory (EI). Not all neurons are binaural, however, and in some cases stimulation of one or the other ear will have no effect. If neurons are classified in terms of whether stimulation of each ear has a predominantly excitatory (E), predominantly inhibitory (I), or no effect (O), nine categories are generated (Table 1). Following Goldberg and Brown, each of these categories is identified in Table 1 by a letter pair specifying the effect of stimulation of each ear (E, I, or O), in the order contralateral–ipsilateral.

The apparent simplicity and comprehensiveness of this classification scheme are unfortunately both illusory, and a number of qualifications and further bases for classification must be considered. Goldberg and Brown (1969) noted that although the neurons in their sample could be classified as EE or EI in terms of the *predominant* effect of stimulation of each ear, many of these neurons could be shown "under appropriate circumstances" to receive both excitatory and inhibitory inputs from one or both ears (Goldberg and Brown 1969, p. 630). As indicated in later sections, the "appropriate circumstances" referred to by Goldberg and Brown include particular conditions of IID and ITD. In a comprehensive study of SOC neurons in the cat, Guinan et al. (1972a) added mixed excitation and inhibition to the three classes of monaural effect used in Table 1, generating a 4×4 classification table. However, they included under this heading only those units that showed excitation and inhibition for different *monaural* stimulus parameters (e.g., units whose excitatory tuning curves had inhibitory sidebands), and the seven categories additional to those in Table 1 together contained only a small proportion of their sample of binaurally classified units (12/235 = 5%). The fact that so few cells fall into these additional categories, combined with the observation that a much higher proportion of cells reveal mixed excitatory and inhibitory input from one or both ears under appropriate *binaural* stimulation conditions, suggests that mixed input as defined by Guinan et al. (1972a) does not provide a reliable basis for classification of binaural units, and it is therefore not included in the present system. The importance of mixed excitatory and inhibitory influences in binaural processing will be emphasized frequently in subsequent sections, however, and reinforces the point made by Goldberg and Brown (1968, 1969) that the E and I designations should be interpreted as indicating *predominant,* not *exclusive,* excitatory or inhibitory input from a given ear.

Although the classification scheme presented in Table 1 is nominally based on the influence of monaural stimulation of each ear, the nature of the input from each ear is not always apparent from monaural stimulation. Most obviously, the low

Table 1. Classification of binaural input patterns based on predominant effect of monaural stimulation of each ear

		Ipsilateral stimulation		
		Excitation	Inhibition	No effect
Contralateral stimulation	Excitation	EE	EI	EO
	Inhibition	IE	II	IO
	No effect	OE	OI	OO

Table 2. Classification of EO and OE cells on basis of relationship between binaural response (Bin) and monaural excitatory response (E)

Bin < E	Bin = E	Bin > E
Inhibition EO/I and OE/I (or EI and IE)	Monaural EO/mon and OE/mon	Facilitation EO/F and OE/F

level or absence of spontaneous activity in many SOC (particularly LSO) neurons in anesthetized preparations has the consequence that monaural inhibition cannot be identified on the basis of extracellular recordings. Most EI (or IE) cells would be classified as EO or OE on the basis of monaural responses, and the inhibition is revealed only by comparison of the response to binaural stimulation with the monaural excitatory response. Table 2 indicates that this comparison can result in monaurally-defined EO or OE cells being assigned to one of three categories. A binaural response smaller than the monaural excitatory response indicates that the input from the other ear is inhibitory. Since this inhibition would presumably have been detected in the presence of spontaneous activity, these units are normally classified as EI (or IE), and this practice will be followed here[16]. If the binaural response does not differ from the monaural excitatory response across a wide range of parameters, the unit is judged to be monaural (EO/mon and OE/mon in Table 2). Monaural inhibitory responses might also be detected in the presence of spontaneous activity and such cells, which appear to have been reported only in DCN, would be designated IO/mon or OI/mon. The third class of EO (or OE) cells shown in Table 2 are those in which the binaural response exceeds the monaural excitatory response, indicating that the ipsilateral input is facilitatory (EO/F and OE/F). Although the presence of neurons of this sort in SOC has been reported (Brownell et al. 1979; Tsuchitani 1977; Watanabe et al. 1968; chinchilla: Langford 1984), they have not been described in detail. In primary auditory cortex (area AI) and in the deep layers of the superior colliculus (SC), however, neurons of this class have been shown to exhibit a distinctive form of sensitivity to IIDs (Irvine and Wise 1983a; Phillips and Irvine 1981b, 1983; Wise and Irvine 1983, 1984, 1985; see Sect. 7.5). Although the facilitatory input from the monaurally ineffective ear in these cases is basically excitatory, the facts that this input does not elicit a response at any intensity and that these neurons have a distinctive form of IID sensitivity (see Sect. 7.5 and Fig. 53), justify their classification as distinct from monaurally defined EE cells.

[16] Note, however, that there could be situations in which a distinction between EI and EO/I units would be useful. This might be the case if the presence or absence of spontaneous activity proved to be an important difference between cell types receiving inhibitory ipsilateral input (cf. Young and Voigt's 1982 distinction between type II and III cells in DCN). Alternatively, it is not inconceivable that binaural inhibition could occur in the absence of monaural inhibition of spontaneous activity, e.g., if the binaurally manifested inhibition were exerted presynaptically on inputs other than those responsible for the spontaneous activity of the cell.

These arguments also apply to the class of cells characterized by the most dramatic form of binaural facilitation. Cells unresponsive to monaural stimulation of either ear (OO in Table 1) would appear to be, by definition, nonauditory. However, there is a class of binaural neurons that are unresponsive to monaural stimulation, but show marked facilitation to binaural stimulation and are hence designated OO/F. Although Galambos et al. (1959) mentioned the existence of such cells in the first reported microelectrode study of SOC, the only subsequent account of them in SOC is in a recent study in the chinchilla by Langford (1984). These neurons constitute a significant subgroup in primary auditory cortical area AI (Kitzes et al. 1980; Phillips and Irvine 1981 b, 1983) and in the deep layers of SC (Irvine and Wise 1983 a; Wise and Irvine 1983, 1984, 1985), and exhibit distinctive forms of IID (see Sect. 7.5 and Fig. 52) and transient ITD (Kitzes et al. 1980) sensitivity. Whether the failure to describe OO/F cells in recent studies of SOC in the cat indicates that they do not appear in significant numbers until higher levels, or whether it is simply an artifact of the use of monaural search stimuli, is not clear. Evidence that the OO/F properties of some midbrain neurons in the barn owl are derived from EI properties at lower levels (Moiseff and Konishi 1983 a, b; see Sect. 7.3) provides support for the former alternative. Conversely, the fact that they were observed by Galambos et al. (1959) and have been reported to comprise a large proportion (46%) of a sample of SOC neurons in the chinchilla (Langford 1984) supports the latter view.

Goldberg and Brown (1968, 1969) also distinguished different forms of binaural interaction in cells receiving EE input. As indicated in Table 3, they employed the terminology previously established in spinal physiology to describe interactions between two excitatory inputs: cells were classified as exhibiting facilitation, summation, or occlusion (EE/F, EE/S, and EE/O), depending on whether the binaural response exceeded, was equal to, or was less than, the sum of the monaural responses. In some recent studies of binaural interaction a different convention has been adopted in which the term "summation" is applied to any case in which the binaural response exceeds the stronger monaural response. Summation so defined encompasses not only true summation, but all cases of facilitation and many cases of occlusion (viz., those in which the binaural response is less than the sum, but greater than the stronger monaural response). Apart from the semantic peculiarity of applying the term "summation" to cases in which the binaural response is not the sum of the monaural responses, this practice obscures important differences. For example, EE/F cells have been found to exhibit sensitivity to IIDs that is not exhibited by nonfacilitatory EE cells (see Sects. 7.2.4 and 7.5). A final point noted by Goldberg and Brown (1968, 1969), and by later

Table 3. Classification of EE cells on basis of relationship between binaural response (Bin) and sum of monaural responses (Sum)

Bin < Sum	Bin = Sum	Bin > Sum
Occlusion (EE/O)	Summation (EE/S)	Facilitation (EE/F)

investigators of SOC (e.g., Caird and Klinke 1983; rat: Inbody and Feng 1981), is that the input from one ear in EE cells (or in EI and IE cells) can have a stronger effect on the cell than that from the other. Caird and Klinke (1983) used capital and lowercase letters to identify the dominant and nondominant inputs (e.g., Ee and eE). This would appear to be a useful convention in cases in which aural dominance is an issue of concern.

The binaural classification scheme elaborated here is purely descriptive and, in many respects, arbitrary. Like the CN response pattern classification scheme, it is not intended to be accepted dogmatically (cf. Godfrey et al. 1975a) or to be regarded as providing formal definitions of "cell types." Like that system, it suffers from the facts that a given unit's binaural input pattern may not be invariant across its response area and that, in some cases, a given unit may be difficult to assign to a particular category. Nevertheless, as will be seen in the following sections, it provides a convenient means of describing and distinguishing the patterns of binaural input to the different subdivisions of SOC (and of higher levels of the auditory pathway). The fact that some cells with different patterns of binaural input distinguished by the scheme have proved to have different forms of sensitivity to IID has also provided a partial functional validation of it.

5.2.2 Response Properties and Tonotopic Organization of Neurons in Superior Olivary Complex Subdivisions

Description of the physiological response characteristics of neurons in the various SOC subdivisions is complicated by the fact that a number of the early reports were anecdotal in nature and assigned units to particular nuclei without adequate histological evidence. As an electrode passes through MSO, click-evoked slow potentials reverse polarity in a stereotyped fashion (e.g., Galambos et al. 1959; Tsuchitani and Boudreau 1964), and this polarity reversal was used in a number of studies to identify recording sites. Data reported by Guinan et al. (1972b) indicate that although polarity reversal is always seen in electrode penetrations through MSO, it is also seen in penetrations that do not pass through this nucleus. Identification of recording sites on the basis of this criterion alone is therefore unreliable.

Guinan et al. (1972a, b) described the physiological response characteristics of a large sample of SOC neurons in a study distinguished by exhaustive classification of response characteristics and great care in the accurate histological localization of recording sites. The following account will rely heavily, but not exclusively, on their data and on the detailed accounts of LSO and lateral periolivary nuclei in the cat by Boudreau and Tsuchitani (1968, 1970), Tsuchitani (1977, 1982), Tsuchitani and Boudreau (1967, 1969) and Tsuchitani and Johnson (1985), and of medial SOC in the dog by Goldberg and Brown (1968, 1969). The basic response characteristics of SOC neurons will be discussed in this section and the mechanisms by which interaural time and intensity differences are coded will be considered in the following two sections. A more detailed account of SOC physiology can be found in Goldberg's (1975) review.

1. Medial Nucleus of the Trapezoid Body. Of 218 units isolated in MNTB of the cat by Guinan et al. (1972a, b), 142 (65%) had complex waveforms in which the action potential was preceded by a positive-going "prepotential." As noted previously (Sect. 4.2.4), Li and Guinan's (1971) evidence establishes that the prepotential reflects presynaptic activity in cells receiving calyces of Held. Units with prepotentials are therefore identified as *principal* neurons in MNTB (Morest 1968a; see Sect. 5.1.2). The majority of units without prepotentials are therefore presumably stellate and elongate neurons.

All units with complex waveforms and the majority of other units in MNTB of the cat were found by Guinan et al. (1972a, b) to be EO/mon. The predominance of contralateral monaural input to MNTB confirms earlier preliminary observations in the cat (Galambos et al. 1959; Goldberg et al. 1963) and more detailed data from the dog (Goldberg and Brown 1968). As would be expected on the basis of the secure transmission afforded by the large-diameter TB fibers and calyceal endings, the response properties of principal neurons in MNTB closely resemble those of the large-diameter TB fibers (Brownell 1975) and of globular/bushy cells in posterior AVCN (see Sect. 4.2.4). Thus, units identified as principal neurons by the presence of prepotentials have primarylike with notch, primarylike, or phase-locked PSTHs, irregular firing patterns, and narrow V-shaped tuning curves (Guinan et al. 1972a, b). Brownell (1975) reported that the excitatory tuning curves of many large-diameter TB fibers were associated with inhibitory sidebands; although Guinan et al. did not present data on this issue, it seems probable that principal cells will prove to have inhibitory sidebands. Similarly, although no data are available on either the phase-locking or rate–intensity functions of principal cells, it can reasonably be anticipated that they will prove to have the secure phase-locking and monotonic saturating (i.e., sigmoid) rate–intensity functions characteristic of TB fibers (Brownell 1975) and AVCN "primarylike" and "primarylike with notch" units. Response latencies to clicks of principal neurons in MNTB of the cat (Guinan et al. 1972a, b) and of unidentified neurons in MNTB of the dog (Goldberg and Brown 1968) are consistently shorter than those of cells in other SOC nuclei, reflecting the rapid transmission afforded by the large-diameter fibers and calyceal terminals.

Principal cells in MNTB are tonotopically organized, with high-to-low CFs arranged in a ventromedial-to-dorsolateral direction (Guinan et al. 1972a, b), as illustrated schematically in Fig. 28. Although the three-dimensional tonotopic organization of MNTB has not been described, it seems likely that isofrequency (or frequency-band) sheets correspond to the structural laminae formed by the principal cell dendrites and peridendritic plexus (Morest 1968a). Because cell density is lower in the dorsolateral part of the nucleus, there is a disproportionate frequency representation in MNTB, with a much larger number of neurons having high than low CF (Guinan et al. 1972a, b).

The properties of cells other than those with prepotentials are less primarylike and more diverse than those of the principal cells. A significant proportion of nonprincipal cells respond only at stimulus offset and have fairly broad tuning curves. Guinan et al. (1972b) summarized a number of observations suggesting that these units might correspond to the stellate neurons described by Morest (1968a). They were unable to find any evidence of tonotopic organization of MNTB units other than those with prepotentials.

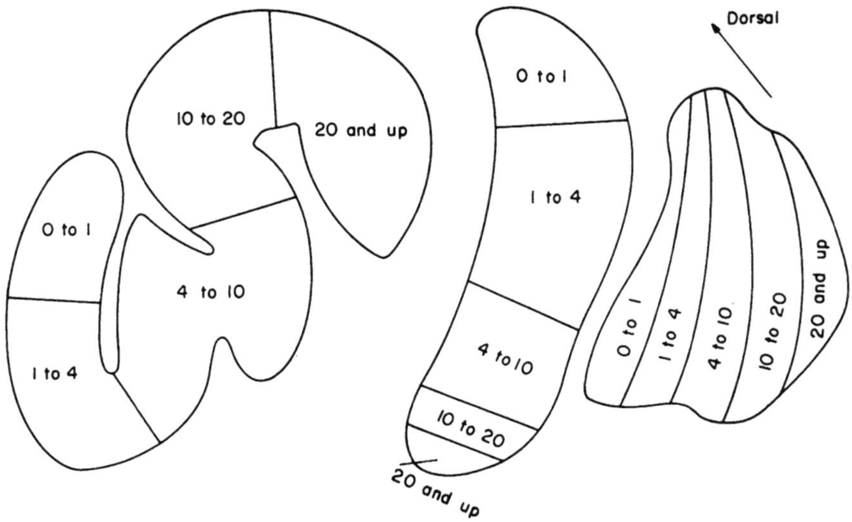

Fig. 28. Schematic illustration of tonotopic organization of lateral superior olivary nucleus (*left*), medial superior olivary nucleus (*center*), and medial nucleus of the trapezoid body (*right*). Outlines of nuclei are subdivided into areas, each of which is labeled with the range of characteristic frequencies (in kHz) of the neurons in that area. (From Guinan et al. 1972b. Copyright by Gordon and Breach Science Publishers Inc.)

2. Medial Superior Olivary Nucleus. Single units in MSO of the cat are extremely difficult to isolate, apparently as a consequence of the thinness of the cell plate (approximately four neurons) and the presence of large-amplitude slow potentials which make it hard to discriminate action potentials. These difficulties are reflected in the fact that of the 432 SOC neurons isolated by Guinan et al. (1972a, b) that could be assigned to response categories, only 11 were located in MSO. The most detailed data are those obtained by Goldberg and Brown (1968, 1969) from a sample of 106 units in the larger, folded MSO of the beagle. Limited data are also available for a sample of 52 units in MSO of the rat (Inbody and Feng 1981). As noted previously, the attribution of units to MSO in a number of other studies must be viewed with suspicion because of the failure to use adequate methods for the histological localization of recording sites.

In both cat and beagle, approximately two-thirds of MSO neurons receive EE binaural input (Guinan et al. 1972a, b; dog: Goldberg and Brown 1968). In the beagle, the remaining neurons were either EI or IE (24%) or EO/mon (11%). Inbody and Feng (1981) reported somewhat different proportions in the rat, for which EE cells comprised 52% and EI cells 48% of the sample. That this might reflect a true species difference, perhaps characteristic of high-frequency animals with less well-developed MSO (see Masterton et al. 1975), is suggested by evidence that a higher proportion of IC neurons receive EI input in the rat than in the cat (Silverman and Clopton 1977; Syka et al. 1983; see Sect. 7.2.3). Data sharply at variance with those for the cat have recently been reported for the chinchilla by Langford (1984): 46% of his sample ($N = 136$) received OO/F binaural input, while 20% received EO/F, and 15% EE input. In the majority (70%) of animals in this study, however, units were assigned to MSO on the grounds that the electrode tracks were confined to "the MSO portion" of SOC. Thus, although the high proportion of OO/F cells in this sample is of great interest, their localization to MSO rather than to adjacent periolivary regions is open to question.

MSO neurons in the cat (Guinan et al. 1972a, b) and in the beagle (Goldberg and Brown 1969) have sharp, V-shaped excitatory tuning curves. In the majority of EE cells in the beagle, the CFs and response areas were similar for the two ears, although in some cells with inhibitory sidebands the excitatory areas for the two ears differed. In EI and IE cells, the excitatory CF was usually near the most effective frequency for the inhibitory ear (Goldberg and Brown 1969). Similar observations on corresponding frequency selectivity in the input from the two ears have been made in the rat (Inbody and Feng 1981). Recently, Caird and Klinke (1983) reported on the properties of a group of units in the cat which they defined as "MSO type" on the grounds that they were located either around MSO or in a region in which the authors assumed they were recording from the axons of MSO neurons. The fact that these units were described as having broad tuning curves, in conjunction with the imprecision in localization, indicates that the claim that these cells can be identified with MSO cells must be viewed with some skepticism.

MSO in both the cat and the beagle is tonotopically organized. Although the data base for the cat is rather small, it indicates that CFs are arranged in an orderly sequence from low to high in the dorsomedial-to-ventrolateral direction (Guinan et al. 1972b; Fig. 28), in accordance with the topographic organization of the projections from rostral AVCN to MSO (van Noort 1969; Warr 1966, 1982). The three-dimensional frequency organization of MSO has not been described, but it seems likely that isofrequency (or frequency-band) planes correspond to the horizontal fibrodendritic laminae described in Sect. 5.1.3. In the folded MSO of the dog, low frequencies are similarly represented in the dorsal limb and high frequencies in the ventral limb (Goldberg and Brown 1968), and similar tonotopic organization has been described in the echolocating bat *Molossus ater* (Harnischfeger et al. 1985). In both the cat and the beagle, there is a disproportionate representation of low frequencies in MSO, the majority of units having CFs below 4 kHz and very few units having CF greater than 11 kHz. As noted previously (Sect. 5.1.3), the major projections to MSO are from bushy/large spherical cells in rostral AVCN via the medium-diameter TB fibers. Osen (1969b) predicted that the frequency representation in MSO would be limited to low-to-middle frequencies, on the grounds that its input is derived from the rostral large spherical cell area (AA) that does not receive innervation from the basal turn of the cochlea. The frequency representation in MSO in fact corresponds closely with that in AA (Bourk et al. 1981; see Fig. 18 and Sect. 4.2.2) and with the range of CFs (0.3 – 12 kHz) found by Brownell (1975) in the medium-diameter TB fibers.

The bushy/spherical cells in rostral AVCN whose axons (comprising the medium-diameter TB fibers) provide the major input to MSO, have response characteristics that are in most respects indistinguishable from those of AN fibers (Brownell 1975; Sect. 4.2.4). What evidence is available suggests that these characteristics are probably preserved in the monaural responses of MSO neurons – presumably as a consequence of the calyx-like properties of the large endings on the principal cells (Schwartz 1972) – but the data are unfortunately fragmentary. Thus, Goldberg and Brown (1969) demonstrated secure phase-locking at low frequencies in a small number of MSO neurons that were studied in detail, and Moushegian et al. (1964, 1967) reported phase-locking in units which they identified (on questionable grounds) as MSO cells. In both the beagle and the cat, MSO neurons have sustained responses to tone bursts and irregular discharge (Guinan et al. 1972a, b; dog: Goldberg and Brown 1968), but there is almost no evidence on their PSTH types. Goldberg and Brown (1968, 1969) did not report observations on response patterns, and data for the cat are fragmentary. Kiang et al. (1973) assert that the available PSTHs for MSO neurons are primarylike with notch, which would indicate a transformation from the primarylike PSTHs of bushy cells in AA, but the size of the sample on which this statement is based is not stated. Guinan et al. (1972a, b) reported PSTHs for only two MSO cells: one was primarylike with notch and the other was phase-locked.

There has similarly been no systematic study of the monaural rate–intensity functions of MSO neurons. If the properties of "primarylike" AVCN units were preserved in MSO, the functions would be expected to be sigmoid (i.e., monotonic and saturating; Sect. 4.2.4 and Brownell 1975). Contrary to this expectation, a number of the monaural rate–intensity functions presented by Goldberg and Brown (1969) are nonsaturating, while others are nonmonotonic. Although it is not clear whether these cells were representative, the existence of such functions is indicative of transformations from the primarylike characteristics of the major input. If the extended dynamic range exhibited in some of these functions were achieved without broader frequency tuning, it would indicate that such dynamic range can be achieved without the integration of information from fibers whose CF does not equal the stimulus frequency. Goldberg and Brown (1969) also reported that the *binaural* dynamic range of EE cells in MSO of the beagle was frequently considerably greater than that for monaural stimulation. This extended dynamic range reflects the prevalence of facilitatory and summative binaural interactions in EE cells and the fact that the ipsilateral and contralateral responses commonly have different thresholds.

It is clear from this account that there is an urgent need for solution of the technical problems of recording from MSO neurons to allow the accumulation of a systematic body of data on their response characteristics. Despite the limitations of the evidence on the basic response characteristics of MSO neurons, Goldberg and Brown's (1969) observations on the sensitivity of beagle MSO cells to interaural time and intensity differences constitute one of the foundations of our understanding of spatial processing in the auditory system. This evidence is examined in the following sections.

One aspect of the responses of MSO neurons that appears not to have received attention concerns the source of inhibitory input to EI and IE cells (and to cells with mixed input). Only the large endings with type 1 characteristics degenerate after CN lesions (Sect. 5.1.3) and it therefore seems unlikely that inhibition is conveyed over the long-axon projections from CN. Short-axon inhibitory interneurons have not been described in MSO, and the origin of the small-vesicle terminals that do not degenerate after CN ablation has not been established with certainty. However, Kiss and Majorossy (1983) have argued that spinous multipolar cells are local interneurons giving rise to one class of such terminal, and Schwartz (1984) has described a class of intrinsic axons in MSO. Marginal cells receive CN afferents from only one side (Lindsey 1975) and have a much higher proportion of terminals with small, flattened vesicles (Schwartz 1980). It is therefore tempting to speculate that they might correspond to EI and IE cells. Whatever the merits of this speculation, the distinctive dendritic morphology of the cell types in MSO suggests correlations with binaural input categories that merit investigation with intracellular labeling techniques.

3. Lateral Superior Olivary Nucleus. There is general agreement that the majority of LSO neurons in the anesthetized cat, other than those with low (<1 kHz) CF, receive IE binaural input (Boudreau and Tsuchitani 1968, 1970; Caird and Klinke 1983; Goldberg et al. 1963; Guinan et al. 1972a, b; Tsuchitani 1977). The remaining cells also receive ipsilateral excitatory input, with the contralateral ear either having no effect (OE/mon; Guinan et al. 1972a, b; Tsuchitani 1977) or an unspecified effect (Tsuchitani 1977). Most LSO neurons with CF below 1 kHz are OE/mon (Boudreau and Tsuchitani 1968; Guinan et al. 1972a, b). Brownell et al. (1979) reported similarly homogeneous ipsilateral excitation of LSO neurons, but more diverse contralateral inputs, in decerebrate cats; only 50% of their cells were IE and a significant proportion received EE input. As noted previously, LSO receives only a small direct projection from contralateral CN (Sect. 5.1.4), and the contralateral inhibitory input to LSO neurons is thought to be derived from MNTB (Sect. 5.1.2). The prevalence of IE binaural input to LSO neurons in the cat has prompted Cant (1984) to suggest that the type 1 terminals on the distal dendrites of LSO neurons are excitatory and are derived from the AVCN projection, while the type 2 terminals on the soma and proximal dendrites are inhibitory and are derived from the MNTB projection. If the MNTB principal cells act chiefly as inhibitory interneurons to LSO cells, their segregation in a separate nucleus is puzzling. It might be that the structural specializations required for rapid transmission require such segregation. Alternatively, the other projections of MNTB whose functions are not yet known, such as that to NLL (see Sect. 6.1), might be derived from principal cells and require their segregation.

The excitatory ipsilateral tuning curves of most LSO neurons are sharp and V-shaped (Boudreau and Tsuchitani 1968, 1970; Caird and Klinke 1983; Guinan et al. 1972a, b; Tsuchitani 1977; Tsuchitani and Boudreau 1966, 1967, 1969). Neurons in LSO exhibit very low levels of spontaneous activity in anesthetized animals, and inhibitory sidebands can therefore not be readily detected. However, Brownell et al. (1979) demonstrated inhibitory sidebands in the ipsilateral tuning curves of 9 of 10 LSO units in decerebrate cats, and sidebands have also been described in units with spontaneous activity in anesthetized cats (Fig. 29; Caird and Klinke 1983). The anatomical substrate for ipsilateral inhibitory input to LSO neurons is not known, although the absence of inhibitory sidebands in the AVCN spherical cells which provide the major ipsilateral input to LSO (Brownell 1975; Sect. 4.2.4 and 5.1.4) suggests that they are generated by mechanisms intrinsic to LSO (see Brownell et al. 1979 for discussion).

Contralateral inhibitory tuning curves for IE neurons are generally reported to be similar or identical to those for ipsilateral excitation (Boudreau and Tsuchitani 1968; Caird and Klinke 1983; Guinan et al. 1972b; Tsuchitani and Boudreau 1969). In the absence of spontaneous activity, inhibitory tuning curves have

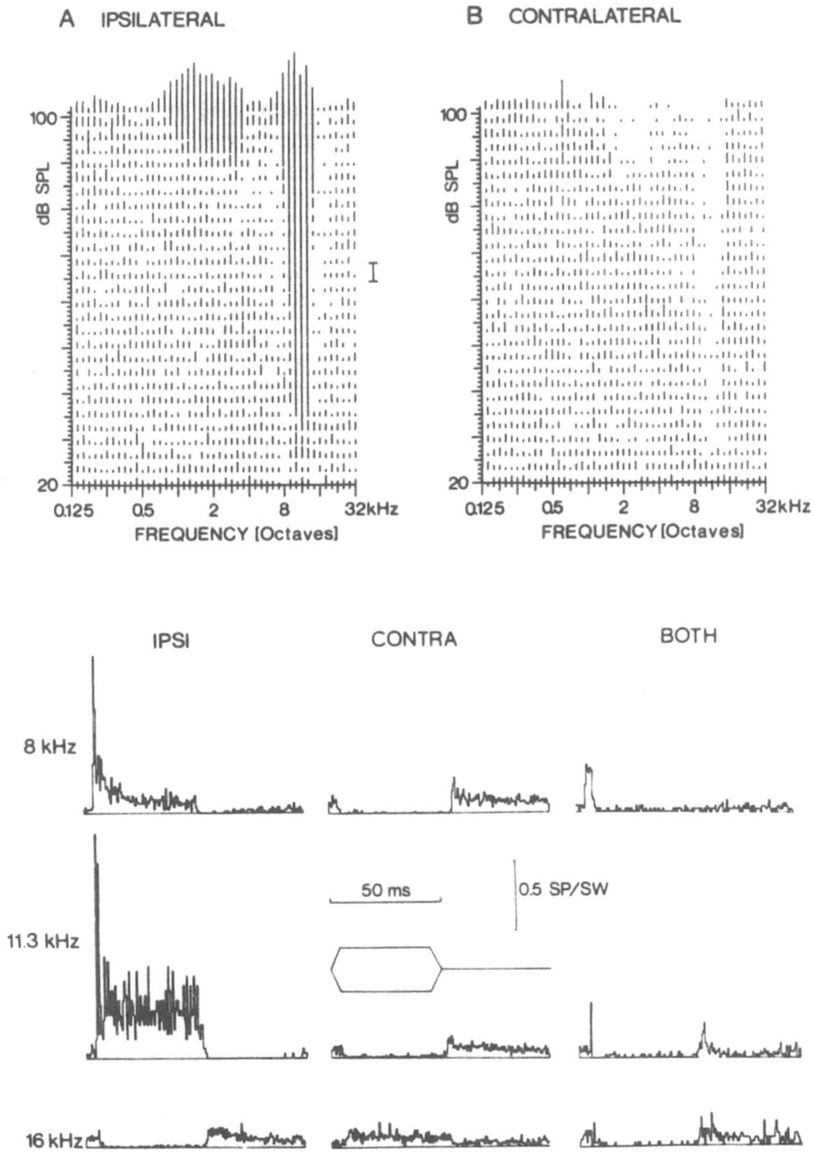

Fig. 29 A, B. Plots of **A** ipsilateral and **B** contralateral frequency responses of an IE cell in the lateral superior olivary nucleus of cat. In each plot, the height of each *bar* shows the number of spikes evoked by a tone burst at that frequency–intensity value. The *calibration bar* to the right of plot **A** respresents 10 spikes per sweep. Peristimulus time histograms below frequency–response plots show responses to monaural ipsilateral (*IPSI*) and contralateral (*CONTRA*) stimulation and to simultaneous binaural stimulation (*BOTH*) at three frequencies indicated. The calibration *SP/SW* indicates the spikes per histogram bin (512 µs binwidth) divided by the number of stimulus presentations. Stimulus envelope is shown beneath time *calibration bar*. (Caird and Klinke 1983)

usually been generated by presenting an ipsilateral CF tone and determining the degree of inhibition produced by a contralateral stimulus as its frequency and intensity are varied (e.g., Boudreau and Tsuchitani 1968; Tsuchitani and Boudreau 1969). The data so obtained could therefore be interpreted as indicating that inhibition occurs at lowest intensities when the frequencies to the two ears are the same, rather than as revealing the absolute shape of the inhibitory tuning curve. However, similar excitatory and inhibitory tuning curves are also obtained when inhibition is assessed by suppression of spontaneous activity. This is beautifully illustrated in a cell described by Caird and Klinke (1983), in which the correspondence extends to the presence of excitatory (or disinhibitory) sidebands in the contralateral curve, corresponding to the inhibitory sidebands in the ipsilateral curve (Fig. 29; note 16-kHz PSTHs).

LSO is tonotopically organized, with high frequencies represented in the ventromedial limb and low frequencies in the dorsolateral limb (Fig. 28; Boudreau and Tsuchitani 1970; Guinan et al. 1972b; Tsuchitani 1977; Tsuchitani and Boudreau 1966), reflecting the topography in the projections from CN (van Noort 1969; Warr 1966, 1982). As in MNTB and MSO, it is likely (but not yet established empirically) that isofrequency (or frequency-band) planes correspond to fibrodendritic laminae. As indicated in Fig. 28, there is a disproportionate representation of high frequencies in LSO (Boudreau and Tsuchitani 1970; Guinan et al. 1972b; Tsuchitani 1977). Guinan et al. (1972b) have pointed out that the relative numbers of high- and low-CF units are more disparate than is suggested by standard frontal representations, because the high-frequency ventromedial limb has a greater rostrocaudal extent than the low-frequency dorsolateral limb. The fact that cells with CF below 1 kHz located in the dorsolateral limb receive OE/mon input is in accord with the report by Glendenning et al. (1985) that this region of LSO receives a much weaker projection from MNTB than does the remainder of the nucleus.

The majority of LSO units in anesthetized cats have chopper PSTHs to ipsilateral CF tone bursts (Guinan et al. 1972a, b; Tsuchitani 1977, 1982; Tsuchitani and Johnson 1985), although primarylike responses with a sharp initial peak (e.g., Fig. 29; Caird and Klinke 1983) and phase-locked responses (Tsuchitani 1977) have also been described. Tsuchitani (1982) and Tsuchitani and Johnson (1985) distinguished a number of types of chopper response pattern, and also argued that at least some of the responses characterized by a sharp initial peak are recorded from units whose responses are affected by the presence of the microelectrode. The discharge regularity characteristic of the chopper pattern is reflected in symmetric (gaussian) ISIHs in most LSO neurons, but some have asymmetric ISIHs (Guinan et al. 1972a, b; Tsuchitani 1982; Tsuchitani and Johnson 1985). LSO units in the echolocating bat *Molossus ater* also exhibit chopper PSTHs (Harnischfeger et al. 1985).

Although some stellate cells in AVCN (which have chopper response patterns) probably project to LSO, the major ipsilateral projection is apparently derived from small spherical/bushy cells with primarylike or primarylike with notch response patterns (see Sects. 5.1.4 and 4.2.4). It therefore appears that the chopper pattern must be generated within LSO, perhaps by temporal integration and membrane properties of the type described in VCN "choppers" (Sects. 4.2.4 and 4.2.5). The chopper pattern in LSO may, however, be partly a consequence of barbiturate anesthesia. Brownell et al. (1979) reported that LSO units in decerebrate cats generally had PSTHs characterized by a sharp inital peak followed by a sustained lower level of discharge (cf. Fig. 29). In two such cells for which recordings were obtained before and after an anesthetic dose of pentobarbital, the PSTHs were transformed to a chopper pattern.

Anesthesia similarly modifies the rate–intensity functions of LSO neurons. In anesthetized cats, monaural rate–intensity functions (both excitatory and inhibitory) for LSO units are sigmoid with limited dynamic range (see monaural ipsilateral – i.e., excitatory – function in Fig. 30A; Boudreau and Tsuchitani 1968, 1970; Tsuchitani 1977; Tsuchitani and Boudreau 1966, 1969). They thus resemble those of AVCN units (Sect. 4.2.4) and medium-diameter TB fibers (Brownell 1975). In decerebrate cats, however, 9 of 16 LSO rate–intensity functions were nonmonotonic, and in one neuron a nonmonotonic function was transformed to a characteristic sigmoid function by barbiturate anesthesia (Brownell et al. 1979). Nonmonotonicity of rate–intensity functions presumably reflects inhibitory inputs to LSO neurons that are suppressed by barbiturate anesthesia.

Although phase-locked PSTHs have been reported in LSO (Guinan et al. 1972a, b; Tsuchitani 1977), there has been no systematic study of phase-locking in LSO neurons. The phase-locking of medium-diameter TB fibers is similar to that of AVCN "primarylike" units (Brownell 1975), and therefore to that of AN fibers (see Sect. 4.2.4). However, if the chopper pattern of LSO units in anesthetized cats is the consequence of temporal integration of many individually subthreshold inputs, this process might result in degradation of phase-locking, as is seen in VCN "choppers" (cf. van Gisbergen et al. 1975c; Sect. 4.2.4).

The role of IE neurons in LSO in the processing of interaural time and intensity disparities is examined in the following sections, but one feature of the contralateral inhibitory input to IE neurons should be noted here. Boudreau and Tsuchitani (1968) reported that the entire response to ipsilateral stimulation, including the first spike at intensities at which first-spike latency had reached an asymptotic minimum, could be inhibited by simultaneous contralateral stimulation. This observation indicates that the total conduction times from the ipsilateral and contralateral ears are approximately equal despite the greater path length and the additional synapse in the contralateral pathway. The rapid contralateral input is achieved by means of the high conduction velocity of the large-diameter globular cell axons, and the rapid and secure transmission provided by the calycoid endings on MNTB principal cells. As noted previously, these factors result in MNTB latencies being consistently shorter, by up to 1 ms, than those in other SOC subdivisions (Guinan et al. 1972a, b; dog: Goldberg and Brown 1969). A further factor contributing to the rapidity and strength of contralateral inhibitory effects might be the arrangement of excitatory and inhibitory synaptic terminals on LSO neurons. As noted previously, Cant's (1984) evidence suggests that the inhibitory terminals from MNTB are located on the soma and proximal dendrites and the excitatory terminals from ipsilateral AVCN on the distal dendrites. If this is the case, the proximity of the inhibitory terminals to the site of impulse initiation would allow them to exert a rapid and powerful effect on the neuron's discharge.

4. Other Superior Olivary Cell Groups. Only limited data are available on the response characteristics of neurons in LNTB, VNTB, and the periolivary nuclei.

Because of the fragmentary nature of this evidence, and the fact that many of these neurons have descending projections whose functions are unkown (see Sect. 5.1.5), these data will be only summarily treated here. Almost all VNTB neurons receive contralateral excitatory input, in accordance with evidence that the major projection to the nucleus is that from contralateral CN. The majority of cells in both dog and cat are EO/mon, but small proportions receive EE or EI input (Goldberg et al. 1963; Guinan et al. 1972a, b; dog: Goldberg and Brown 1968). Conversely, and also in agreement with the anatomical evidence, almost all LNTB neurons in the cat receive ipsilateral excitatory input, the majority being OE/mon, with smaller proportions of EE and IE cells (Boudreau and Tsuchitani 1968; Goldberg et al. 1963; Guinan et al. 1972a, b; Tsuchitani 1977). In both areas the majority of units have V-shaped tuning curves, but significant proportions (almost 40% in LNTB; Tsuchitani 1977) have wide and irregular tuning curves (Guinan et al. 1972a, b; Tsuchitani 1977). The greater proportion of LNTB units have been reported to exhibit chopper patterns to CF tones (Guinan et al. 1972a, b; Tsuchitani 1977), whereas in VNTB chopper and primarylike patterns are equally common and a variety of other response patterns are seen (Guinan et al. 1972 a, b).

Patterns of binaural input similar to those in VNTB and LNTB are seen in medially and laterally located periolivary cell groups, respectively. In general, neurons in periolivary cell groups located medial to MSO receive contralateral excitatory input and diverse ipsilateral input, the majority being EO/mon and smaller proportions being EE and EI (Guinan et al. 1972a, b; dog: Goldberg and Brown 1968). Conversely, neurons in periolivary cell groups located lateral to MSO receive ipsilateral excitatory input, the majority being OE/mon. These characteristics of medially and laterally situated periolivary neurons are also in agreement with the anatomical evidence that the major projections to these cell groups are from contralateral and ipsilateral CN, respectively. In all of the periolivary nuclei for which data have been reported, the majority of tuning curves are V-shaped, but significant proportions of wide and irregular tuning curves are seen (Guinan et al. 1972b; Tsuchitani 1977). Guinan et al. (1972b) were unable to discern any tonotopic organization in the periolivary nuclei, but Tsuchitani (1977) has claimed that all of the lateral periolivary groups are tonotopically organized, with high-to-low frequencies represented from medial to lateral.

5.2.3 Binaural Interaction: Coding of Interaural Intensity Differences

Goldberg and Brown (1969) systematically compared the sensitivity of EE and EI cells in MSO and VNTB of the beagle to IIDs in dichotically presented tones. They found that EE cells were generally insensitive to IIDs, but were sensitive to changes in the average binaural intensity (ABI), or overall intensity, of the stimuli. As noted previously, the binaural rate–intensity functions of EE cells were generally steeper and had wider dynamic range than the monaural functions. Conversely, EI cells were relatively insensitive to ABI, but were sensitive to IIDs. These observations supported an earlier report by Hall (1965) that EI cells in cat SOC were sensitive to IIDs in click stimuli. The general sensitivity of EI and IE cells to IIDs has been confirmed in studies of LSO in the cat (Boudreau and Tsuchitani 1968, 1970; Caird and Klinke 1983; Tsuchitani and Boudreau 1969) and bat (Harnischfeger et al. 1985), MSO of the rat (Inbody and Feng 1981), and higher levels of the auditory pathway in a number of mammals (see Sects. 7.2.4 and 7.5). This association is further supported by the sensitivity to IIDs of EI cells in lower brainstem nuclei of the barn owl (Moiseff and Konishi 1983a, b; see Sect. 7.3), in the midbrain of the domestic fowl (Coles and Aitkin 1979), and in the superior olivary nucleus of the green tree frog (Feng and Capranica 1978).

Data from Boudreau and Tsuchitani (1970) for an IE neuron in LSO of the cat are presented in Fig. 30A. These data clearly illustrate the sensitivity of such cells to IIDs: the response to an excitatory ipsilateral stimulus at any given intensity declines progressively with increases in the intensity of the contralateral inhibitory stimulus. Presentation of the data in this form does not, however, readily allow appreciation of the cell's sensitivity in spatial terms. The data have therefore been replotted in Fig. 30B to show the variation in the cell's response as a function of IID at each of the ipsilateral intensities employed by Boudreau and Tsuchitani. Each of the replotted curves indicates the response generated as the contralateral intensity is varied above and below the fixed ipsilateral intensity indicated in the key (see Sect. 7.2.4 for discussion of alternative methods of generating IIDs). This form of presentation makes it clear that, at each ipsilateral intensity, the cell responds over a range of IIDs for which the ipsilateral stimulus is more intense (negative IIDs by the convention adopted here), and is totally inhibited over a range of IIDs for which the contralateral stimulus is more intense (positive IIDs); at each ipsilateral intensity, total inhibition of the response occurs in the region of zero IID. As indicated by the horizontal line at the top of Fig. 30B, zero IID is produced by sound sources located in the median sagittal plane, and IIDs defined as positive and negative correspond to contralateral and ipsilateral azimuths (see Sect. 2.3, and Fig. 6). This cell therefore exhibits a systematically varying response over a range of IIDs corresponding to ipsilateral azimuths and is totally inhibited over a range of IIDs corresponding to contralateral azimuths. Similar sensitivity is exhibited in the functions for another IE neuron from LSO in Fig. 31, which is discussed in more detail in Sect. 5.2.4. The form of sensitivity revealed in these figures is characteristic of IE cells in LSO (Boudreau and Tsuchitani 1970; Caird and Klinke 1983) and, in mirror symmetric form, of EI cells in MSO and in higher auditory centers (see Sects. 7.2.4 and 7.5, and Figs. 43 and 52). At higher levels, it has been shown that the slope of the IID sensitivity function, and the IID range over which it occurs, can be different for different neurons (Irvine and Wise 1983a, b; Phillips and Irvine 1981b, 1983; Wise and Irvine 1983, 1985). These differences, and their implications for the neural representation of IID and of auditory azimuthal location, will be considered in Sects. 7.5 and 8.2.4.

Although EI and IE cells constitute the major classes of IID-sensitive neurons in the mammalian auditory system, evidence from higher centers has revealed that different forms of IID sensitivity are exhibited by some cells characterized by facilitatory rather than inhibitory binaural interactions. These forms of IID sensitivity will be considered in Sects. 7.2.4 and 7.5.

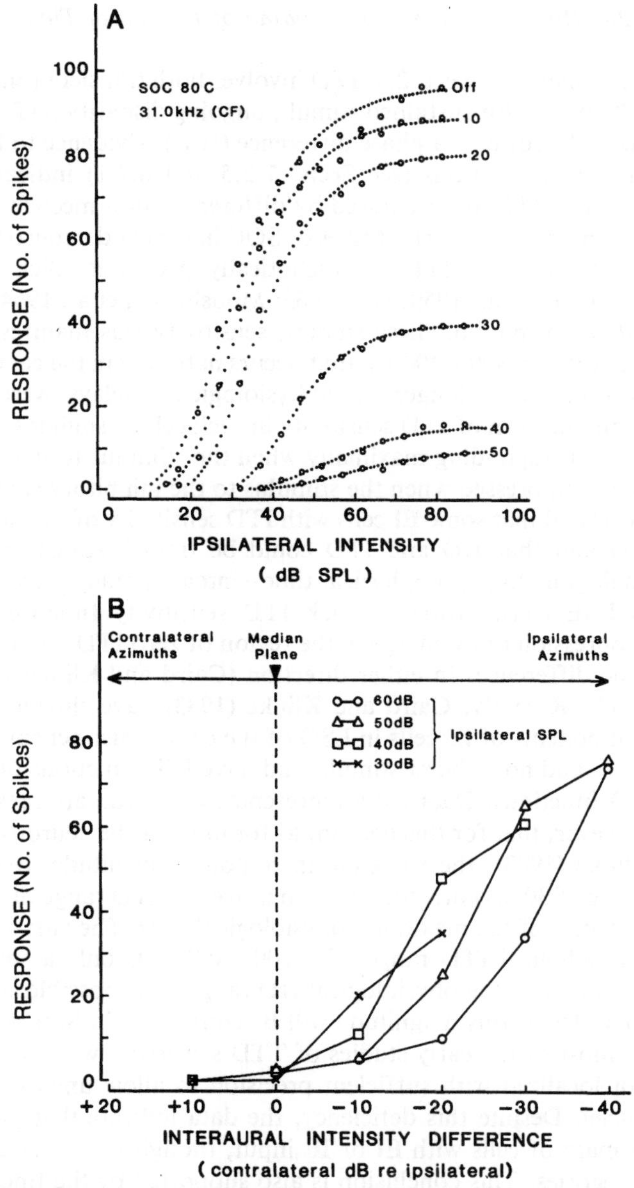

Fig. 30. A Family of binaural intensity functions for an IE cell in the lateral superior olivary nucleus of the cat, showing sensitivity to interaural intensity difference (IID). Each function was generated by presenting a contralateral tone at the characteristic frequency of the neuron (31.0 kHz) at the fixed level indicated (in dB SPL) on the function, and varying the level of the simultaneously presented ipsilateral tone of the same frequency in 5-dB steps. Function marked *Off* is for monaural ipsilateral stimulation. It is apparent that for any given level of stimulus to the ipsilateral (excitatory) ear (on abscissa), the response declines with increases in the level of the stimulus to the contralateral (inhibitory) ear. (Modified from Boudreau and Tsuchitani 1970) **B** Data in **A** have been replotted (as described in text) to show variation in response as a function of IID (contralateral intensity relative to ipsilateral intensity) at different levels of the ipsilateral stimulus (as indicated in *key*). In this and subsequent figures of this sort, the *horizontal line* at the top of the figure indicates the broad azimuthal ranges corresponding to the IID ranges on the abscissa. But note that, as discussed in Sect. 2.3, the IID – azimuth relationship in the cat is nonmonotonic

5.2.4 Binaural Interaction: Coding of Interaural Time Differences

As explained in Sect. 2.3, ITD involves both transient (onset and offset) disparity (TTD) and, for sustained stimuli, ongoing disparity (OTD), which for pure tones takes the form of a phase difference (IPD). Evidence to be examined in this and subsequent sections (see Sects. 7.2.5 and 8.2.1) indicates that these different forms of ITD are processed by different neural mechanisms.

A number of the early studies of SOC in the cat demonstrated that EI and IE cells are sensitive to ITDs in dichotically presented click stimuli (i.e., to TTD; Galambos et al. 1959; Hall 1965; Moushegian et al. 1964, 1967; Watanabe et al. 1968). In most of these studies, sensitivity was manifested to ITDs within the range (up to 300 – 400 µs) that occurs naturally in the cat (see Sect. 2.3), although in some cases longer "nonphysiological" delays were employed. The most common form of ITD sensitivity in such cells is similar to their sensitivity to IID, the cell responding maximally when the stimulus to the excitatory ear leads and being suppressed when the stimulus to the inhibitory ear leads. Hall (1965) demonstrated that some EI cells with ITD sensitivity of this sort were also sensitive to IID and that IID and ITD could be traded against each other in a manner analogous to psychophysical time – intensity trading (see Sect. 2.3). In other EI and IE cells, however, click ITD sensitivity functions are characterized by suppression of discharge in the region of zero ITD and increasing discharge with time differences in either direction (Caird and Klinke 1983; Moushegian et al. 1967). Recently, Caird and Klinke (1983) have shown that the onset response components of IE cells in LSO of the cat are also sensitive to onset disparities in tone- and noise-burst stimuli, and have ITD functions of the same form as their IID functions. Data for a representative neuron are presented in Fig. 31. Note, however, that for this neuron, as for most of the neurons described by Caird and Klinke (1983), the variation in response magnitude over the physiological IID range of 30 dB corresponds to that over an ITD range of approximately 2 ms, far in excess of the maximum physiological ITD. The variation in response over the physiological ITD range of ± 300 – 400 µs is only a small fraction of that occurring over the physiological IID range. The possible significance of sensitivity to TTDs of this magnitude will be considered in Sect. 8.2.2.

In most of the early studies of TTD sensitivity with clicks, recording sites were not localized with sufficient precision to allow units to be assigned to specific nuclei. Despite this deficiency, the data indicate that processing of TTD is the domain of cells with EI or IE input, though not necessarily of *all* cells in these categories. This conclusion is also supported by the finding of Harnischfeger et al. (1985) that most of the cells in SOC and IC of the echolocating bat *Molossus ater* that are sensitive to onset disparities in high-frequency tone bursts receive EI or IE input. There is, however, a clear need for a systematic study of TTD sensitivity in a large sample of histologically localized neurons. Since EI and IE cells are also sensitive to IIDs, and a high-frequency sound located off the median plane generates both an IID and a TTD, it would also be of considerable interest to examine the responses of EI and IE cells to *physiologically appropriate* combinations of these disparities. Some preliminary observations on such combinations were reported by Harnischfeger et al. (1985); their data indicate that sensitivity to ITDs is enhanced in the presence of appropriate IIDs.

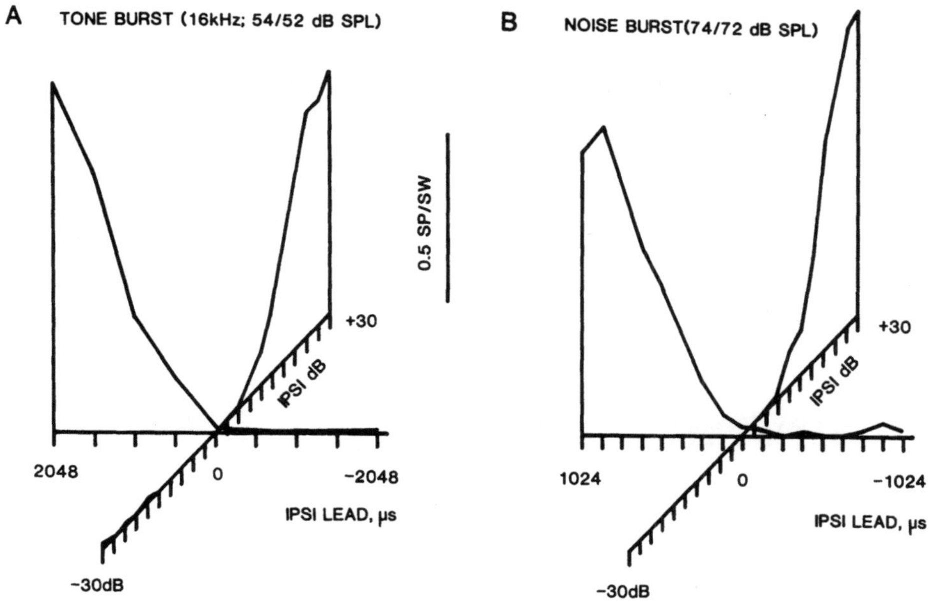

Fig. 31 A, B. Three-dimensional plots showing variation in response of an IE neuron in lateral superior olivary nucleus (LSO) of the cat as a function of interaural time difference (ITD; horizontal plot with abscissa labeled *IPSI LEAD*) and interaural intensity difference (IID; diagonal plot with abscissa labeled *IPSI dB*, expressed relative to contralateral level). Stimulus was **A** tone burst at characteristic frequency of neuron (16 kHz), and **B** noise burst. Response measure is the "peak" response (the response in the peristimulus time histogram bin containing the highest peak plus those in the four neighboring bins – see representative histograms in Fig. 29). Clearly, the neuron is sensitive to both IID and ITD, responding maximally when the stimulus at the ipsilateral ear is more intense or leads in time. The IID and ITD curves are complementary around zero disparity, but note that the response variation that occurs over the near-physiological IID range of 30 dB is seen only over an ITD range greatly in excess of that occurring naturally in the cat. (Redrawn from Caird and Klinke 1983)

The nature and mechanisms of IPD sensitivity were elucidated by Goldberg and Brown (1969) in a detailed analysis of a number of low-frequency neurons in MSO and VNTB of the beagle. They generated ITDs by means of a delay line, but controlled for any contamination by TTD sensitivity by analyzing data for only the last 900 ms of the responses to 1-s tone bursts. A representative delay sensitivity function from their data is presented in Fig. 32 I (full circles). Delays well beyond the physiological range were used to demonstrate that the function is periodic, with period equal to that of the stimulating sinusoid (viz., 2254 µs), indicating that the discharge rate varies as a function of interaural phase rather than of delay per se. Note, however, that the neuron's response varies from its maximum to its minimum values over the physiological range of $\pm 300-400$ µs. Moushegian et al. (1964, 1967) had previously reported phase sensitivity in neurons described as being "within or near" MSO, and Rose et al. (1966) had described periodic delay functions in IC (see Sect. 7.2.5 for detailed account of IPD processing in IC).

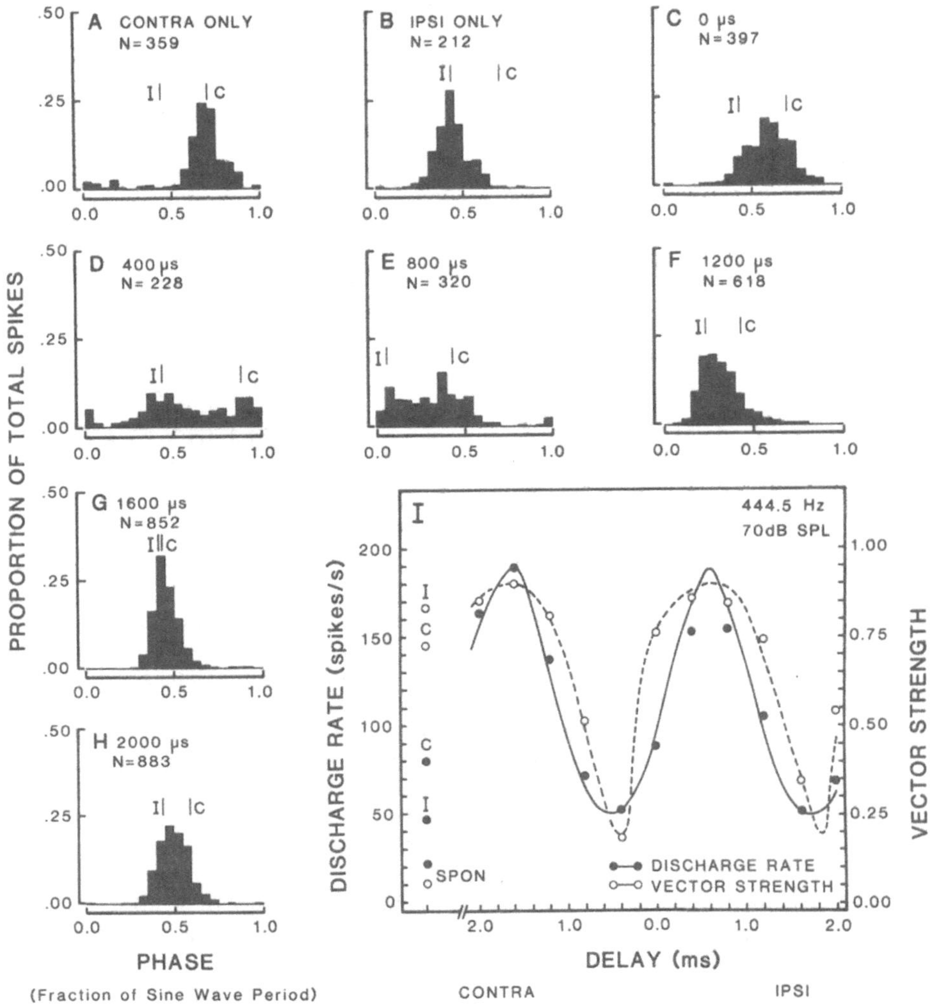

Fig. 32 A–I. Delay sensitivity function **I** and phase (i.e., period) histograms **A–H** for neuron (in MSO of dog) sensitive to interaural phase difference (IPD). Stimulus was 444.5 Hz tone at 70 dB SPL. In **I**, discharge rate (*full circles* and *full line*; left ordinate) and vector strength (*open circles* and *broken line*; right ordinate) are plotted as a function of interaural delay. *Points* on left indicate responses to contralateral (*C*) and ipsilateral (*I*) monaural stimulation, and in the absence of stimulation (*SPON*). In **A–H**, phase (period) histograms show proportion of total spikes as a function of sine wave period for **A** monaural contralateral and **B** monaural ipsilateral stimulation, and **C–H** for binaural stimulation at contralateral delays indicated. In each case, *N* = total number of spikes in histogram (based on five stimulus presentations in all panels except **H**, which is based on six). Discharge rates plotted in **I** are derived by dividing *N* values by the product of the number of sweeps and the duration of the stimulus for which data were analyzed (viz., 0.9 s; see discussion in text). *Vertical bars* above each histogram indicate the mean phase for contralateral (*C*) and ipsilateral (*I*) ears expected on the basis of monaural stimulation. In **D–H**, the *bar* marked *C* is shifted when necessary in accordance with the delay of the contralateral stimulus. (Modified from Goldberg and Brown 1969)

Detailed analysis of a small sample ($N = 8$) of IPD-sensitive cells in MSO and VNTB enabled Goldberg and Brown (1969) to elucidate the mechanisms underlying IPD sensitivity. Each of the IPD-sensitive cells in MSO exhibited phase-locking to monaural stimulation of either ear. This is illustrated by the monaural phase (i.e., period) histograms for the cell in Fig. 32A and B, which show the proportion of total spikes bearing a specific phase relation to the stimulating sinusoid. Clearly, both the contralateral (Fig. 32A) and ipsilateral (Fig. 32B) responses are securely phase-locked, but the phase relation between stimulus and discharge differs for the two ears, as revealed by the different positions of the vertical bars indicating the mean phase for each ear. In the remaining panels (Fig. 32C – H), the binaural phase histograms and the number of spikes (N) for various delays are presented. Note that the histogram and value of N in Fig. 32H are based on six stimulus presentations, while those in the other panels are based on five presentations. In each of these panels the position of the vertical bar indicating the mean phase for the contralateral ear is shifted in accordance with the delay of the contralateral stimulus. It is apparent that the maximum discharge occurs at the delay that brings the phase-locking points for the two ears into coincidence (1600 μs contralateral delay; Fig. 32G), and the minimum discharge occurs when they are most separated (400 μs contralateral delay; Fig. 32D). The most favorable delay (that at which the cell fires most strongly) therefore corresponds to the difference between the monaural phase-locking points for the two ears. In a sample of eight IPD-sensitive units (six in MSO and two in VNTB) which were phase-locked to stimulation of each ear, the correlations between these two measures were 0.92 and 0.99 for data expressed in absolute time and in degrees, respectively. Despite the small sample size, these elegant data therefore provide strong support for a "coincidence detector" model of IPD sensitivity, of the general type originally proposed by Jeffress (1948), and discussed in more detail in Sect. 8.2.2. Crow et al. (1978) have reported a similar agreement for a sample of SOC neurons (presumed to be in MSO) in the kangaroo rat. This evidence clearly establishes a major function of the faithful preservation of AN fiber discharge characteristics provided by the end-bulbs on bushy cells in anterior AVCN. The precise relaying of phase-locked information to MSO by these cells via the medium-diameter TB fibers is the basis of IPD processing by MSO neurons.

The discharge rate at the most favorable delay for the unit in Fig. 32 is greater than the sum of the monaural responses, and that at the least favorable delay is smaller than the contralateral response alone. At other intensities, the response of this unit at the least favorable delay was smaller than either of the monaural responses. These observations, and similar data in other cells, provide evidence for binaural facilitation when the excitation from the two ears arrives in phase, and inhibition when it arrives out of phase (Goldberg and Brown 1969). These data therefore not only indicate that the input from a given ear to an MSO neuron can be both excitatory and inhibitory, but that the inhibition must also be phase-locked (see Yin and Kuwada 1984 for discussion of the need for phase-locked inhibition in these cells). A final point to be noted in Fig. 32 is that binaural phase-locking is most precise at the most favorable delay (Fig. 32G) and least precise at the least favorable delay (Fig. 32D), the vector strength curve (open circles) paralleling the discharge rate function.

Goldberg and Brown (1969) also reported that while some ($N = 4$) IPD-sensitive cells in VNTB were like those in MSO in exhibiting phase-locked responses to each ear, others ($N = 5$) had more diverse binaural input patterns, including predominantly inhibitory input from the ipsilateral ear. Moushegian and his colleagues (Moushegian et al. 1964, 1967, 1975; Crow et al. 1978) have also reported IPD sensitivity in cells (in the region of MSO) with EI input in the cat and kangaroo rat, and have noted that in some cells the vector strength function is not periodic like the discharge rate function. IPD sensitivity in IC is also associated with diverse binaural input patterns (e.g., Kuwada et al. 1984; see Sect. 7.2.5). Goldberg and Brown (1969) suggested that such cells in VNTB are not the primary site of binaural convergence and that their IPD sensitivity reflects binaural convergence at some other site. Although this is a plausible suggestion, it must be acknowledged that the available data do not establish that phase-locked excitatory responses to stimulation of each ear are a *necessary* condition for the primary processing of IPD. It is conceivable that a periodic delay function could be generated by binaural convergence in a neuron receiving, for example, phase-locked excitatory and inhibitory input from one ear and phase-locked inhibitory input from the other.

The distinction between sensitivity to IPD (or, more generally, to any feature of the acoustic stimulus) as a manifestation of the initial processing of that feature and sensitivity that reflects initial processing elsewhere is an important one. In any situation in which the known or potential sources of input to a neuron include neurons that are themselves sensitive to the feature in question, the distinction will not be easily made. In the case of IPD, the delay sensitivity of the MSO cells described by Goldberg and Brown (1969) undoubtedly reflects the initial binaural processing of IPD by these cells, but it is not clear that this is the *only* form the initial processing can take. Whatever might be the outcome of this question, it is clear that *sensitivity* to IPD is not the preserve of cells with a particular form of binaural input. Further discussion of IPD processing and of the means by which ongoing time disparities are coded in the nervous system by populations of neurons will be deferred until the considerably larger body of data on IPD sensitivity in IC neurons has been considered (see Sect. 7.2.5).

5.3 Behavioral/Lesion Evidence on the Role of the Superior Olivary Complex in Sound Localization

The anatomical and physiological evidence of SOC's specialization for binaural processing has prompted a number of studies of the effects of brainstem lesions on sound localization performance. Most of these studies have employed a test situation originally devised by Neff and his colleagues (Neff 1962; Neff et al. 1956; Fig. 33A and B), in which an acoustic stimulus is presented behind one of two goal-boxes located on the perimeter of a semicircular enclosure. The animal (usually a cat) is then released from the centrally situated starting cage and is rewarded for approaching the correct goal-box (Fig. 33B). This task is intended to measure the animal's ability to localize the sound, in the sense of determining its azimuthal location, on the basis of acoustic information alone. However, the use of two symmetrically placed goal-boxes makes it possible for the animal to deter-

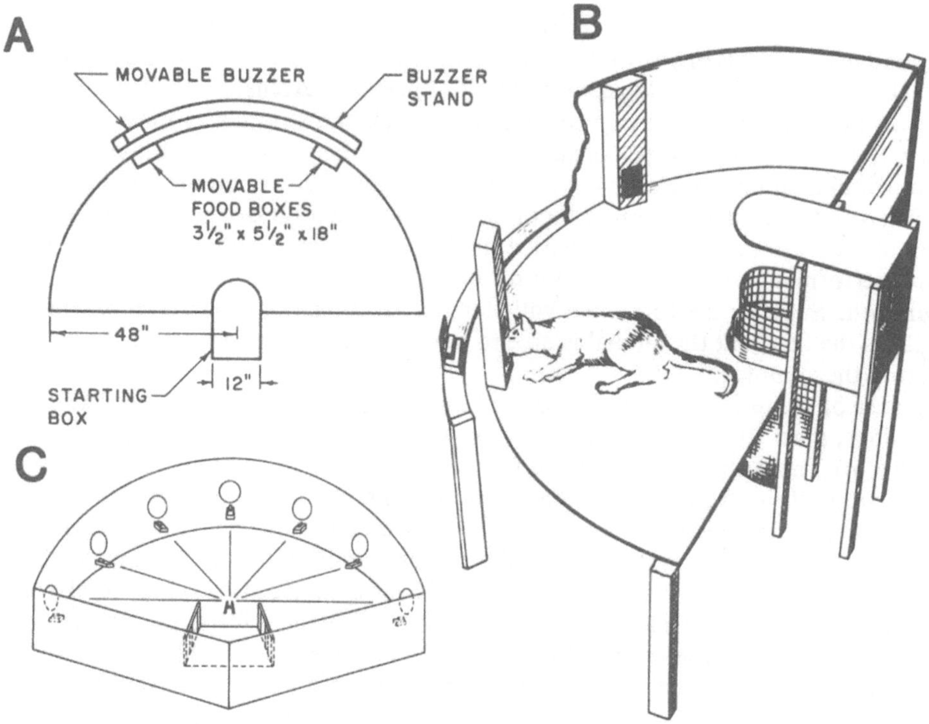

Fig. 33 A – C. Plan **A** and perspective drawing **B** of apparatus used by Neff and colleagues to test auditory localization. See text for description of task. In **B**, the cloth walls are shown partly cut away and the starting box doors are shown in the lowered (open) position. (Moore et al. 1974) **C** Schematic diagram of seven-speaker version of Neff situation used by Jenkins and colleagues. (Masterton et al. 1981)

mine only the *laterality* of the signal on the basis of acoustic information and to make a visually guided approach to the goal-box on that side. In Neff's original experiments (e.g., Neff 1962), this possibility was controlled for by training and testing the animals in darkness, but it is not clear that this procedure has been observed in all subsequent studies with this apparatus. More recently, Jenkins and his colleagues (Jenkins and Masterton 1982; Jenkins and Merzenich 1984; Masterton et al. 1981) have used a situation in which seven goal-boxes are placed at 30° intervals around the perimeter of the enclosure (Fig. 33C). Although this procedure does not allow thresholds to be determined with as much precision as in the movable-speaker situation, it controls for the possibility of visual guidance and has also allowed the detection of important features of the effects of lesions at levels above the SOC (Jenkins and Masterton 1982; Jenkins and Merzenich 1984; Masterton et al. 1981; see Sect. 7.6).

Normal cats tested in the Neff situation have thresholds (angular separation of speakers at which 75% correct performance is achieved) of the order of 5° or less when relatively long-duration (e.g., 500 ms), spectrally complex (buzzer or noise) signals are used (Casseday and Neff 1975; Moore et al. 1974; Nauman, cited by Neff 1968). If an animal trained to criterion on this task is subjected to unilateral destruction of the cochlea, an extensive period of retraining is required and post-operative thresholds are elevated by 10° – 12° (Moore et al. 1974; Nauman, cited by Neff 1968; Neff and Casseday 1977). The need for extensive retraining is indicative of the animal's having to learn a new task and strongly supports the interpretation that normal performance is based on the use of binaural cues and that the animal has to relearn the task using monaural information. Jenkins and Masterton (1982) demonstrated that monaural localization depends critically on stimulus duration: performance with a single click was barely above chance level, and performance with a white-noise burst decreased with decreases in burst duration. Superior performance with longer-duration noise bursts apparently reflects the fact that the animal is able to make scanning movements of the head during the stimulus.

Because of the technical difficulty of making lesions restricted to SOC or its subdivisions, most studies at this level have investigated the effects of lesions of TB. There are, nevertheless, marked differences between the lesions in different animals, both in the extent of damage to TB and in the extent to which other structures are involved. Moore et al. (1974) investigated the role of the various auditory commissural systems in auditory localization. Whereas section of the corpus callosum or the commissure of the inferior colliculus (ICO) had no effect on localization performance, TB lesions had significant effects, the nature of which depended on the extent of the lesion. Very large lesions which effectively isolated CN on one side not surprisingly resulted in a need for retraining and post-lesion performance similar to that of animals rendered monaural by cochlear destruction. In a subsequent study using similar stimuli and procedures, Casseday and Neff (1975) reported that animals with large TB lesions were totally unable to relearn to localize. Given the performance of monaural animals, this is a surprising result unless the lesions were so large as to include all striae and effectively isolate CN on both sides. Despite this conflict, both studies indicated that partial lesions of TB without involvement of other structures resulted in small increases (2° – 5°) in angular thresholds with little or no need for retraining. Jenkins and Masterton have shown that the localization decrements consequent on TB lesions are present either throughout both quadrants of the frontal field or on the side ipsilateral to the lesion.

Although these data establish that the TB is essential to transmission of the information on which binaural localization is based, they do not establish that the processing of this information takes place in SOC. Evidence supporting the latter assertion is, however, provided by studies in which the effects of lesions to the input and output pathways of SOC have been contrasted. The most compelling data were obtained in a study by Masterton et al. (1967), using what was essentially a lateralization task with cats. In these experiments, click trains could be presented independently to the left or right ears via small transducers held within the pinnae by means of a flexible leather helmet. Cats were originally trained in a shuttle-box shock-avoidance situation, using a 1 kHz click train to the right ear as the warning signal (discriminative stimulus) and silence as the safe signal. They were subsequently tested for transfer from this task to two other tasks. In task 2, the warning and safe signals were click trains to the right and left ears, respectively. In task 3, trains of click pairs were used: in the warning signal, clicks to the right ear preceded those to the left ear by an interval (ITD) of 0.5 ms, while in the safe signal

ITD was reversed. By analogy with human lateralization data, it would be expected that the latter stimuli would be perceived as a single click train lateralized to the side of the leading ear and would thus be perceptually equivalent (or at least similar) to the stimuli in task 2. In further observations made with the task 3 stimulus, ITD was systematically varied to determine the threshold.

In normal animals there was rapid transfer from task 1 to tasks 2 and 3, confirming the perceptual similarity of the stimuli in the latter tasks, and the ITD threshold was approximately 50 µs. In two cases with complete TB transections, performance on tasks 1 and 2 was near normal, but the animals were unable to perform the ITD-based task 3 and had ITD thresholds of the order of 500 µs. In animals with subtotal TB lesions, transfer to and performance on task 3 (with standard ITD of 0.5 ms) was normal, but ITD thresholds were elevated. The nature of the deficit in animals with subtotal lesions of the lateral lemniscus (LL) was quite different from that following TB lesions. These animals had normal ITD thresholds and were able to relearn task 3 to normal levels, but they failed to show normal transfer from task 2 (right vs left discrimination) to task 3 (right leading vs left leading). Thus, while these tasks were qualitatively (perceptually) similar for normal animals, this similarity was lost in the animals with LL lesions. The marked difference in the nature of the effects of TB and LL lesions provides strong evidence that SOC is critical for the processing of ITD information.

The importance of binaural integration in SOC is also indicated by Jenkins and Masterton's (1982) evidence that whereas TB lesions result in impaired localization bilaterally or in the ipsilateral half of space, unilateral lesions at higher levels result in a *contralateral* localization deficit. Evidence for the contralateral representation of auditory spatial location and on the mechanisms underlying this representation will be discussed in detail in later sections (see Sect. 8.2.5 for general discussion). Total bilateral transection of LL in this and other studies (Casseday and Neff 1975; Jenkins and Masterton 1982) abolishes localization performance totally. The effects of unilateral and bilateral lesions at levels above SOC (see Sect. 7.6.1) indicate that the information processed in SOC must reach higher levels to be utilized behaviorally.

These behavioral/lesion data indicate that the critical integration of binaural information for azimuthal localization takes place in SOC. Further support for this view is provided by a study in which Masterton et al. (1975) took advantage of species differences in the size of the MSO to perform what they suggested might be termed a "natural ablation" experiment. They tested cats, rats, tree shrews, and hedgehogs in the Neff situation (with a fixed 60° separation between the speakers) using pure-tone pulses with duration and rise time such that localization depended on IPD at low frequencies and on IID at high frequencies. Whereas all species performed well at high frequencies (at which IID provided the cue), performance at low frequencies (IPD as cue) varied directly as a function of the relative size of MSO. Most dramatically, the hedgehog, lacking an MSO, was totally unable to localize at frequencies below 8 kHz. Although these data demonstrate a relationship between the relative size of an animal's MSO and its ability to utilize IPD information for low-frequency localization, they do not allow of the same interpretation as would abolition of IPD processing by MSO lesion in an animal normally capable of utilizing this cue. In some other species (e.g., mouse and porpoise) in which MSO is small or absent, the large spherical cell area of AVCN is also small or absent (mouse: Martin 1981; Webster and Trune 1982; porpoise: Osen and Jansen 1965). Although no data are available on the neuronal architecture of CN in the hedgehog, or on phase-locking in AN fibers in any of these species, it seems likely that the absence of MSO is merely *one* feature of auditory systems in which mechanisms for the preservation and central processing of phase information are lacking or poorly developed. The inability of the hedgehog to localize low-frequency sounds on the basis of IPD would then be attributable to its having an auditory system of this sort rather than to the absence of an MSO per se.

5.4 Hypotheses Concerning Parallel Processing in the Cochlear Nucleus and Superior Olivary Complex

The evidence presented in the preceding sections on the physiological characteristics and connections of neurons in the various subdivisions of SOC allows further elaboration and examination of the notion that different subdivisions of CN give rise to parallel processing pathways in the auditory brainstem. Some of the qualifications of these notions that were emphasized in Sect. 4.2.7 will be mentioned but not reiterated in detail here.

The projections of neurons in the threee major subdivisions of CN to the subdivisions of SOC provide evidence of largely segregated pathways (cf. Warr 1982), within which further subdivisions are possible on the basis of the projections of particular cell types in CN. DCN projections terminate in VNLL and IC, almost completely bypassing SOC, a fact reflected in the absence of pauser and buildup response patterns in SOC. This pattern of projections allows certain functions for the DCN projection system to be excluded (e.g., a significant role in the processing of interaural disparities or in the control of descending systems in the lower brainstem), but does not permit any more precise definition of its function than the highly qualified account given in Sect. 4.2.7.

Although neurons in both AVCN and PVCN project to NLL and IC, their projections to SOC are largely segregated, AVCN projecting mainly to the major SOC nuclei and PVCN mainly to the (broadly defined) periolivary cell groups. The fact that the periolivary nuclei have major descending projections to the cochlea and CN indicates that one of the functions of the pathways originating in PVCN is the control of centrifugal auditory systems in the lower brainstem. However, our ignorance about the function of these centrifugal systems prohibits any clearer definition of this function. Moreover, the fact that cells in PVCN also have major projections to NLL and IC indicates that this is not the *only* function of these cells.

The major projections of neurons in AVCN to SOC are to the major nuclei involved in the processing of the interaural disparities that serve as cues to azimuthal sound location. Within this system, a further (partial) segregation is supported by physiological and comparative anatomical evidence. Both MSO and the CN region containing large bushy/spherical cells receiving end-bulbs (viz., AA or Osen's large spherical cell area) have a disproportionate representation of low frequencies. Neurons in AA are characterized by secure phase-locking, and the processing of IPD apparently takes place predominantly in MSO. Both of these regions are well developed in species whose low-frequency hearing permits the use of IPD, and this system appears to be *necessary* for the preservation of precise phase information and the processing of IPD. Conversely, the MNTB/LSO system has a disproportionate high-frequency representation and is specialized for the processing of IIDs; this system is hypertrophied in animals with high-frequency hearing for which IPD is not available as a cue to azimuthal location. Although this evidence indicates a segregation of IPD and IID processing, it must be emphasized that this segregation is incomplete and that it cannot be characterized in terms of either a general ITD – IID or low frequency – high frequency segregation. Both the incompleteness and the difficulties of characterization are indicated by the fact that MSO also contains IID-sensitive EI cells, and that fine IID discrimination is possible in species (such as humans) in which MNTB/LSO are small or absent, suggesting that MSO might be *sufficient* for IID coding at moderately high frequencies. Both qualifications are also emphasized by evidence on the processing of TTD. This processing appears to be carried out by EI or IE cells, which are present in both LSO and MSO, and occurs in species in which one or the other nuclei is poorly developed or absent. In summary, LSO appears to be specialized for the processing of both IID and TTD, while MSO is specialized for the processing of IPD, but is capable of processing both IID and TTD.

This discussion has been restricted to the role of SOC in the processing of interaural disparities. As noted in Sect. 4.2.7, however, there are also grounds for supposing that the preservation of temporal information in the projections from AA to MSO is also an important element in the central processing of frequency spectrum information. Unfortunately, there are no data bearing on the way in which spectral information is represented in SOC, although models that ascribe a major role in spectral coding to SOC will be considered in Sect. 8.1.

There is a danger in accounts of this sort that qualifications, although important, might be allowed to obscure functionally significant differences. The converse danger, of course, is that ignoring important qualifications can allow the adoption of an oversimplified and misleading point of view. In attempting to steer a path between these extremes, it is necessary to emphasize that the connections of different cell groups in CN with the different subdivisions of SOC strongly indicate broad functional segregation, but that there remain many uncertainties as to both the degree of segregation of these systems and their proper functional characterization.

6 Nuclei of the Lateral Lemniscus: Anatomy and Physiology

6.1 Neuronal Architecture and Connections

The axons of neurons in CN and SOC ascend to the midbrain on each side in LL (see Fig. 1). Embedded within the fibers of LL is a band of cells, extending from the region immediately rostral to LSO to just ventral to IC. Traditionally, two nuclear groups have been distinguished within this cell band – a dorsal nucleus (DNLL) ventral and slightly lateral and caudal to IC, and a ventral nucleus (VNLL) extending from rostral SOC to DNLL (e.g., Berman 1968; Ramon y Cajal 1909; Stotler 1953; Taber 1961). Within the traditionally defined VNLL a number of regions can be distinguished on cytoarchitectonic grounds (Adams 1979; van Noort 1969; Roth et al. 1978; Taber 1961). Although there is a considerable overlap between these regions, the cytoarchitectonic differences, in conjunction with differences in connections, have prompted a redefinition of the cell groups ventral to DNLL. In the revised scheme, VNLL comprises a ventral region of oval cells and a vertically extended region of multipolar and horizontal cells – termed by Adams (1979) the ventral and medial zones of VNLL. The region of sparsely distributed multipolar cells between VNLL so defined and DNLL – termed by Adams (1979) the dorsal zone of VNLL – is designated the intermediate nucleus (INLL; Brunso-Bechtold et al. 1981; Glendenning et al. 1981; Kane and Barone 1980). Hypertrophy of INLL and VNLL has been described in some mammals with ultrasonic hearing, viz., the dolphin and some bats (Zook and Casseday 1982a; Zvorykin 1964).

The only modern study devoted explicitly to the cellular morphology of NLL is Kane and Barone's (1980) account of DNLL neurons in the cat, but brief descriptions of neurons in the other nuclei have been provided incidentally to accounts of the connections of NLL by Adams (1979) and Glendenning et al. (1981). Kane and Barone (1980) classified Nissl-stained DNLL neurons into nine categories on the basis of soma size and shape, but the great majority of cells in the nucleus (83% in their sample) are medium- to large-sized and round or ovoid (Adams 1979; Kane and Barone 1980). In Golgi preparations, the somas and dendritic trees of most cells are oriented horizontally, perpendicular to the fibers of LL (Kane and Barone 1980; see cells in area designated DL in Fig. 35C). Brief accounts of Golgi material from other species indicate similar horizontal organization of DNLL dendrites in the mouse (Ramon y Cajal 1909 – see his Fig. 360; Willard and Ryugo 1983) and human (Geniec and Morest 1971). Kane and Barone (1980) reported that although the axons of the majority of cells of most types ran dorsally, some cells of most types had axons that ran ventrally in LL. Some medium-sized cells had medially directed axons that presumably contribute to the commissure of the lateral lemniscus (CLL; commissure of Probst), which interconnects DNLL on the two sides.

Both INLL and VNLL also contain a variety of cell types (Adams 1979; Taber 1961), and only the predominant types will be briefly described here. Cells in INLL are sparsely distributed and the predominant cell type is multipolar in shape (Adams 1979). In Golgi preparations in the mouse, INLL neurons have horizontally oriented dendrites like those in DNLL (Willard and Ryugo 1983), but there appears to have been no Golgi study in the cat. In both the cat (Adams 1979) and the mouse (Willard and Ryugo 1983), VNLL can be subdivided into medial and ventral (lateral in the mouse) zones. The principal cell types in the medial zone in both species are multipolar cells and horizontal cells with dendrites perpendicular to the ascending auditory fibers. The predominant cell type in the ventral zone of VNLL in the cat bears a close resemblance to the principal cells of MNTB. In protargol material, cells in the ventral zone in the cat and the lateral zone in the mouse are seen to receive large axosomatic calyceal endings similar in form to, but smaller than, those in MNTB (Adams 1979; van Noort 1969; Stotler 1953; mouse: Willard and Ryugo 1983).

The afferent and efferent connections of NLL have been extensively studied with both degeneration and axoplasmic tracer techniques. DNLL and VNLL are distinguished by very different patterns of ascending input. The major sources of ascending afferents to DNLL are ipsilateral MSO, and ipsilateral and contralateral LSO (Elverland 1978; Glendenning et al. 1981; Glendenning and Masterton 1983; Henkel and Spangler 1983; van Noort 1969). After HRP injections restricted to DNLL, more than 70% of labeled cells are located in SOC (Glendenning et al. 1981); smaller projections are derived from contralateral VCN (Glendenning et al. 1981; Warr 1966, 1982), contralateral DNLL (via CLL), and ipsilateral VNLL (Glendenning et al. 1981). The projections to DNLL from CN and SOC are summarized schematically in Fig. 34. In marked contrast to this pattern, the major source of ascending afferents to VNLL is contralateral VCN (Fig. 34; Glendenning et al. 1981; van Noort 1969; Warr 1966, 1969, 1972, 1982; see also Fig. 26). After HRP injections restricted to VNLL, the overwhelming majority of labeled cells are located in contralateral VCN, with a higher proportion in AVCN than in PVCN (Glendenning et al. 1981). A weak projection from contralateral DCN is indicated by degeneration and tracer studies (Adams and Warr 1976; Fernandez and Karapas 1967; Glendenning et al. 1981). There is also a weak projection to VNLL from ipsilateral SOC (Elverland 1978; Glendenning et al. 1981; Henkel and Spangler 1983), but this appears to be derived from MNTB and periolivary cell groups rather than from the major binaural nuclei (Glendenning and Masterton 1983; Glendenning et al. 1981). A small dorsomedial region of VNLL in both cat (Glendenning et al. 1981; Warr 1969) and rhesus monkey (Strominger and Strominger 1971) is unique in receiving a dense projection from *ipsilateral* VCN.

The pattern of afferent connections to INLL is in many respects intermediate between those to DNLL and VNLL, in that it receives substantial projections both from contralateral VCN and from ipsilateral SOC and VNLL (Fig. 34; Glendenning et al. 1981; van Noort 1969). Although this pattern of input suggests that INLL can be regarded as transitional between DNLL and VNLL, it is in fact distinguished from these nuclei by the fact that it receives a substantial projection from MNTB (Fig. 34). After HRP injections centered on INLL, 20% – 25% of labeled auditory cells are located in MNTB (Glendenning et al. 1981).

Although the evidence reviewed here has established the major sources of ascending afferents to NLL and important differences in the patterns of input to the various nuclei, one question concerning the nature of their afferent supply re-

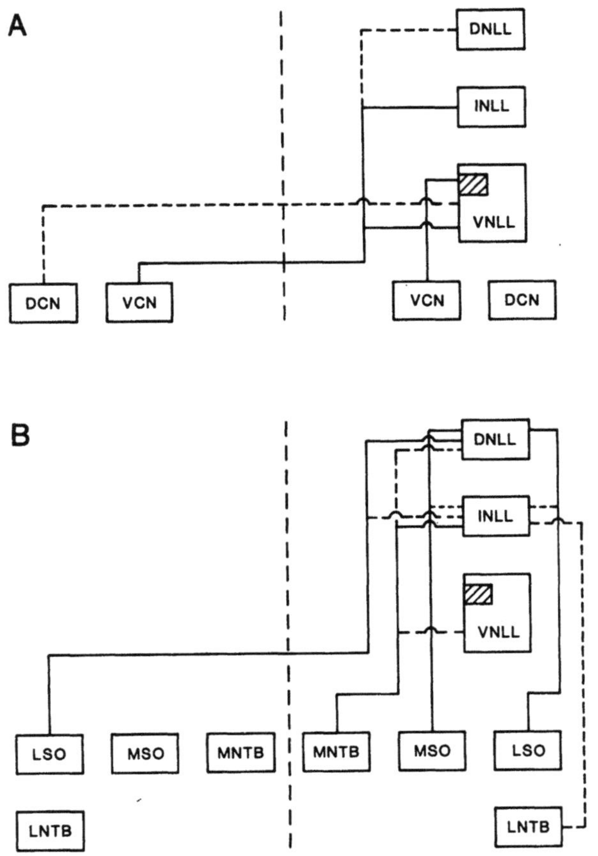

Fig. 34 A, B. Schematic diagrams summarizing ascending projections to the nuclei of the lateral lemniscus from ipsilateral and contralateral (shown to left of *bold broken vertical line*) subdivisions of **A** the cochlear nucleus and **B** the superior olivary complex. Major and minor projections (defined in terms of relative magnitude) are shown by *full* and *broken lines*, respectively. *Cross-hatched area* in ventral nucleus of lateral lemniscus (*VNLL*) indicates dorsomedial region receiving input from ipsilateral ventral cochlear nucleus (*VCN*). Although all projections have been represented as terminating in nuclei of lateral lemniscus, it is uncertain to what extent they are derived from collaterals of fibers proceeding to the inferior colliculus. Note also that each of the major nuclei depicted schematically here contains neurons of different morphological characteristics, and that these diagrams therefore present an extremely simplified picture of the ascending auditory pathways

quires clarification. It is unclear to what extent the innervation of NLL neurons is provided by collaterals of fibers proceeding to IC as opposed to fibers terminating solely in NLL. Only in the ventral zone of VNLL, in which large-diameter axons terminate as calyceal endings, does it seem certain that the endings are not derived from collaterals. Van Noort (1969) described the fibers terminating elsewhere in NLL as thin tufts detaching from LL at right angles, but asserted that, although they had the appearance of collaterals, some were the terminals of single fibers. In this connection, it would be of interest to determine the proportions of single- and double-labeled neurons in CN and SOC subdivisions after injections of different retrograde tracers into IC and one or other of the NLL. It should be emphasized that in the schematic summary of the ascending projections to NLL in Fig. 34 the projections have been shown as terminating in these nuclei rather than as deriving from collaterals solely as a matter of illustrative convenience.

The marked differences between the sources of afferents to DNLL and VNLL are associated with equally marked differences in their ascending projections to the auditory midbrain. DNLL has strong projections to ipsilateral IC and (via the commissure of Probst) to contralateral DNLL and IC, whereas VNLL projects almost only to ipsilateral IC and DNLL (Adams 1979; Brunso-Bechtold et al. 1981; Elverland 1978; Goldberg and Moore 1967; Kudo 1981; van Noort 1969; Roth et al. 1978; Whitley and Henkel 1984). Brunso-Bechtold et al. (1981) reported that INLL projected bilaterally to IC, but Zook and Casseday (1982b) reported that INLL projected only ipsilaterally to IC in the mustache bat, in which the general pattern of ascending projections to IC conforms to that in most mammals. The nature of the ascending projections of this recently distinguished division of NLL therefore remains uncertain. Consideration of the detailed organization of the projections of NLL to auditory midbrain nuclei will be deferred to the sections examining the target structures (see Sects. 7.1.3 and 7.1.5). In addition to these major projections, NLL have ascending projections via extralemniscal pathways that bypass the IC. Thus, both DNLL and VNLL project to the medial and dorsal divisions of the medial geniculate body (MBG; Henkel 1983; Kudo 1981; Whitley and Henkel 1984), projections that form part of Morest's (1965b) lateral tegmental system. DNLL also projects to the deep layers of the superior colliculus (Kudo 1981; see Sect. 7.1.5). Projections from NLL to the midbrain reticular formation (Irvine and Jackson 1983; Stotler 1953; rat: Davis et al. 1982), and descending projections to SOC and CN (Kane and Conlee 1979; Kane and Finn 1977; Kudo 1981; van Noort 1969; Whitley and Henkel 1984), have also been described.

6.2 Physiology of Lateral Lemniscus Neurons and Functional Considerations

There has been only a single study devoted explicitly to examination of the response properties of NLL neurons (Aitkin et al. 1970; Brugge et al. 1970), although Guinan et al. (1972a, b) obtained data for a sample of VNLL neurons in the course of their study of SOC. Aitkin et al. (1970) reported both sustained and onset discharge patterns for neurons in both DNLL and VNLL, but did not indicate the relative proportions of these two types. Some sustained responses were characterized by discharges throughout the stimulus, while others had a pattern in which an onset response was followed by a period of silence and then a sustained discharge for the duration of the stimulus. Phase-locking was frequently observed at frequencies below 1 kHz. Guinan et al. (1972a, b) similarly reported that their small sample of VNLL units exhibited either sustained responses (with chopper pattern predominant) or onset responses.

Aitkin et al. (1970) determined single unit CFs and response areas, but did not report detailed observations on FTCs, although all of the response areas they presented are indicative of sharp frequency tuning. Guinan et al. (1972a, b) reported that the majority (15 of 22) of VNLL units in their sample had V-shaped FTCs, the remainder being broadly tuned or having FTCs with various departures from a simple V shape. Aitkin et al. (1970) reported that both DNLL and VNLL were tonotopically organized with high-to-low frequency units arrayed from dorsal to ventral in each case. Tonotopic organization was consistently observed in DNLL, and was characterized by an unusually rapid transition from low to high frequencies. Tonotopy was a less consistent observation in VNLL (being seen in only four of ten penetrations), and no systematic frequency progressions could be discerned in the interstitial region which has subsequently been defined as INLL. In apparent conflict with Aitkin et al., Guinan et al. (1972b) reported that they could detect no tonotopic organization in VNLL. However, the dorsomedial-to-ventrolateral orientation of their penetrations was oblique to the tonotopic axis indicated by the data of Aitkin et al. and might have made the detection of tonotopy difficult. Some degree of frequency topography in VNLL is suggested by the report by Ryan et al. (1982) of differential (dorsal and ventral) 2-DG uptake following low- and high-frequency stimulation in the Mongolian gerbil, although Nudo and Masterton (1984) failed to find 2-DG evidence of tonotopy in either VNLL or DNLL of a number of species. The uncertainty of the physiological and 2-DG evidence concerning tonotopy in VNLL is reinforced by anatomical evidence. Whereas the afferent and efferent connections of DNLL with tonotopically organized structures (viz., MSO and IC) are topographic (Adams 1979; Brunso-Bechtold et al. 1981; Henkel and Spangler 1983; Roth et al. 1978), the connections of VNLL with IC are reported not to be topographically organized (Adams 1979; Brunso-Bechtold et al. 1981; Roth et al. 1978; Whitley and Henkel 1984). The uncertainty concerning the tonotopic organization of VNLL cannot be resolved on the basis of the available data, but may reflect the fact that this nucleus comprises at least two cytoarchitectonically different zones.

Physiological evidence on the laterality of input to neurons in the different nuclei is in good agreement with the anatomical data on their afferent connections. Aitkin et al. (1970) reported that the majority of units in DNLL (88% of a sample of 121) were binaural and the remaining 12% were monaural (mainly EO/mon), in accordance with the predominant projections to DNLL being derived from SOC. Conversely, the majority of VNLL units (89% of a sample of 57) were EO/mon and only 11% were binaural, in accordance with the predominant projections to this nucleus from contralateral VCN. Units in the interstitial region (INLL) had intermediate properties, 62% being binaural and 38% EO/mon. The latter proportions are almost identical to those reported for VNLL by Guinan et al. (1972a, b; viz., 12 of 32 EO/mon; 18 EE or EI), suggesting that the latter's sample might have included units from both VNLL and INLL.

Binaural neurons in DNLL were shown by Brugge et al. (1970) to be sensitive to ITD and IID. Neurons responsive to low frequencies were sensitive to IPD, with periodic delay functions of the form described previously for MSO neurons (see Sect. 5.2.4). In contrast to such cells in MSO, however, not all IPD-sensitive cells showed sustained, phase-locked responses to monaural stimulation of either ear. Some were EE, but exhibited onset responses to monaural stimuli and sustained responses only to binaural stimuli at optimal delays, whereas others were EO to monaural stimulation. As in VNTB (Goldberg and Brown 1969; see Sect. 5.2.4), these characteristics suggest that IPD sensitivity in DNLL reflects binaural integration taking place elsewhere (presumably in MSO), but are not incompatible with the interaction's taking place in DNLL itself. In two cases examined in detail, the delay producing maximal discharge was that bringing the monaural discharges into coincidence, as previously demonstrated in MSO by Goldberg and Brown (1969).

DNLL neurons receiving EI binaural input, like units of this class in MSO (Goldberg and Brown 1969; Sect. 5.2.3) are sensitive to IID, but relatively insensitive to ABI (Brugge et al. 1970). The IID sensitivity functions of these cells are generally mirror symmetric with those of IE cells in LSO (Boudreau and Tsuchitani 1968, 1970, 1980; see Sect. 5.2.3 and Fig. 30). That is, response is maximal over a range of positive IIDs (i.e., when the stimulus at the contralateral ear is more intense, corresponding to contralateral azimuths) and is totally or near-totally suppressed over some range of negative IIDs (corresponding to ipsilateral azimuths); cf. functions for IC cells in Fig. 43. These characteristics suggest that the sensitivity of EI cells in DNLL to IIDs reflects binaural integration by EI cells in ipsilateral MSO and/or IE cells in contralateral LSO. Given that ipsilateral LSO also projects to DNLL, it is of interest that Brugge et al. (1970) did not report the presence of the IE cells that this projection might be expected to produce. These cells are also very uncommon in IC, and discussion of this issue will be deferred to Sect. 7.2.3.

An interesting feature of IID sensitivity in DNLL neurons reported by Brugge et al. (1970) is that the initial spikes of sustained responses were never suppressed, even at large IIDs favoring the ipsilateral ear. This is also a common observation in IC, in which onset response components in cells with some form of sustained discharge are relatively insensitive to IIDs, and further consideration of this issue will also be deferred (see Sect. 7.2.4).

The available physiological and anatomical data unfortunately throw little light on the functional importance of NLL. All of the structures that project to NLL also project to IC (see Sect. 7.1.3). Indeed, as indicated in the preceding section, it is possible that the major innervation of NLL is by collaterals of fibers proceeding to IC. The major projections of VNLL and DNLL are, in turn, to IC (e.g., Adams 1979; Kudo 1981). The fact that VCN and SOC thus have indirect projections to IC via NLL in addition to their direct projections strongly suggests that the function of the indirect pathways must be to permit some form of transformation or integration of the information being transmitted. The available physiological data, however, provide no evidence as to the nature of any such transformations. There is thus an urgent need for systematic study of the response properties of NLL neurons to determine the respects in which they differ from those of the cells from which their input derives. As discussed in the preceding section, DNLL and INLL also have significant extralemniscal projections, namely those to medial MGB, the deep layers of SC, and midbrain reticular formation. However, NLL are by no means the *only* source of auditory input to these structures (see Sects. 7.1.4 and 7.1.5), and these projections do little to clarify the functional significance of NLL. In the rat, a projection from NLL to ventral regions of the nucleus reticularis pontis caudalis has been identified as a link in the circuit subserving the acoustic startle reflex (Davis et al. 1982).

7 The Auditory Midbrain: Anatomy and Physiology

7.1 Neuronal Architecture and Connections of Auditory Midbrain Structures

7.1.1 Preliminary Considerations: Parcellation of Auditory Midbrain

The major component of the auditory midbrain is the inferior colliculus (IC), the posterior quadrigeminal body of the classical anatomists, which is visible as a convex protuberance on the dorsal surface of the caudal midbrain (see Fig. 1). The IC is the site of termination of almost all LL fibers (Aitkin and Phillips 1984a; Goldberg and Moore 1967; Morest 1965b), and the source of axons that ascend via the brachium of the inferior colliculus (BIC) to the auditory thalamus (see Fig. 1). Ramon y Cajal (1909) identified three major divisions of IC in Golgi-impregnated material from a number of mammals, and a similar (though terminologically different) tripartite parcellation based on the appearance of the cat IC in Nissl-stained sections (e.g., Berman 1968; Rockel and Jones 1973a) has provided the context for most modern research on the connections and physiological response properties of IC neurons.

This parcellation, illustrated in Fig. 35B, identifies a large, centrally located ovoid mass of densely packed cells as the central nucleus of the inferior colliculus (ICC). On the free surfaces of the IC, the central nucleus is covered by a layer of cells, divided here into a pericentral nucleus (ICP) covering the dorsal and posterior surfaces of ICC, and an external nucleus (ICX), covering the lateral and anterior surfaces. Within ICC so defined, a further division (indicated approximately by the broken line in Fig. 35B) was made by Rockel and Jones (1973a) between dorsomedial and ventrolateral divisions, on the grounds of neuronal architecture and fiber connections. The ventrolateral division is defined by the pronounced laminar organization of cell bodies, dendrites, and the afferent axonal plexus (see Sect. 7.1.2 for detailed account), whereas the dorsomedial division is defined by a concentration of large cells and the absence of lamination. The distinction between these regions is supported by the fact that the ventrolateral division receives most of its afferent input from LL, whereas the dorsomedial division receives its input primarily from the auditory cortex and from the contralateral IC via the commissure of the inferior colliculus (ICO; see Sect. 7.1.3 for details).

Although Ramon y Cajal (1909) had stressed the *cortical* (layered or stratified) structure of the surface layers dorsal and lateral to the central nucleus, stratification is not apparent in Nissl material (e.g., Berman 1968) and was explicitly denied by van Noort (1969) and by Rockel and Jones (1973c). In accord with Ramon y Cajal's view, however, Morest and his colleagues have argued that the dorsal region of IC in a number of species has a clearly cortical structure (Morest 1966; Morest and Oliver 1984; human: Geniec and Morest 1971; tree shrew: Oliver and Hall 1978). Recently, Morest and Oliver (1984) and Oliver and Morest (1984) have stressed this point in putting forward a subdivision of IC in the cat, based on the distribution of cell types and the organization of the neuropil in

Golgi-impregnated material, that differs significantly from the standard (Nissl-based) parcellation. Their parcellation is illustrated in Fig. 35A. The central nucleus[17] is defined by the presence of fibrodendritic laminae, and thus corresponds approximately to the ventrolateral division of ICC in the standard parcellation. However, Morest and Oliver divide the central nucleus into central, medial, ventral, and lateral subdivisions (C, M, V, and L in Fig. 35A) on the basis of differences in laminar pattern and in the distribution of cell types (see Sect. 7.1.2). The regions dorsal and caudal to the central nucleus are described as having a four-layered cortical structure, and are therefore defined as the dorsal cortex (I – IV in Fig. 35A) and caudal cortex, respectively. The successively deeper layers of the cortex contain predominantly larger cells and have distinctive patterns of neuropil. The dorsal cortex so defined (together with the dorsomedial nucleus – labeled DM in Fig. 35A) corresponds approximately to the dorsal part of ICP and the dorsomedial division of ICC in the standard parcellation. The caudal cortex corresponds approximately to the caudal part of ICP.

The central nucleus and the cortex comprise two of three major components of IC that are classified by Morest and Oliver (1984) as "tectal," on the basis of the fundamental pattern of dendritic branching of the constituent neurons. "Tectal" neurons are described as typically multipolar, with four or more dendrites having a number of branches, usually with several higher-order collaterals and conspicuous dendritic appendages. The third tectal component comprises a number of regions that are termed "paracentral"; some of these are shown in Fig. 35A, namely, the dorsomedial nucleus (DM), the lateral part of the commissural nucleus (CL), and the ventrolateral and lateral nuclei (VL and LN, respectively). As discussed in more detail in the following section, some of these paracentral nuclei (viz., LN and the rostral pole nucleus, which is not visible at the level shown in Fig. 35A) correspond to some parts of ICX as defined by Berman (1968). Morest and Oliver (1984) provide a detailed comparison of their subdivision of IC with previous subdivisions, and no further attempt to establish equivalences will be made here.

The existence of conflicting parcellations of IC presents a dilemma for the reviewer. On the one hand, almost all of the studies of the connections and physiological response properties of IC neurons that are to be reviewed have been done in the context of the traditional three-division parcellation of IC. On the other hand, recent anatomical evidence in a number of species supports both the cortical structure of the area designated ICP in the standard parcellation and the heterogeneity of the areas considered to comprise ICX. Furthermore, recent connectionist and physiological evidence has indicated that within conventionally defined ICC there is at least a partial segregation of the terminations of ascending afferents from different brainstem sources and of cells with different response characteristics (e.g., Brunso-Bechtold et al. 1981; Roth et al. 1978; Semple and Aitkin 1979, 1981; see Sects. 7.1.3 and 7.2.3). It seems probable that subdivisions so defined might prove to correspond to those defined by Morest

[17] Morest and Oliver (1984) use the abbreviation "CN" to refer to the central nucleus in their parcellation (e.g., Fig. 35C). This term has been used previously in this review as the standard abbreviation for "cochlear nucleus" and will be retained in that sense in the text.

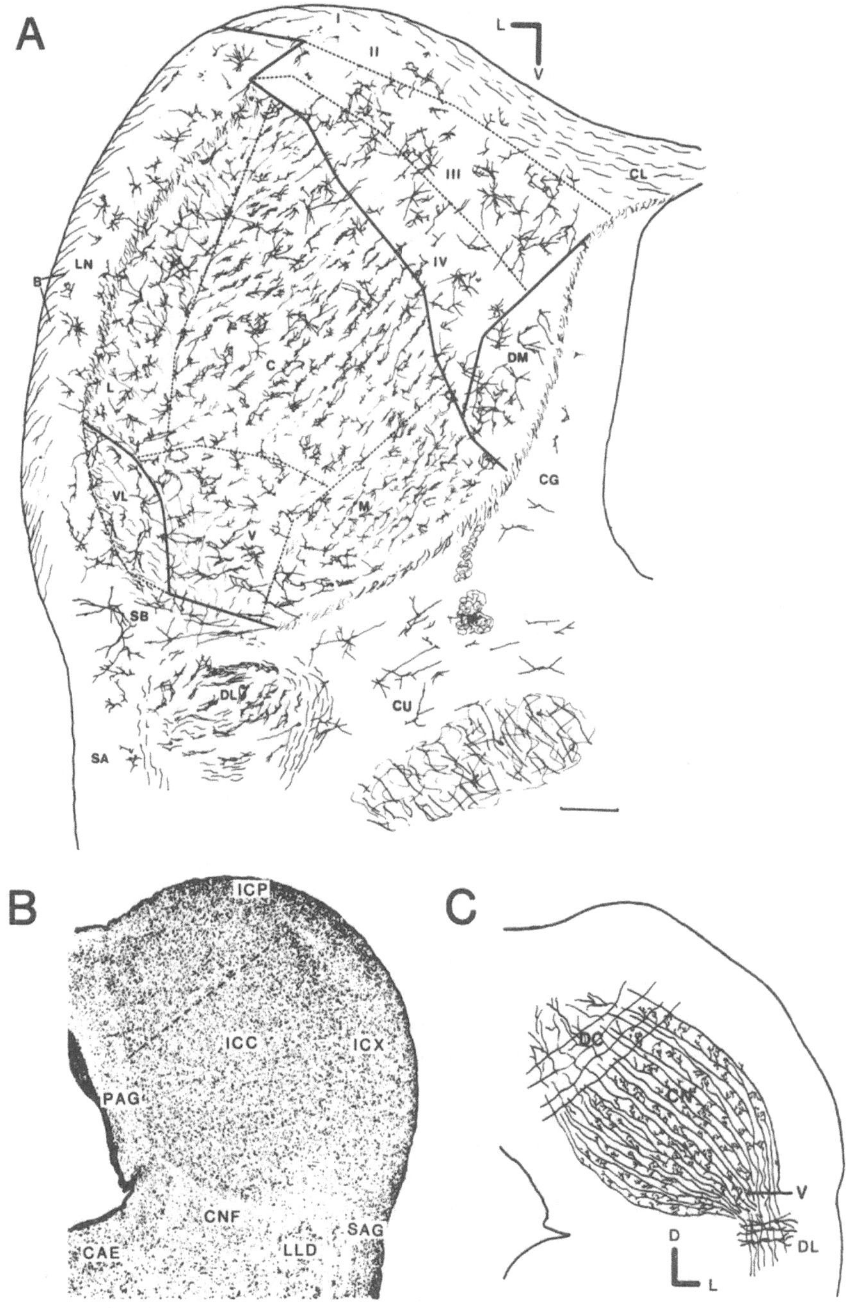

Fig. 35. A Major divisions of Morest and Oliver's (1984) parcellation of inferior colliculus (IC) of the cat shown on drawing of coronal section (Golgi-Cox method) through caudal IC in cat aged 2 months. Abbreviations: ***B***, brachium of the inferior colliculus; *C* central part of central nucleus; *CG* central gray; *CL* lateral part of commissural nucleus; *CU* cuneiform area; *DL* dorsal nucleus of lateral lemniscus; *DM* dorsomedial nucleus; *L* lateral part of central nucleus; *LN* lateral nucleus; *M* medial part of central nucleus; *SA* sagulum; *SB* subcollicular area; *V* ventral (interstitial) part of central nucleus; *VL* ventro-

and Oliver (1984), and that the next decade will see the establishment of correlations between morphologically and physiologically defined subdivisions (and cell types) in IC of the type that followed the precise definition of the neuronal architecture and parcellation of CN by Osen (1969a) and Brawer et al. (1974). Given the poor definition in Nissl material of many of the IC subdivisions identified by Morest and Oliver (1984), and the example provided by the CN data reviewed in Sect. 4.2.5, it is almost certain that progress in this direction will require the intracellular labeling of physiologically characterized neurons. Only the most preliminary data of this sort (Kuwada et al. 1980) are so far available for IC, however, and attempts to establish any but the simplest correlations between currently available data and the finer subdivisions of Morest and Oliver's (1984) parcellation would require unwarranted speculation. In the following account of the neuronal architecture and connections of IC, the term "laminated central nucleus" will be used to refer to the (approximately equivalent) ventrolateral ICC of the standard parcellation and central nucleus of Morest and Oliver's (1984) parcellation. In referring to areas outside the laminated central nucleus, an attempt will be made to indicate the region's identification according to both schemes.

Surrounding the IC are a number of areas receiving auditory input that Morest and Oliver (1984) classified as "tegmental," on the basis of the dendritic branching pattern of the predominant neural type. "Tegmental" neurons have fewer dendrites than tectal neurons, often only two or three, with long, tapering trajectories and fewer secondary branches with sparse appendages. Among the tegmental components of the auditory midbrain are the cuneiform area and sagulum (identified as CU and SA in Fig. 35A, and as CNF and SAG in Fig. 35B), the subcollicular area (SB in Fig. 35A), and the nucleus of BIC. Two other tegmental regions, termed the rostral and dorsal intercollicular areas, may correspond in part to the rostral part of ICX in Berman's (1968) parcellation. Because little is known about the characteristics of these intercollicular areas, or about the nature of their contribution to auditory brainstem processing, they will not be considered in any detail.

Fig. 35 (*continued*) lateral nucleus; I, II, III, IV layers of dorsal cortex. (Morest and Oliver 1984) Scale = 0.5 mm. **B** Major divisions of standard parcellation of inferior colliculus of the cat, shown on Nissl-stained coronal section. *Broken line* indicates approximate position of border between ventrolateral and dorsomedial divisions of central nucleus. Abbreviations: *CAE* nucleus caeruleus; *CNF* cuneiform nucleus; *ICC* central nucleus of inferior colliculus; *ICP* pericentral nucleus of inferior colliculus; *ICX* external nucleus of inferior colliculus; *LLD* dorsal nucleus of lateral lemniscus; *PAG* periaqueductal gray; *SAG* sagulum. (Modified from Berman 1968). **C** Some axons of the central nucleus and dorsal cortex of the inferior colliculus. Axons characteristic of the central nucleus (here designated *CN*) enter from the lateral lemniscus and course dorsomedially, parallel to the disk-shaped dendrites of the principal cells (not shown). Within the central nucleus, they emit side branches which form grape-like clusters of terminal boutons. In this drawing the ventral part (*V*), the interstitial zone of the nucleus, is distinguished by less elaborately branched endings. The other subdivisions of the central nucleus are not distinguished here. In contrast, axons within the dorsal cortex (*DC*) consist of fibers running orthogonal to the lemniscal fibers as well as fine collateral continuations of the lemniscal axons. Other abbreviations: *DL* dorsal nucleus of lateral lemniscus. Freehand drawing from a rapid Golgi preparation from a young kitten. (Morest and Oliver 1984)

One further midbrain structure of significance for the brainstem processing of auditory information does, however, merit attention here. The intermediate and deep layers of the superior colliculus (SC), in contrast to the exclusively visual superficial layers, receive projections from a number of auditory brainstem structures (e.g., Edwards et al. 1979; see Sect. 7.1.5) in addition to projections from visual and somatosensory sources. The efferent projections of deep SC (SCD) are to premotor brainstem areas involved in the control of saccadic eye movements (e.g., Graham 1977; Edwards and Henkel 1978) and, in animals with mobile ears, movements of the pinnae (Henkel 1981; Henkel and Edwards 1978). These and other data from electrical stimulation and lesion studies (see Sects. 7.6.2 and 7.6.3) indicate that SCD is involved in the control of orientation movements of the eyes and ears to targets of all modalities. The accuracy of saccades to auditory targets (monkey: Whittington et al. 1981; human: Zahn et al. 1978; Zambarbieri et al. 1981, 1982) indicates that SCD must either contain or have access to a precise representation of auditory space. There has consequently been considerable recent interest in the nature of the auditory input to neurons in SCD and in the manner in which the spatial location of auditory targets is represented by these neurons (e.g., Gordon 1973; Hirsch et al. 1985; King and Palmer 1983; Middlebrooks and Knudsen 1984; Wise and Irvine 1983, 1984, 1985). The structure and auditory connections of SCD will therefore be described in Sect. 7.1.5, and evidence on auditory responses and spatial representation in this nucleus will be examined in Sect. 7.5.

7.1.2 Neuronal Architecture of Major Subdivisions of Inferior Colliculus

The ventrolateral division of ICC in the cat – Morest and Oliver's (1984) central nucleus – is defined by the presence of laminae formed by the flattened dendritic fields of the principal cells and the reciprocal planar organization of afferent axons from LL (Morest 1964a; Morest and Oliver 1984; Oliver and Morest 1984; Rockel and Jones 1973a). A similar laminar organization is apparent in the central nucleus of all mammals for which IC has been examined in Golgi material (e.g., human: Geniec and Morest 1971; squirrel monkey: Fitzpatrick 1975; rat: Faye-Lund and Osen 1985; mouse: Willard and Ryugo 1983; brush-tailed possum: Aitkin and Kenyon 1981; American opossum: Willard and Martin 1983; tree shrew: Oliver and Hall 1978).
Rockel and Jones (1973a) and Oliver and Morest (1984) used different criteria to distinguish varieties of principal cell (those whose dendrites conform to the laminar pattern) in the cat, and consequently generated different classifications. The studies are in agreement, however, that the principal cells have flattened, or *disk-like*, dendritic fields (Figs. 35A and 36), the height of which (70 – 100 μm) determines the thickness of the lamina. Both studies described the principal cells as having dendritic appendages, characterized as spines (Rockel and Jones 1973a) or as either short protrusions or pedunculated appendages (Oliver and Morest 1984), on the distal dendrites. Rockel and Jones (1973a) estimated that the principal cells compromised 60% of ICC neurons, whereas Oliver and Morest (1984) give estimates of approximately 75% – 85% for different subdivisions of the central nucleus as defined in their parcellation. Since Rockel and Jones's (1973a)

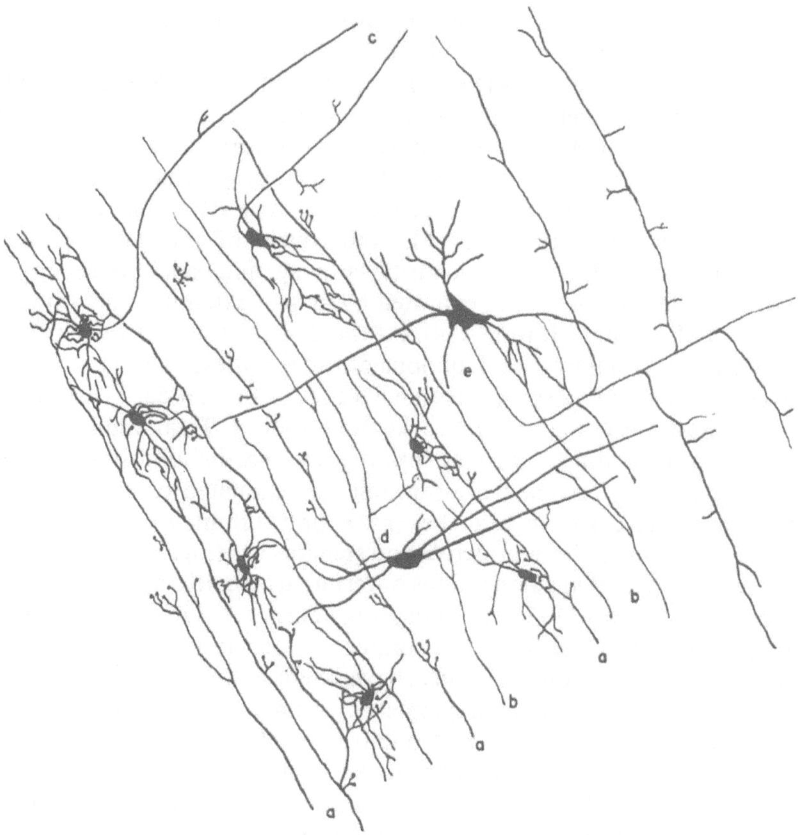

Fig. 36. Scheme of the intrinsic organization of the central nucleus of the cat inferior colliculus. The major cell types in the central nucleus are distinguished by the arrangement of their dendrites with respect to the laminar afferent fibers. Neurons with disk-shaped dendritic fields have their long axes parallel to thick and thin laminar afferents (*a, b*) and receive inputs from narrow sectors of the afferent fiber spectrum. Disk-shaped neurons emit axons with sparse collaterals within the central nucleus (*c*). Stellate cells have dendrites crossing many strata of laminar afferents and can receive inputs from wide sectors of the afferent fiber spectrum. Axons emerging from simple stellates (*d*) and complex stellates (*e*) also contribute to the axonal plexus of the central nucleus and may play a role in the modulation of local activity. (Oliver and Morest 1984)

figure is based on the total cell population of ICC, including the dorsomedial division in which disk-shaped cells are absent, these estimates are in reasonable agreement. In both studies, the major class of nonprincipal cells comprised stellate (Oliver and Morest 1984) or multipolar (Rockel and Jones 1973a) cells of various kinds, whose spherical dendritic fields – particularly in the case of the larger cells – extend across several laminae (Fig. 36).

A number of classes of axon can be distinguished in the laminated central nucleus in Golgi impregnations (Oliver and Morest 1984; Rockel and Jones 1973a). The major class of afferent fibers comprises the axons of the lateral lemniscus, which enter IC ventrolaterally and course dorsomedially parallel to the dendritic layers of the principal cells (Figs. 35C and 36). In parasagittal sections, the axons are seen to give rise to a number of branches forming a longitudinally extended slab or sheet. Rockel and Jones (1973a) described all lemniscal axons (in adult cats) as being of large diameter (5 μm or more), but Oliver and Morest (1984) described both coarse (up to 3 μm) and thin (less than 1 μm) lemniscal axons in kittens. It seems likely that the thin axons of unknown origin that Rockel and Jones described as conforming to the laminae (their type III axons) are in fact thin LL axons. After large lesions of LL, the course of the degenerating fibers recapitulates the pattern of the dendritic laminae, and bands containing coarse- and thin-fiber degeneration are separated by areas containing only thin-fiber degeneration (Oliver and Morest 1984). At the dorsal border of the laminated central nucleus, fine collateral continuations of the lemniscal axons extend into the deep layers of the dorsal cortex (dorsomedial ICC of the standard parcellation) (Oliver and Morest 1984; Fig. 35C). Rockel and Jones (1973a) similarly described the laminae as extending some way into the dorsomedial division.

As they course dorsomedially through the laminated central nucleus, the lemniscal axons give off side branches that supply terminals to the surfaces, especially the dendrites, of both principal and stellate cells (Figs. 35C and 36). In the case of the principal cells, aggregations of terminals are described as forming "clasps" (Rockel and Jones 1973a) or "grape-like clusters" (Oliver and Morest 1984) on the proximal dendrites and, less commonly, on the soma (Fig. 36). EM study of the terminals of lemniscal axons shows them to contain large spherical vesicles and to make one or more asymmetric contacts with the postsynaptic element (Rockel and Jones 1973b; see also Oliver 1984b). The type 1 cytological characteristics of these terminals makes it tempting to speculate that lemniscal axons have predominantly excitatory effects on IC neurons. However, these characteristics have not been established for the terminals of *all* lemniscal axons and, as noted previously, cytological features can suggest, but not establish, the functional nature of synapses (see footnote on p. 48).

The other major class of axons in the laminated central nucleus comprises the efferent axons of the principal and stellate cells and their intrinsic collaterals. The main segments of many of the axons projecting from the central nucleus course rostrolaterally toward the BIC, almost perpendicular to the fibrodendritic laminae (Fig. 36). Principal cells form relatively few intrinsic collaterals, but some large stellate cells can have very complex collateral branching patterns (see axon e in Fig. 36; Morest and Oliver 1984; Oliver and Morest 1984). Some small stellate cells may be Golgi type II cells with axon collateral systems restricted to the central nucleus. Rockel and Jones (1973b) described several types of axonal terminal in ventrolateral ICC in addition to those of the lemniscal axons, some of which were characterized by flattened or pleomorphic vesicles and symmetric synaptic contacts, and might therefore be inhibitory. However, these types of terminal could not be associated with particular classes of axon.

The final aspect of the neuronal architecture of the laminated central nucleus to be considered concerns the pattern of the fibrodendritic laminae, of which Rockel and Jones (1973a) and Oliver and Morest (1984) have given somewhat different accounts. The two studies agree that in the central part of the laminated central nucleus – Oliver and Morest's (1984) division C – the laminae in transverse sections run from dorsomedial to ventrolateral at about 40° – 60° to the vertical plane (cf. orientation of dendrites and of lemniscal axons in Figs. 35A and C, respectively), and that the anteroposteriorly extended laminae are seen in horizontal sections to be oriented from posterolateral to anteromedial (at an angle of about 20° to the sagittal plane). In other respects, however, the two accounts differ. Rockel and Jones (1973a) described the laminae as curving as they approached the edges of ICC, particularly at the lateral edge where they swept dorsally at a steep angle. They therefore argued that the laminae, which appear as flat sheets in the central region, in fact form a series of concentric curves, somewhat analogous to the layers of an onion. In contrast to this account, Morest and Oliver (1984) have argued that the orientation (and some other features) of the laminae in the four subdivisions of the central nucleus are different, and that – in the three subdivisions in which they have been reconstructed (viz., the central, medial, and lateral) – they are relatively flat. The most significant difference in orientation is in the lateral subdivision, in which the laminae are oriented from ventromedial to dorsolateral, almost at 90° to those in the central subdivision. This description is not, in fact, markedly different from Rockel and Jones's (1973a) description of the lateral edges of the laminae as sweeping dorsally at a steep angle. The crux of the disagreement lies in the conflict between Oliver and Morest's claim that the laminae are *not continuous* across the boundaries of their proposed subdivisions of the central nucleus and Rockel and Jones's implicit assumption that the laminae are *continuous* throughout an undifferentiated ventrolateral division.

The neuronal architecture of the other major divisions of IC has been described in less detail than that of the laminated central nucleus, and discussion is complicated by the different parcellations of these regions. There is also very much less physiological data on the properties of neurons outside the laminated central nucleus, and the neuronal architecture of these regions will consequently be treated only briefly.

As noted previously, the dorsomedial division of ICC in the standard parcellation (Rockel and Jones 1973a) corresponds approximately to layer IV of the dorsal cortex and the dorsomedial nucleus of Morest and Oliver's (1984) parcellation. This region is characterized by the absence of disk-shaped cells and dendritic lamination, and by the presence of a high proportion of large multipolar/stellate cells. Fine collateral extensions of lemniscal axons combine with non-lemniscal afferents (see Sect. 7.1.3) to form a dense meshwork of neuropil that is characterized by a conspicuous horizontal plexus in layer IV (see horizontally oriented fibers in area designated DC in Fig. 35C).

The three superficial layers of Morest and Oliver's (1984) dorsal cortex correspond approximately to the part of ICP dorsal to ICC in the standard parcellation. Rockel and Jones (1973c) described a variety of predominantly multipolar cell types in ICP, and – in agreement with Morest and Oliver's (1984) account – observed that the smallest cells were found most dorsally and progressively larger cells more ventrally. However, Rockel and Jones (1973c) stated that although dorsal ICP consequently has a "superficial appearance of lamination," they did not find it useful to regard it as having a cortical structure. Conversely, Morest and Oliver's (1984) and Morest's (1966) characterization of this region as cortical is in line with Ramon y Cajal's (1909) description and with later descriptions in other species (e.g., tree shrew: Oliver and Hall 1978; rat: Faye-Lund and Osen

1985; mouse: Willard and Ryugo 1983; American opossum: Willard and Martin 1983). It should also be noted that one of the features cited by Rockel and Jones (1973c) in arguing against a cortical organization was the fact that the dendrites of cells at a given depth extend into other levels. As Morest and Oliver (1984) point out, this is in fact a common feature of cortical structures. The superficial layers of the dorsal cortex (dorsal ICP) are characterized by a fine, horizontally oriented axonal plexus. Ascending lemniscal axons do not reach this region, and the plexus is composed of both intrinsic axon collaterals and nonlemniscal afferents (predominantly from auditory cortex; see Sect. 7.1.3; Morest and Oliver 1984; Rockel and Jones 1973c).

The area designated as ICX in the standard parcellation comprises a number of heterogeneous regions, and details of neuronal architecture have been described for only some of these areas. The wedge-shaped rostroventral extension of IC that underlies the caudal end of SC was designated part of ICX by Berman (1968) and is so labeled in Fig. 41. This region corresponds to Morest and Oliver's (1984) nucleus of the rostral pole, which they described as having a tectal structure, comprising medium-sized stellate neurons in tightly knit neuropil. This description is in accord with that of van Noort (1969), who considered this region to be a rostroventral extension of the central nucleus (cf. Osen 1972). In contrast to the clearly tectal structure of the rostroventral process, the more dorsal parts of traditionally defined ICX covering the rostral surface of the central nucleus appear to correspond to some parts of the ventral and dorsal intercollicular areas, which Morest and Oliver (1984) characterized as tegmental in structure.

The part of ICX lateral to ICC corresponds approximately to Morest and Oliver's (1984) lateral nucleus (Fig. 35A and B), which they describe as containing neurons of both tectal and tegmental types. The neurons are of smaller average size than those of the central nucleus and cortex, and have lower packing density because of the many myelinated fibers of passage that dominate the axonal plexus of the region. These fibers include efferent axons from other collicular divisions projecting into BIC and descending afferent fibers from BIC. The fact that this region of traditionally defined ICX comprises neurons of both tectal and tegmental types is of interest in connection with the fact that it receives both auditory and somatosensory afferents (see Sect. 7.1.3).

7.1.3 Afferent Connections of Inferior Colliculus Subdivisions

Projections to IC from almost all of the major divisions of the brainstem auditory system have been described in previous sections of this review. The highly convergent nature of these projections is most clearly demonstrated by experiments in which large injections of HRP have been made into conventionally defined ICC. Such injections in the cat result in large numbers of labeled neurons in DNLL, INLL, and LSO bilaterally, VNLL and MSO ipsilaterally, and AVCN, PVCN, and DCN contralaterally (Adams 1979; Brunso-Bechtold et al. 1981; Roth et al. 1978). In addition to these major projections, smaller numbers of labeled cells are found ipsilaterally in all three divisions of CN, MNTB and sagulum, contralaterally in MSO, and bilaterally in the periolivary cell regions (see Fig. 27 for locations of periolivary projection cells). The projections to IC

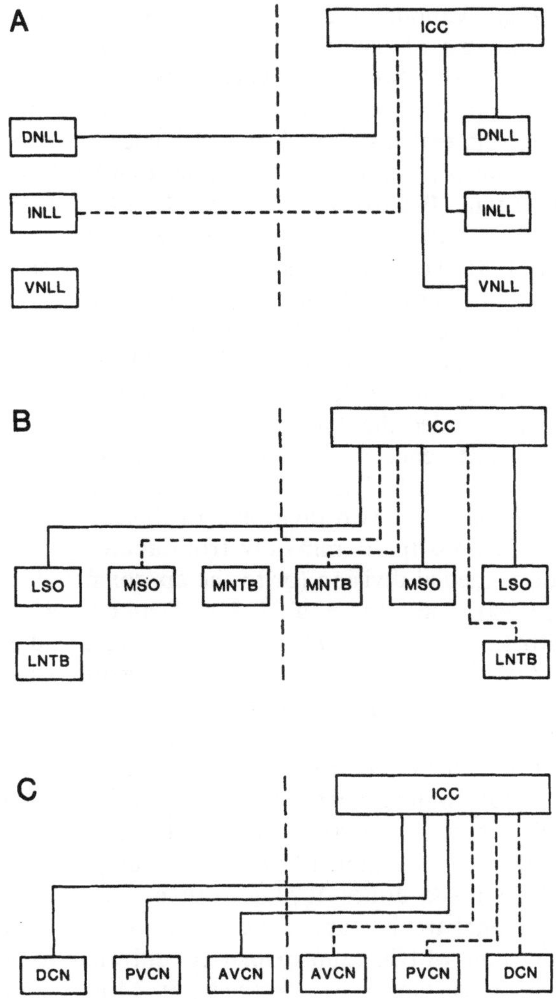

Fig. 37A–C. Schematic summaries of major (*full lines*) and minor (*broken lines*) ascending projections to ICC from ipsilateral and contralateral (shown to left of *bold vertical broken line*) nuclei of **A** the lateral lemniscus, **B** superior olivary complex, and **C** cochlear nucleus. ICC also receives sparse projections from a number of other brainstem structures, as described in text. Note also that each of the major nuclei depicted schematically here contains neurons of different morphological characteristics, and that these diagrams therefore present an extremely simplified picture of the ascending auditory pathways

from CN, SOC, and NLL are summarized schematically in Fig. 37. Similar patterns of labeling after large HRP injections have been seen in a variety of other mammals (mouse: Ryugo et al. 1981; Willard and Ryugo 1983; rat: Beyerl 1978; Druga and Syka 1984a; gerbil: Nordeen et al. 1983; brush-tailed possum: Aitkin and Kenyon 1981; American opossum: Willard and Martin 1983, 1984; bat: Schweitzer 1981; Zook and Casseday 1982b).

The large injections used in these studies result in uptake in both the laminated ventrolateral and large-celled dorsomedial divisions of the standard parcellation of ICC, and hence do not distinguish between the sources of ascending afferents to the two divisions. Smaller injections of HRP into the ventrolateral division, and anterograde degeneration and tracer studies, indicate that all of the major brainstem structures labeled after large injections (and summarized in Fig. 37) project to the ventrolateral division, though not to all parts of it (see later discussion; Brunso-Bechtold et al. 1981; Elverland 1978; Glendenning and Masterton 1983; Henkel and Spangler 1983; Kudo 1981; Oliver 1984b; Osen 1972; Warr

1982). Although Golgi and anterograde degeneration studies indicate that fine collaterals of lemniscal axons reach the dorsomedial division (Goldberg and Moore 1967; Morest and Oliver 1984; Rockel and Jones 1973a), these projections are apparently derived from only a subset of the structures projecting to the ventrolateral division. Thus, evidence from both anterograde and retrograde axoplasmic tracer studies indicates that LSO does not project to the dorsomedial division (Brunso-Bechtold et al. 1981; Elverland 1978; Glendenning and Masterton 1983), while evidence on MSO is contradictory (Brunso-Bechtold et al. 1981; Henkel and Spangler 1983). The ascending projections to the dorsomedial division appear to be derived mainly from contralateral CN (Brunso-Bechtold et al. 1981; Oliver 1984b; Osen 1972) and from NLL (Brunso-Bechtold et al. 1981; Kudo 1981; Whitley and Henkel 1984). Brunso-Bechtold et al. (1981) reported that labeled cells were seen in ipsilateral CN *only* after HRP injections that included the dorsomedial division.

The difference in the strength of ascending projections to ventrolateral and dorsomedial ICC is reversed by the pattern of descending and commissural projections to the two divisions. In the cat and most other mammals, descending projections (predominantly from auditory cortex) to ICC are restricted to the dorsomedial division (or to the approximately equivalent layer IV of the dorsal cortex) (Andersen et al. 1980b; Diamond et al. 1969; Morest and Oliver 1984; Rockel and Jones 1973a; mouse: Willard and Ryugo 1983; American opossum: Willard and Martin 1983). Two puzzling exceptions to this pattern are Fitzpatrick and Imig's (1978) report of cortical projections to both the dorsomedial and the laminated portions of ICC in the owl monkey, and reports of cortical projections to both divisions of ICC in the rat (Beyerl 1978; Druga and Syka 1984b). In the cat, axons in the ICO arise from neurons throughout ventrolateral ICC, but their midbrain terminations are largely restricted to the contralateral dorsomedial region (Aitkin and Phillips 1984b; Brunso-Bechtold et al. 1981; Rockel and Jones 1973a; Roth et al. 1978). The marked differences in the neuronal architecture of the laminated ventrolateral and large-celled dorsomedial divisions of ICC (approximately equivalent to the central nucleus and layer IV of the dorsal cortex, respectively, in Morest and Oliver's parcellation) are thus associated with equally marked differences in the sources of their afferents. The ventrolateral division receives massive ascending projections from a large number of brainstem nuclei and (in the cat) little or no descending projections; the dorsomedial division, conversely, receives only sparse projections from a more restricted set of brainstem nuclei, but receives major descending and commissural projections.

Thus far, ascending projections to ICC have been described only in terms of the brainstem nuclei from which the projections originate. This level of description, and block diagrams such as that presented in Fig. 37, ignore the structural and functional heterogeneity of many brainstem "nuclei" that has been emphasized in previous sections. Evidence on the topographic organization of ascending projections to ICC and on the cell types giving rise to these projections is reviewed in the following paragraphs.

Evidence from degeneration and axonal tracer studies indicates that projections to ICC from almost all brainstem nuclei are topographically organized (Adams 1979; Brunso-Bechtold et al. 1981; Henkel and Spangler 1983; van Noort 1969;

Oliver 1984b; Osen 1972; Roth et al. 1978). Projections to dorsal ICC consistently derive from neurons in the low-frequency regions of brainstem structures, while projections to ventral ICC consistently derive from neurons in the high-frequency regions of these nuclei. Small injections of tritiated amino acids into DCN (Oliver 1984b) or MSO (Henkel and Spangler 1983) have been shown to result in bands of dense labeling in ventrolateral ICC, the orientation of which corresponds to that of fibrodendritic laminae. This topographic organization of projections is in accordance with physiological evidence of a dorsal-to-ventral, low-to-high-frequency tonotopic organization in ventrolateral ICC (see Sect. 7.2.1). The fact that physiological evidence reveals only a single frequency/cochlear representation in ICC indicates that ascending afferents from multiple brainstem sources converge in a highly organized fashion onto the laminated central nucleus. The only brainstem nucleus whose projections to ICC do not appear to be topographically organized is VNLL (Adams 1979; Brunso-Bechtold et al. 1981; Roth et al. 1978; Whitley and Henkel 1984). As noted in Sect. 6.2, the physiological evidence for tonotopic organization of VNLL is equivocal.

A complex organization of LSO's projections to ICC in the cat is indicated by Glendenning and Masterton's (1983) double-retrograde tracing study. Although each LSO projects in a topographically organized fashion to ICC on each side, the neurons projecting ipsilaterally and contralaterally are partially segregated and only a small percentage send collaterals to both sides. The projection to the ipsilateral ICC is greater from the lateral (low-frequency) arm of each LSO, while that to contralateral ICC is greater from the medial (high-frequency) arm of each LSO (Fig. 38A and B). The lateral-to-medial (low-to-high-frequency) gradient in the percentage of LSO neurons projecting contralaterally is clearly indicated in the mean data for a number of animals in Fig. 38B. Glendenning and Masterton (1983) and Masterton et al. (1982) have pointed out that the bilateral projections of LSO result in an organization analogous to that of the optic chiasm (Fig. 38C), and have argued for the importance of this chiasm-like organization in the representation of auditory spatial location. This notion is examined in Sect. 8.2.5. As noted in Sect. 5.1.4, a similar study in the American opossum by Willard and Martin (1984) indicated that a larger proportion (approximately a third) of LSO central neurons projecting to IC were double-labeled, and that the topographic organization of the projections was somewhat different from that in the cat. Species differences in the organization of these projections therefore require examination.

Small injections of retrograde tracers into the ventrolateral division result in labeling in only a subset of the nuclei labeled after large injections, indicating at least a partial segregation of afferents from different sources within ICC (Aitkin and Schuck 1985; Brunso-Bechtold et al. 1981; Roth et al. 1978; Semple and Aitkin 1981) – what Brunso-Bechtold et al. (1981) have termed a "nucleotopic" organization. Thus, Roth et al. (1978) compared the patterns of labeling produced by two small injections of HRP into the high-frequency region of ICC. A posteriorly located injection resulted in labeling only in contralateral CN (all three divisions) and in ipsilateral VNLL, while a centrally located injection resulted in labeling restricted to ipsilateral MSO, LSO, and periolivary nuclei. Brunso-Bechtold et al. (1981) confirmed that the projections from CN and VNLL were heaviest to caudal ICC and those from SOC heaviest to rostal ICC.

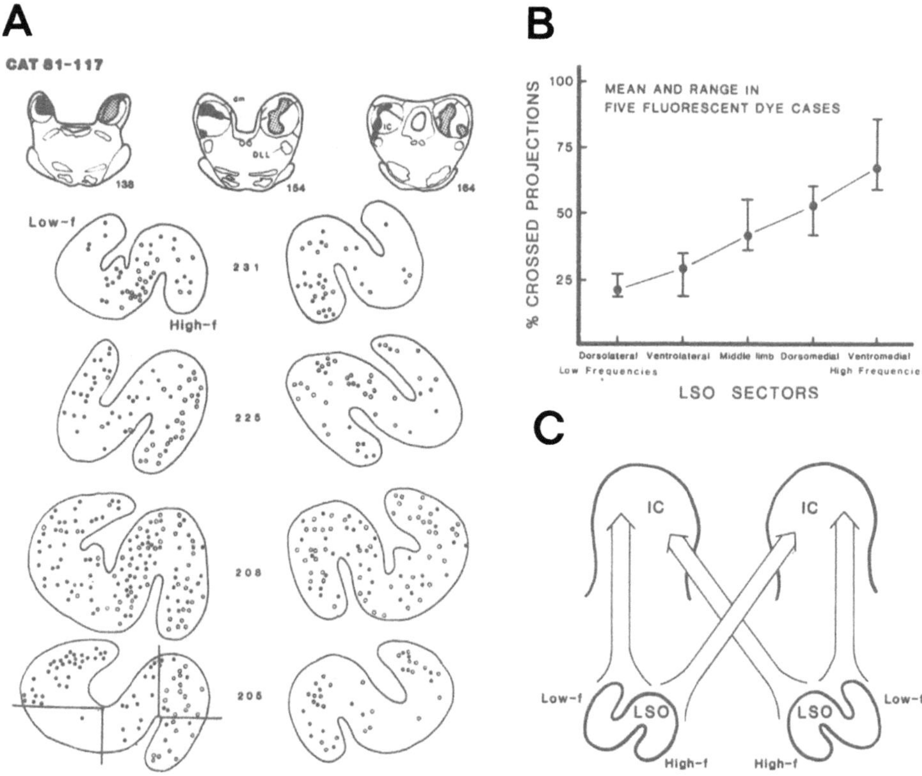

Fig. 38. A Distribution of labeled cells in left and right lateral superior olive (*left- and right-hand columns*) following injection of fast blue in the left inferior colliculus and nuclear yellow in the right inferior colliculus. Approximate extent of injection sites is indicated in tracings of midbrain sections at top of figure. *Black dots*, cells labeled with fast blue; *open circles*, cells labeled with nuclear yellow. Sections are 40 μm thick. Bottom left-hand section of LSO shows lines used to divide it into five sections (used in **B**). Note that the *black dots* in the left LSO represent cells giving rise to uncrossed fibers, but in the right LSO they represent cells giving rise to crossed fibers. Conversely, the *open circles* in the left LSO give rise to crossed fibers, but in the right LSO they give rise to uncrossed fibers. **B** Lateral-to-medial gradient in the percentage of labeled lateral superior olivary cells giving rise to crossed and uncrossed projections. Mean percentage of crossed projections and entire range (*vertical bars*; based on five cases with largest fluorescent dye injections) are plotted as a function of the five sectors defined in **A**. **C** Schematic drawing illustrating chiasm-like arrangement of projections from lateral superior olivary nucleus (*LSO*) to inferior colliculus (*IC*). *Low-f* and *High-f*: low- and high-frequency regions of tonotopic sequence in LSO. (Glendenning and Masterton 1983)

Both anterograde and retrograde labeling evidence indicate that MSO projects predominantly to the dorsolateral and rostral region of ventrolateral ICC (Aitkin and Schuck 1985; Brunso-Bechtold et al. 1981; Henkel and Spangler 1983). Semple and Aitkin (1981) described a case in which a small iontophoretic injection of HRP into a physiologically characterized region of ventrolateral ICC resulted in labeled cells only in contralateral LSO and DCN. These data remain preliminary, but in conjunction with physiological evidence on the partial segregation of binaural response types in ICC to be reviewed in Sect. 7.2.3, they strongly suggest a degree of functional segregation in the projections to the laminated central nucleus. Knowledge of the zones of termination of projections from different brainstem regions, and of the relationship of these zones to the subdivisions identified by Morest and Oliver (1984), is an important requirement for the understanding of the functional organization of the laminated central nucleus. In the first study directed to these issues, Oliver (1984b) has presented anterograde tracer evidence that DCN projects only to the central and medial divisions of the laminated central nucleus of contralateral IC.

The structural definition of labeled neurons provided by the HRP method has allowed the neuron types projecting to ICC in the cat to be identified in some studies. In AVCN, many type I stellate cells and small cells, but few (if any) bushy (spherical or globular) cells or type II stellate cells are labeled after IC injections (Adams 1979, 1983a; Cant 1982; Roth et al. 1978). The absence of projections from bushy/large spherical cells to IC is supported by Bourk's (1976) observation that "primarylike" units with large prepotentials could not be antidromically activated from IC. Adams (1983a), using double-retrograde tracers, found that almost all the stellate (multipolar) cells projecting to IC also send collaterals to DCN. In PVCN, many multipolar and small cells are labeled; although a few octopus cells may be labeled, their identification is uncertain and labeling is generally restricted to regions containing few octopus cells (Adams 1979; Roth et al. 1978). In DCN, fusiform cells, giant cells, and a few small cells in the superficial layers are labeled (Adams 1979; Brunso-Bechtold et al. 1981; Oliver 1984b; Roth et al. 1978). With respect to SOC, there is general agreement that both bipolar and multipolar central cells in MSO are labeled after IC injections, but some disagreement as to whether marginal cells participate in this projection (Adams 1979; Aitkin and Schuck 1985; Henkel and Spangler 1983; Kiss and Majorossy 1983; see Sect. 5.1.3). Adams's (1979) data also indicated the following patterns of labeling in other nuclei: large multipolar and small bipolar cells in LSO; a few, apparently stellate, cells in MNTB; all cell types in DNLL and VNLL.

Many of the cell types projecting to ICC have been shown in previous sections to have different physiological properties, some of which can be related to reasonably well-defined functional roles. More generally, there can be little doubt that the differences in soma size and shape and in dendritic structure that form the basis of the definition of morphological cell types are associated with differences in electrophysiological properties and functions (see Brawer et al. 1974; Morest et al. 1973). The diversity of cell types projecting to ICC thus indicates a degree of convergence even more dramatic than that suggested by the number of gross nuclear subdivisions from which projections are derived.

In contrast to the detailed information as to the cell types from which ICC's afferents are derived, almost nothing is known about the mode of termination of particular projections. Recently, however, Oliver (1984b) reported that terminals containing round vesicles and making asymmetric synaptic contacts on the dendrites of disk-shaped and stellate cells in the medial and central regions of the central nucleus were labeled after small injections of tritiated proline or leucine into DCN (see also Oliver and Morest 1979). The procedures used in this study –

combined LM and EM autoradiography after small injections of tritiated amino acids – offer a powerful means of determining both the locus and mode of termination of particular projections to ICC.

Afferent projections to the regions comprising the other major IC subdivisions in the standard parcellation (viz., ICX and ICP) have been described in much less detail than those to ICC. The major afferents to ICP (corresponding approximately to layers I–III of the dorsal and caudal cortex of Morest and Oliver's parcellation) are descending projections from auditory cortex (Andersen et al. 1980b; Diamond et al. 1969; Morest and Oliver 1984; Rockel and Jones 1973c) and commissural projections from the contralateral IC (Kudo and Niimi 1980; Moore and Goldberg 1963; Rockel and Jones 1973c). Evidence from degeneration and tracer studies indicates that there are no ascending auditory projections to ICP from lower brainstem centers (Glendenning and Masterton 1983; Kudo 1981; Rockel and Jones 1973c; mouse: Ryugo et al. 1981; Willard and Ryugo 1983). Rockel and Jones (1973c) suggested the possibility of projections from DNLL to ICP, but no evidence of such projections was found in an autoradiographic study by Kudo (1981).

As noted previously, ICX in the standard parcellation is approximately equivalent to a rather diverse set of subdivisions in Morest and Oliver's (1984) parcellation (viz., the lateral nucleus, the rostral pole nucleus, and parts of the intercollicular area). The rostral pole nucleus (or rostroventral process), which was described by Morest and Oliver (1984) as having a tectal structure and by van Noort (1969) as part of the central nucleus, receives a major projection from contralateral CN (van Noort 1969; Oliver 1984b; Osen 1972). The other regions comprising ICX receive only sparse lemniscal projections, the origins of which have not been determined in detail, but include contralateral DCN (Aitkin et al. 1981; Goldberg and Moore 1967; Kudo 1981; Morest and Oliver 1984; Oliver 1984b; mouse: Ryugo et al. 1981; Willard and Ryugo 1983). There is, however, a strong projection to these parts of ICX from ICC (Aitkin et al. 1981; Andersen et al. 1980a; Kudo and Niimi 1980; Rockel and Jones 1973a). A distinctive feature of these parts of ICX (and of the lateral nucleus, at least, in Morest and Oliver's parcellation) is that, in a number of species, it also receives somatosensory projections from the dorsal column nuclei and trigeminal complex (Aitkin et al. 1981; Hand and van Winkle 1977; Morest and Oliver 1984; mouse: Willard and Ryugo 1979, 1983; American opossum: RoBards 1979; RoBards et al. 1976). In the mouse, Willard and Ryugo (1979, 1983) have described a laminar segregation of input to the "external cortex" (approximately equivalent to lateral ICX), somatosensory afferents terminating in the superficial layer, and auditory afferents (from DCN) terminating in the deep layer. There appears to have been no corresponding investigation in the cat.

7.1.4 Efferent Connections of Inferior Colliculus Subdivisions

The IC has both ascending projections to auditory thalamus and descending projections to midbrain and brainstem structures. Because of the restriction of this review to brainstem processing, the ascending projections will be only briefly summarized here. The roles in auditory brainstem processing of projections to other midbrain or lower brainstem structures are in most cases little understood, and these projections will therefore also be described only briefly.

Description of the ascending projections of IC is complicated by the fact that different investigators have used different parcellations of the medial geniculate body (MGB) and posterior thalamus (see Aitkin et al. 1984b; Calford 1983 for reviews). The following summary uses Morest's (1964) subdivision of MGB, and is a simplified integration of the results of recent tracer studies by Andersen et al. (1980a), Calford and Aitkin (1983), and Kudo and Niimi (1978, 1980), to which the reader is referred for more detailed accounts.

All three divisions of the traditional parcellation of IC project bilaterally to MGB, the contralateral projections (via ICO) being weaker, but having the same termination pattern. The laminated ventrolateral division of ICC projects predominantly, and in a topographically organized fashion, to the laminated ventral division of MGB, which is the major source of input to primary auditory cortex (area AI) and to the other tonotopic cortical fields (see Aitkin et al. 1984b; Imig and Morel 1983 for recent reviews of the thalamocortical auditory system). The ventrolateral division also has weaker projections to the medial and deep dorsal divisions of MGB and to the lateral division of the posterior nuclear group. The ascending projections of the ventrolateral division are derived from disk-shaped neurons and from stellate neurons (Oliver 1984a).

ICP and the dorsomedial division of ICC project most strongly to the dorsocaudal division of MGB, which also receives a major projection from sagulum. The fact that the areas designated ICP and dorsomedial ICC in the standard parcellation have common thalamic targets supports Morest and Oliver's contention that they should be considered components (laminae) of a single dorsal cortex. The major ascending projection from ICX is to the medial and deep dorsal divisions of MGB. Thus, although the projections of the major divisions overlap in some regions of auditory thalamus (notably in the medial division of MGB), the major projection of each division is to a different thalamic division.

IC also has projections to lower brainstem auditory nuclei and to regions outside the auditory pathway, notably to the dorsolateral pontine nucleus and to the deep layers of SC (e.g., Andersen et al. 1980a; Conlee and Kane 1982; Edwards et al. 1979; Hashikawa 1983; Hashikawa and Kawamura 1983; Kawamura 1975; van Noort 1969; Powell and Hatton 1969; Rasmussen 1964). Projections to the dorsolateral pontine nucleus comprise a component of one of the pathways over which auditory input reaches the cerebellum (e.g., Aitkin and Boyd 1975, 1978; Boyd and Aitkin 1976; Kawamura 1975). Recent tracer studies indicate a considerable degree of segregation of the IC neurons giving rise to projections to lower auditory nuclei and to structures outside the auditory pathway. Thus, descending projections to SOC and CN are derived predominantly from ICC, although ICP and ICX also project, less strongly, to CN (Conlee and Kane 1982; Hashikawa 1983). Conversely, the projections to the dorsolateral pontine nucleus and to deep SC are derived almost entirely from ICX and ICP, with little or no contribution from ICC (Edwards et al. 1979; Hashikawa 1983; Hashikawa and Kawamura 1983; see following section for more detailed account of auditory afferents to deep SC). A high degree of segregation in the sources of IC's ascending and descending projections to auditory structures, and of its projections outside the auditory pathway, is also indicated by preliminary data obtained with double-retrograde tracers (Hashikawa and Kawamura 1983). When different fluorescent tracers were injected into MGB and CN, cells labeled with one or the other tracer were found in ICC and, to a lesser extent, in ICX, but no double-labeled cells were seen. Similarly, injections into the pontine nuclei and SC both resulted in labeled cells predominantly in ICX and ICP, but most cells were single-labeled, with only a few cells in ICX showing weak double-labeling.

As noted in the previous section, some divisions of IC have substantial intrinsic projections to other divisions and commissural projections to the contralateral IC. Thus, the laminated ventrolateral division of ICC has substantial projections to ipsilateral ICX, and also projects through the commissure to the contralateral dorsomedial division of ICC and to ICP (Aitkin and Phillips 1984b; Andersen et al. 1980a; Kudo and Niimi 1980; Rockel and Jones 1973a). The dorsomedial division and ICX project to the contralateral dorsomedial division (Kudo and Niimi 1980), and ICP projects to ICC (Rockel and Jones 1973a).

The evidence reviewed in this and the preceding section establishes a considerable degree of segregation in the afferent and efferent connections of the major traditional divisions of IC, and suggests strongly that this segregation extends to subdivisions or cell types within these divisions. These data therefore suggest a segregation of functions within IC. Functional segregation within IC, and in relation to that previously described in the lower brainstem (Sects. 4.3 and 5.4), will be examined after consideration of the physiological evidence on processing in the auditory midbrain (see Sect. 7.7).

7.1.5 Neuronal Architecture and Auditory Connections of Deep Layers of Superior Colliculus

The superior colliculus (SC), which lies rostromedial to IC and dorsomedial to MGB (see Fig. 1), is a highly laminated structure comprising three cell layers interdigitated between four fiber layers (Kanaseki and Sprague 1974; Fig. 39B). A distinction is commonly made between the superficial layers of SC (SCS; laminae 1 – 3), which have exclusively visual connections, and the deep layers (SCD; laminae 4 – 7), which have diverse afferent and efferent connections (see Edwards 1980; Huerta and Harting 1984 for reviews). Although Edwards (1980) described SCD as having an essentially reticular morphology, the cell types seen in Golgi preparations in fact seem to be a mixture of "tectal" and "tegmental" types, as defined by Morest and Oliver (1984) (see Sect. 7.1.2). Thus, Norita (1980) described many of the neurons in SCD, particularly those with large- and medium-sized soma, as multipolar with many thick dendrites having a number of branches with higher-order collaterals and a large number of spines (see large and medium-sized multipolar cells, labeled L and MM, in Fig. 39A). This description coincides almost exactly with Morest and Oliver's definition of "tectal" neuronal morphology. Other neurons, particularly the "small vertical" and "horizontal" cells (see cells labeled SV and H in Fig. 39A) have fewer dendrites, with straight trajectories and few secondary branches, and thus conform to the definition of "tegmental" neurons.

One undoubtedly reticular characteristic of SCD, however, is the multiplicity and diversity of its afferent connections. This is most clearly demonstrated when

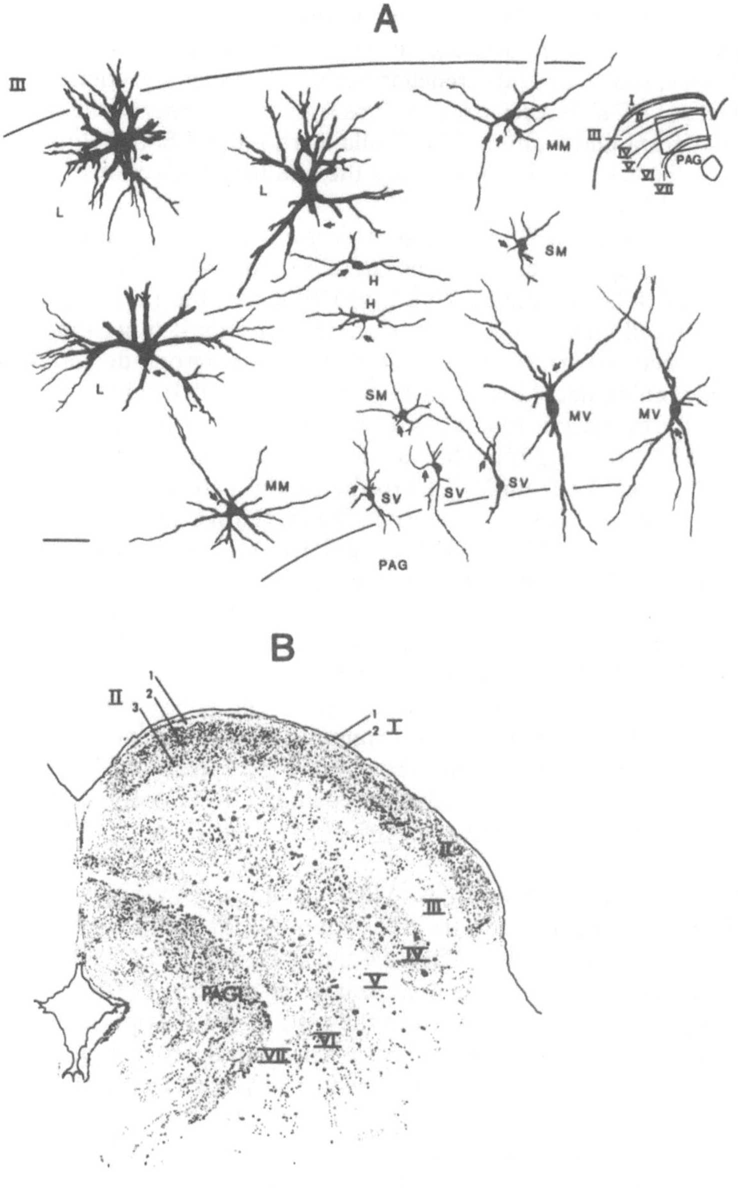

Fig. 39. A Composite drawing from coronal sections stained by rapid Golgi method, showing cell types in deep layers of cat superior colliculus. Area represented by drawing is indicated by *rectangle* in *insert*. Abbreviations for cell types: *L* large; *MM* medium-sized multipolar; *MV* medium-sized vertical; *SM* small multipolar; *SV* small vertical; *H* horizontal. *Arrows* indicate axon initial segments. Other abbreviations as in **B**. *Bar* represents 100 µm. (Modified from Norita 1980). **B** Projection drawing of coronal section stained with cresyl violet through superior colliculus of cat, illustrating laminar organization (laminae labeled I – VII). Arabic numerals indicate sublaminae of laminae I and II. *PAGl* periaqueductal gray, pars lateralis. (Kanaseki and Sprague 1974)

large injections of HRP are made into SCD. Edwards et al. (1979) reported labeled cells in over 40 subcortical structures, including visual, auditory, somatosensory, reticular, and cerebellar nuclei, after such injections in the cat. The deep layers of SC also receive projections from many areas of the cerebral cortex (e.g., Kawamura and Konno 1979; Paula-Barbosa and Sousa-Pinto 1973; Stein et al. 1983; Tortelly et al. 1980). Only the auditory projections to SCD will be summarized here.

Edwards et al. (1979) reported projections to SCD from the following auditory brainstem structures: ICP (ipsilaterally), nucleus of BIC (NBIC) (contralaterally), and ICX, sagulum, VNLL, DMPO, and the periolivary region medial to MNTB (all bilaterally). Most of these projections, other than that from VNLL, have been confirmed in other studies using anterograde and retrograde tracers. Glendenning and Masterton (1983) reported HRP labeling in the same nuclei, with the exceptions that labeled cells were seen in the cuneiform nucleus, but not in VNLL. Kudo (1981) obtained autoradiographic evidence of projections from DNLL (but not from VNLL) to the caudal half of SCD, and projections from DNLL have also been described in the rat (Druga and Syka 1984c) and tree shrew (Casseday et al. 1979). Whitley and Henkel (1984) have presented autoradiographic evidence that extralemniscal projections from VNLL pass through SCD en route to medial MGB, and it is therefore possible that HRP uptake by damaged fibers of passage might have given rise to VNLL labeling after the large injections used by Edwards et al. (1979). Projections from ICP, ICX, and NBIC have been confirmed in both anterograde and retrograde tracer studies (Grofová et al. 1978; Hashikawa and Kawamura 1983; Kudo and Niimi 1980; Kudo et al. 1984). Finally, projections from more restricted sets of auditory brainstem structures have been described after large HRP injections in rodents: ICX and NBIC in the hamster (van Buskirk 1983); DNLL and ICX in the rat (Druga and Syka 1984c).

Auditory cortical projections to SCD are apparently derived predominantly from "belt" areas rather than from the tonotopic auditory cortical fields (Kawamura and Konno 1979; Paula-Barbosa and Sousa-Pinto 1973; see Aitkin et al. 1984b for a recent review of the organization of auditory cortex). SCD also receives a strong bilateral projection from the anterior ectosylvian sulcus (Stein et al. 1983; Tortelly et al. 1980), a region in which auditory, visual, and somatosensory response fields have been described (Clarey and Irvine 1986; Clemo and Stein 1983; Mucke et al. 1982).

Despite the convergence of afferents from many sources to SCD, many of the projections are characterized by restricted zones of termination and by the concentration of terminals into dense "patches" (see Huerta and Harting 1984 for review). Some degree of topography of auditory projections to SCD is indicated by reports that projections from IC, DNLL, and periolivary nuclei are restricted to the caudal half of the nucleus (Edwards et al. 1979; Kudo 1981; Kudo and Niimi 1980). The smaller number of auditory brainstem structures projecting to SCD in rodents is apparently associated with the fact that auditory input is restricted to the caudal half of the nucleus in these species (see Sect. 7.5). Topographic organization of auditory afferents to SCD is of great interest because of evidence for topographic organization of sensitivity to IID in SCD that is to be discussed in Sect. 7.5.

The efferent connections of SCD are also diverse (see Huerta and Harting 1984 for review). Of particular interest are direct and indirect projections to extra-ocular motor nuclei (Edwards and Henkel 1978; Graham 1977), and projections to a paralemniscal zone in the lateral midbrain that projects to the subdivision of the facial nucleus that innervates the pinna muscles (Henkel 1981; Henkel and Edwards 1978). These projections are in accord with electrical stimulation and lesion evidence for the involvement of SCD in the control of saccadic eye movements and of orienting movements of the pinnae and head (see Sects. 7.6.2 and 7.6.3). The role of SCD in the control of orienting of receptors to sensory targets of all modalities has prompted recent interest in the representation of the location of auditory targets by SCD neurons. These issues are considered in detail in Sect. 7.5.

7.2 Physiology of Inferior Colliculus Neurons

All of the physiological investigations of IC neurons to be considered in the following sections have been carried out in the context of the traditional division of IC into central, external, and pericentral nuclei. While the data must therefore, of necessity, be discussed in this context, attempts will be made to relate the findings to the subdivision proposed by Morest and Oliver (1984) where this can be done without unwarranted speculation.

7.2.1 Frequency Tuning and Tonotopic Organization

The majority of neurons in the laminated ventrolateral division of ICC are sharply tuned for frequency, with V-shaped tuning curves that may or may not have low-frequency tails (Aitkin et al. 1975; Merzenich and Reid 1974; Rose et al. 1963). In contrast to the predominantly sharp tuning of ICC neurons, the tuning curves of most neurons in ICX (Aitkin et al. 1975, 1978b, 1981) and in ICP (Aitkin et al. 1975; Merzenich and Reid 1974) are broad and irregular, extending over several octaves. The responses of neurons in the latter divisions are generally less secure and more labile than those of ICC neurons.

Almost all descriptions of frequency tuning and organization of ICC neurons have been based on responses to monaural contralateral stimulation. Merzenich and Reid (1974) mentioned that ipsilateral and contralateral CFs were approximately the same in units that could be driven monaurally from either ear (EE cells), and Kuwada et al. (1984) have confirmed this observation in IPD-sensitive EE cells. Recently, Semple and Kitzes (1985) made a detailed quantitative comparison of the ipsilateral and contralateral tuning characteristics of EE neurons in ICC of the gerbil. They found that there were small but consistent differences in ipsilateral and contralateral CF and tuning curve shape, and large differences in threshold. Ipsilateral CFs were slightly (but statistically significantly) lower and ipsilateral tuning curves generally sharper; mean ipsilateral threshold at CF was 14 dB higher than contralateral. Although these differences are important, they do not require any qualification of the account of ICC's frequency organization provided by the use of contralateral stimulation.

In one of the earliest electrophysiological studies of the auditory midbrain, Erulkar (1959) reported that some IC neurons (the location of which was not further specified) with narrow excitatory tuning curves also had inhibitory sidebands. There appears to have been no subsequent account of inhibitory sidebands in IC, and the absence or low level of spontaneous activity in IC neurons in anesthetized preparations (e.g., Bock and Webster 1974b) precludes their detection. Bock et al. (1972) recorded from IC neurons in unanesthetized (awake restrained) cats, and reported that suppression of spontaneous activity was a common feature of at least some parts of the response area of many IC neurons. However, they did not obtain threshold tuning curves and were therefore unable to make any statement on the presence of inhibitory sidebands. Most of the neurons in their sample responded over a very wide frequency range at suprathreshold intensities. Their procedures did not allow them to localize recording sites histologically, and it is consequently uncertain whether most of their recordings were made from ICX/ICP, or whether ICC neurons exhibit broader frequency tuning in the absence of anesthesia. The former alternative seems the more probable, in view of Ryan and Miller's (1978) report that the majority of units (28 of 43) isolated in IC of alert rhesus monkeys had excitatory isorate functions that were sharply tuned for frequency, while the remaining 15 units were broadly tuned. In 11 of the sharply tuned units, the excitatory region was associated with inhibitory sidebands. The sample from which these data were obtained comprised a majority of cells histologically localized to ICC and a smaller proportion of cells whose precise location in IC could not be specified. Willott and Urban (1978) have also reported sharp tuning and inhibitory sidebands in ventrolateral ICC of mice that were described as "tranquilized." The data from these two species suggest that sharp tuning and inhibitory sidebands are common properties of ICC units in the absence of anesthesia.

In electrode penetrations from dorsal to ventral through ventrolateral ICC of the cat, successively encountered units or unit clusters have progressively higher CFs (Aitkin et al. 1970, 1975; Merzenich and Reid 1974; Rose et al. 1963; Fig. 40). This dorsal-to-ventral, low-to-high-frequency tonotopic organization is in accordance with the topographic organization of lower brainstem projections to ICC discussed in Sect. 7.1.3. Sharp frequency tuning and dorsal-to-ventral tonotopic organization have been shown to be characteristic of ventrolateral ICC in a variety of mammals (rhesus monkey: Ryan and Miller 1978; squirrel monkey; Fitzpatrick 1975; rabbit: Aitkin et al. 1972a; Syka et al. 1981; guinea pig: Popelar and Syka 1982; rat: Clopton and Winfield 1973; brush-tailed possum: Aitkin et al. 1978a; mouse: Harnischfeger 1978; Willot and Urban 1978; various echolocating bats: e.g., Grinnell 1963a; Jen and Schlegel 1982; Vater et al. 1979).

The frequency organization of neurons in the dorsomedial division of ICC and in ICX and ICP is less clear. Merzenich and Reid (1974) stated that auditory responses were not seen in the dorsomedial division in pentobarbitone-anesthetized cats and that the tonotopic organization they described was restricted to the ventrolateral division, but most other investigators have made no explicit statement on the characteristics of the dorsomedial division. Semple and Aitkin (1979), who studied the frequency organization of IC using rostrocaudally oriented horizontal electrode penetrations (see details later in this section), reported that they were unable consistently to establish borders between the ventrolateral and dorsomedial divisions. They noted, however, that there were no discontinuities in tonotopic organization in their data, and that if their maps in fact included data from the dorsomedial division, it would have to be concluded that the tonotopic organization extended into this division. In the squirrel monkey, Fitzpatrick (1975) reported that ICC is characterized by a single tonotopic organization which extends across both the laminated region and a region which she suggested is homologous to the dorsomedial division in the cat. The implications of the squirrel monkey data are unclear, however, given the uncertainty of this homology and the fact that corticofugal projections to ICC in this species extend throughout the central nucleus. Recent 2-DG data on the frequency organization of IC bear on the question of the frequency organization of the dorsomedial division, and this issue will be considered further in the light of these data later in this section.

Tonotopy is difficult to discriminate in ICX and ICP because of the broad tuning of neurons in these regions in cats (as discussed previously) and in other mammals (squirrel monkey: Fitzpatrick 1975; rabbit: Aitkin et al. 1972a; brush-tailed possum: Aitkin et al 1978b). In ICP of the cat, a dorsal-to-ventral, high-to-low-frequency tonotopic organization is apparent in some penetrations (Aitkin et al. 1975; Merzenich and Reid 1974). A similar high-to-low-frequency topography has been observed in a proportion of penetrations through ICX in some studies (Aitkin et al. 1975; Rose et al. 1963), but not in others (Aitkin et al. 1978a). The uncertainty on this issue may be related to the fact that the area designated ICX in the standard parcellation includes a number of areas of very different neuronal architecture (Morest and Oliver 1984; Sect. 7.1.2).

In contrast to the extensive evidence for the basic dorsal-to-ventral tonotopic organization of ICC, there are only limited physiological data that bear on the *three-dimensional* tonotopic organization of the nucleus. Merzenich and Reid (1974) made multiple, parallel, dorsoventrally oriented penetrations through ICC and reconstructed isofrequency contours by joining the points within individual penetrations at which units or unit clusters with particular CFs were located. Representative data from four penetrations in one experiment, and the reconstructed frequency contours, are shown in Fig. 40. The resultant contours were described as flat, and as lying approximately horizontal in the medial part of ICC (as illustrated in Fig. 40) but tilting downward from caudal to rostral and from medial to lateral elsewhere in the nucleus. Merzenich and Reid (1974) stressed that the changes in CF within a given penetration were stepwise rather than continuous, and that constructed isofrequency contours therefore represent the orientation of what they termed "frequency-band laminae." Although the fact that a frequency band rather than a single frequency is represented in a slab or sheet of tissue is an important one, there are obvious dangers in using the term "lamina" to refer to both a structural feature (the fibrodendritic laminae described previously) and a functional feature. These dangers are the more significant when the relationship between the structural and functional features is an empirical question of importance. To avoid these dangers, but to emphasize the importance of the point made be Merzenich and Reid (1974), the term "frequency-band contour" will be used here. Although the frequency-band contours inferred by Merzenich and Reid were flat rather than curved, their orientation corresponded approximately to that of the fibrodendritic laminae in ICC as

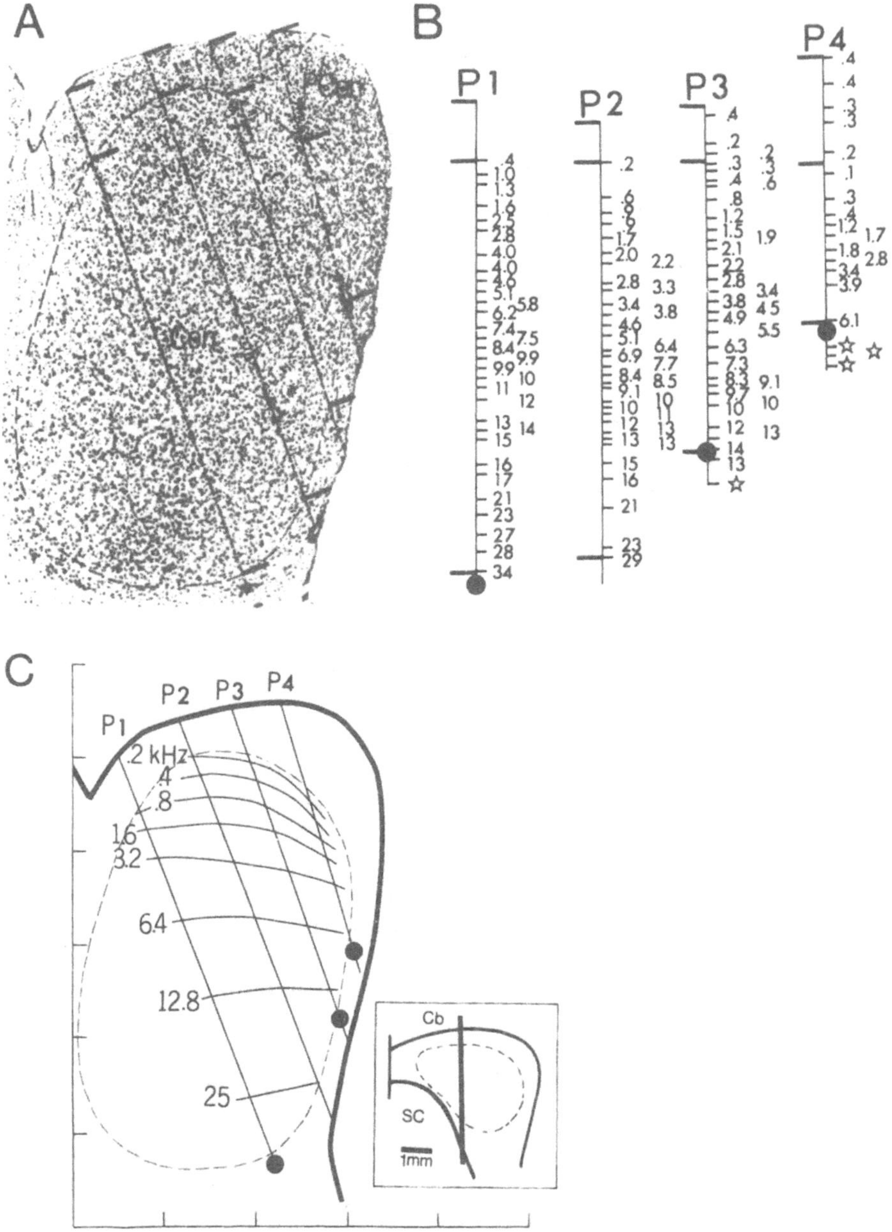

Fig. 40. A Sagittal section through IC at the approximate level indicated in the *inset* in **C.** The caudal apect of the nucleus is to the right. The paths of four microelectrode penetrations through ICC (here marked *Cen*) are indicated by *lines. Horizontal bars* mark the approximate boundaries of ICP (here designated *PCen*) and ICC along electrode tracks. The approximate boundary of ICC is outlined by the *light broken line.* **B** Reconstruction of the four electrode penetrations (*P*1 – *P*4) shown in **A.** The *bold horizontal bars* mark the boundaries of ICP and ICC and correspond with those in **A.** Each number is the CF in kHz, defined for a unit or a small cluster of units isolated at the site denoted by the *light horizontal bar.* At points marked with *stars,* units responded only to auditory stimuli of

described by Rockel and Jones (1973a). The data therefore support the view that structural lamination in ICC (and in other nuclei) is the substrate of tonotopic organization (e.g., Morest 1964b, c; Rockel and Jones 1973a).

Roth et al. (1978) and Semple and Aitkin (1979) also obtained data in the cat in accordance with this view. Semple and Aitkin (1979) made caudorostrally oriented penetrations of ICC in the horizontal plane; as illustrated in Fig. 41A and B, respectively, they obtained constant-CF sequences in the medial part of ICC, but progressively decreasing CF sequences more laterally. This pattern of change is in accordance with Merzenich and Reid's (1974) description of frequency-band contours as being horizontal in medial ICC, but as tilting rostrally in lateral IC, and with Rockel and Jones's (1973a) account of the laminae as tilted in this plane. Of particular interest is the report by Roth et al. (1978) that penetrations from lateral to medial through ICC in the cat produced a short sequence of decreasing CFs followed by a long sequence of increasing CFs (Fig. 42A). This sequence is exactly what would be expected if frequency-band contours corresponded to fibrodendritic laminae, as both Rockel and Jones (1973a) and Oliver and Morest (1984) described the laminae as oriented from ventrolateral to dorsomedial in central ICC, but as rising steeply from ventromedial to dorsolateral in the lateral part of ICC. A similarly good correspondence between the orientation of reconstructed frequency-band contours and that of fibrodendritic laminae has been reported in squirrel monkey ICC by Fitzpatrick (1975).

The correlations between frequency-band contours and fibrodendritic laminae established by the data reviewed here are indirect, the electrophysiological and anatomical data being obtained in different animals, and most often in different studies. More direct data have been obtained from experiments in which frequency mapping of the low-frequency region of ICC was followed by Golgi processing of the midbrain. Although only preliminary reports of these experiments have so far appeared (Chan and Yin 1982; Yin et al. 1983), the data indicate a good correspondence between frequency-band contours and fibrodendritic laminae.

Further evidence on the frequency organization of ICC is provided by recent 2-DG studies. In the anesthetized cat, binaural tone-burst stimulation at a given frequency results in a band of 2-DG labeling in each IC sloping from dorsomedial to ventrolateral at 45° – 60° in the frontal plane (Servière and Webster 1981; Webster et al. 1978). Stimulation with alternating high- and low-frequency tone bursts results in two parallel bands, whose dorsoventral position is in accordance with the electrophysiologically determined tonotopic organization of ICC (Fig. 42B; Servière et al. 1984). The orientation of these bands corresponds

Fig. 40 (*continued*) high intensity. The sites of lesions marking positions along electrode tracks are indicated by the *full circles.* **C** Approximate isofrequency contours, drawn for frequencies representing octave separations. These contours were drawn by determining the points within individual penetrations reconstructed in **B** at which the specified frequencies were represented. The axes are the vertical and horizontal stereotaxic planes; divisions are 1 mm in length. Note that at this relatively medial level, the lines joining isofrequency points are approximately horizontal. (Merzenich and Reid 1974)

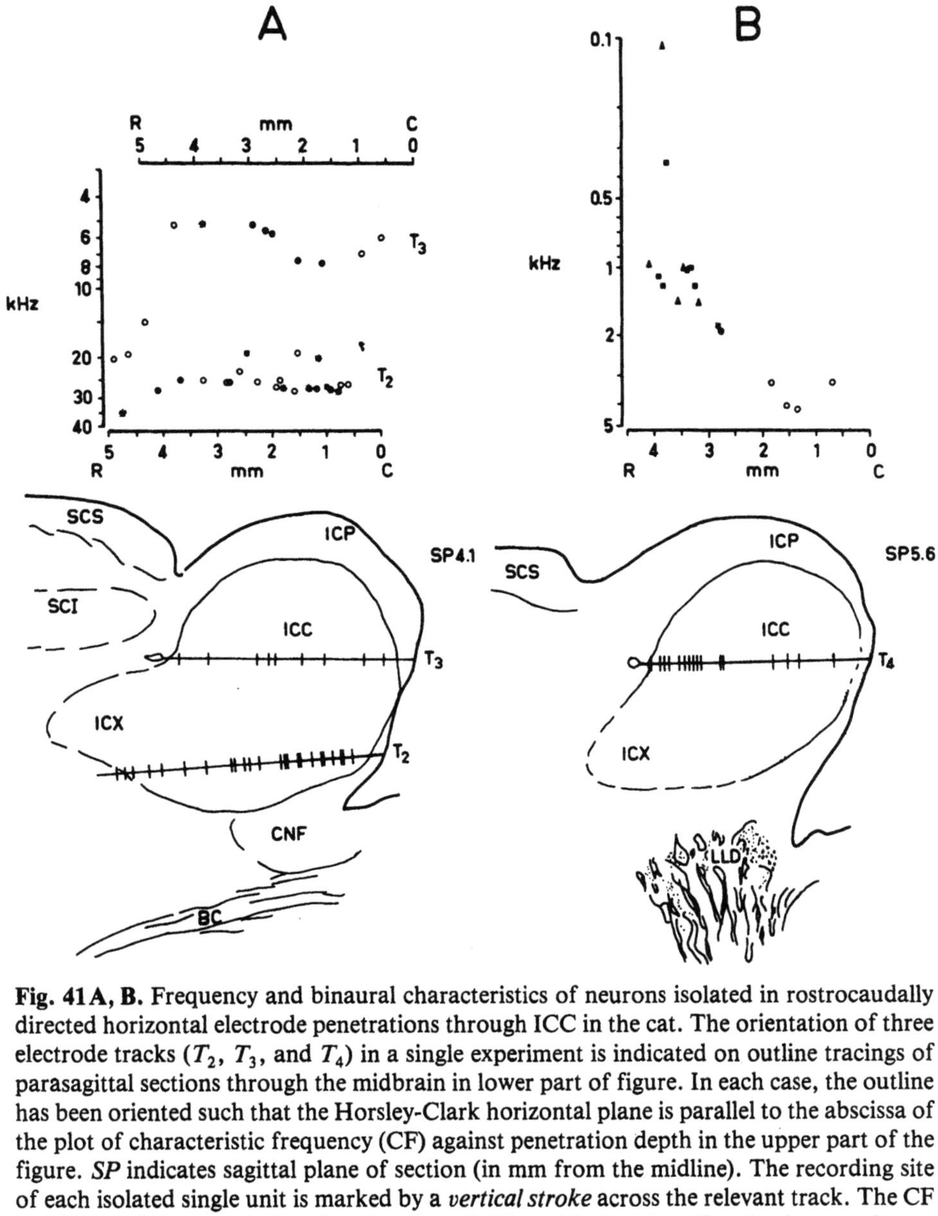

Fig. 41 A, B. Frequency and binaural characteristics of neurons isolated in rostrocaudally directed horizontal electrode penetrations through ICC in the cat. The orientation of three electrode tracks (T_2, T_3, and T_4) in a single experiment is indicated on outline tracings of parasagittal sections through the midbrain in lower part of figure. In each case, the outline has been oriented such that the Horsley-Clark horizontal plane is parallel to the abscissa of the plot of characteristic frequency (CF) against penetration depth in the upper part of the figure. *SP* indicates sagittal plane of section (in mm from the midline). The recording site of each isolated single unit is marked by a *vertical stroke* across the relevant track. The CF and binaural characteristics of each unit are indicated in the plots directly above each outline. CF (in kHz) is plotted as a function of the penetration depth (in mm from the caudal border of IC) at which the unit was isolated (note separate abscissae for penetrations T_2 and T_3 in **A**). Binaural characteristics of unit are indicated by symbol used to represent it: *square* EE; *full circle* EI; *open circle* EO/mon; *triangle* "delay-sensitive"; *star* "other" (see definitions in text). Other abbreviations: *BC* brachium conjunctivum; *C* caudal; *CNF* cuneiform nucleus; *ICC* central nucleus of inferior colliculus; *ICP* pericentral nucleus of inferior colliculus; *ICX* external nucleus of inferior colliculus; *LLD* dorsal nucleus of inferior colliculus; *R* rostral; *SCI* intermediate layers of superior colliculus; *SCS* superficial layers of superior colliculus. (Semple and Aitkin 1979)

A

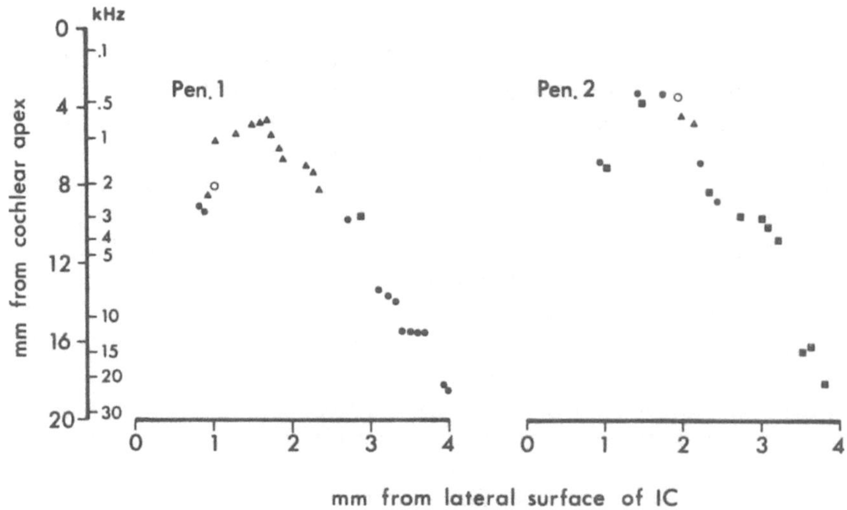

B

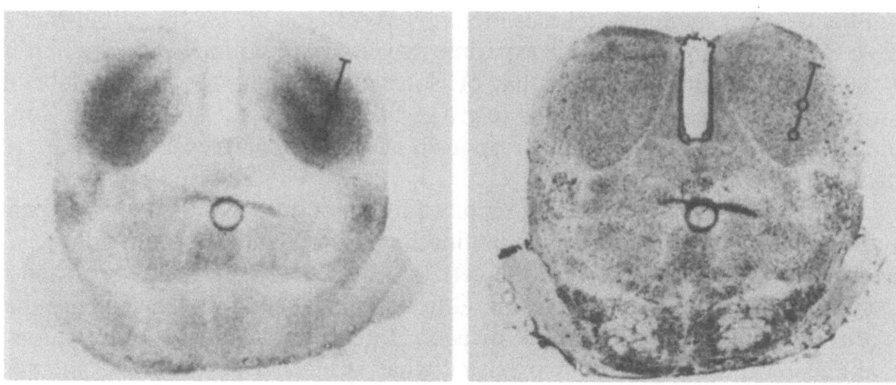

Fig. 42. A Frequency tuning and binaural characteristics of individual neurons isolated in two lateromedially oriented horizontal penetrations (*Pen.* 1 and *Pen.* 2) through ICC in a single cat. For each neuron, characteristic frequency (in kHz) and represented cochlear place (mm from cochlear apex according to Greenwood's (1974) frequency-position function) is plotted as a function of the neuron's location (in mm) from the lateral surface of the IC. Binaural characteristics of neuron are indicated by symbol: *open circles* EO/mon; *full circles* "IID-sensitive" (EI); *full triangles* "delay-sensitive"; *squares* EE; *stars* "other" (see definitions in text). (Roth et al. 1978). **B** Autoradiograph of midbrain (*left*) and Nissl stain of section producing autoradiograph (*right*) from cat exposed to binaural tone burst stimulation (alternating bursts of 2.0 and 13.5 kHz). *Oblique line* indicates position of electrode penetration in which tonotopic sequence was recorded prior to 2-DG exposure, and two *circles* interrupting line indicate points at which dye deposits were made to mark locations of units with CFs of 2.0 and 3.0 kHz (*upper circle*) and 13.4 kHz (*lower circle*). (Servière et al. 1984)

closely to that of the fibrodendritic laminae in the central part of ICC. The bands of increased 2-DG labeling are presumed to reflect the location of those neural elements in which the stimulus causes an increase in activity, and consequently in glucose utilization, sufficient to be detected by the method. As discussed later in this section, there is considerable uncertainty as to the nature of the neural structures (cell bodies, axonal terminals, or dendrites) and of the processes (spike activity, synaptic potentials, etc.) in which increased activity leads to 2-DG accumulation (see, e.g., Auker et al. 1983; Lippe et al. 1980; Nudo et al. 1984; Schwartz et al. 1979). Nevertheless, if the basic assumption is correct, those neural elements within whose response area a pure tone stimulus falls should be labeled, and neurons with CF equal to the stimulus frequency should be located within the band of increased label produced by the stimulus. Servière et al. (1984) and Servière and Webster (1981) confirmed this prediction, by partially mapping IC electrophysiologically prior to 2-DG injection and marking (by means of dye spots) the locations of units whose CF matched the subsequently used exposure frequency. In each case, the dye spots were located within the 2-DG band produced by stimulation of the contralateral ear with that frequency (Fig. 42B). A similar correspondence between electrophysiologically determined and 2-DG-labeled frequency contours in the neostriatum of chicks has been described by Theurich et al. (1984). Frequency-specific 2-DG labeling in IC that is in broad agreement with electrophysiologically demonstrated tonotopic organization has also been demonstrated in the cat by Nudo and Masterton (1984) and in a number of other mammals (rhesus monkey: Webster et al. 1984b; gerbil: Melzer 1984; Ryan et al. 1982). In some experiments with awake animals, the orientation of 2-DG bands corresponds to that of isofrequency contours, but the labeling pattern comprises multiple bands (e.g., gerbil: Ryan et al. 1982; rabbit: Jones and Disterhoft 1983). The factors responsible for these complex banding patterns have not been established.

Two features of the 2-DG labeling reported by Servière et al. (1984) are apparently at variance with evidence on the lamination pattern and electrophysiologically determined tonotopy of ICC in the cat. Labeled 2-DG bands were reported to extend into the dorsomedial division, although the labeling was less intense than that in the laminated ventrolateral division. Interpretation of this observation is qualified by uncertainty as to precisely what is labeled by the 2-DG method. Although fibrodendritic laminae do not extend into the dorsomedial division (layer IV of the dorsal cortex), fine collaterals of lemniscal axons do extend beyond the dorsal border of the ventrolateral division (Morest and Oliver 1984; Rockel and Jones 1973a; see Sect. 7.1.2). If 2-DG labeling mainly reflects activity in synaptic terminals rather than cell bodies (see discussion below), then labeling would be expected to extend beyond the region of lamination. If the terminals of these fine collaterals did not provide secure drive to dorsomedial division neurons, tonotopic organization based on electrophysiological recordings from cell somas would not extend into this region.

The second disparity arises from the fact that the bands of 2-DG label reported by Servière et al. (1984) maintain their dorsomedial-to-ventrolateral orientation to the lateral edge of ICC. Given the agreement between the Golgi studies (Morest and Oliver 1984; Rockel and Jones 1973a) and the electrophysiological evidence (Roth et al. 1978; Fig. 42A), there would appear to be little doubt that both the laminae and frequency-band contours run dorsolaterally in the lateral region of ICC, almost oblique to those in the central and medial parts of the central nucleus. The reasons for the failure of this feature to emerge in the 2-DG data are unclear; it may reflect limited spatial resolution provided by the technique or some as yet unrecognized feature of the nature of the activity underlying increased 2-DG labeling.

A monaurally presented tone results in a characteristic band of 2-DG label in ICC contralateral to the stimulated ear, but no sign of selective labeling in ipsilateral IC (Servière and Webster 1981; Servière et al. 1984). The interpretation of this differential labeling is unclear, but it presumably reflects both the binaural properties of IC units and the nature of the activity labeled by 2-DG.[18] If, however, broad-band noise bursts are presented simultaneously to the other ear, IC ipsilateral to the ear to which the tone is presented exhibits a band of reduced (relative to surrounding areas) 2-DG uptake, the position of which corresponds to that of the band of increased uptake in the contralateral IC (Webster et al. 1984a; Webster et al. 1985). The interpretation of these "inhibitory bands" will be considered in the context of the binaural properties of IC neurons in Sect. 7.2.3. Webster et al. (1985) have also reported that if a combined noise-tone stimulus is presented to a single ear, the band of increased 2-DG uptake in the contralateral IC is bordered by a band of decreased uptake. They have suggested that this band of decreased uptake might reflect the inhibitory sidebands of IC neurons with CF near the tone frequency.

The great advantage of the 2-DG method, notwithstanding the important qualifications concerning the nature of the activity underlying increased (or decreased) labeling, is that it presumably labels that activity in *all* of the neural elements activated by a given stimulus. This is in marked contrast to even the most exhaustive electrophysiological mapping study, in which only a minute fraction of activated elements is identified. The 2-DG technique therefore offers the possibility of obtaining detailed three-dimensional reconstructions of the fre-

---

[18] If the activity reflected in increased 2-DG uptake is an increase in spike activity in cell somas and/or an increase in excitatory synaptic activity (e.g., Auker et al. 1983), then selective contralateral labeling after monaural stimulation would be readily explicable in terms of the fact that almost all IC neurons are excited by monaural contralateral stimulation, whereas a much smaller proportion (approximately 30%) are excited by monaural ipsilateral stimulation (see Sect. 7.2.3 and Table 4A). If, however, increased 2-DG uptake reflects increased activity in axonal terminals (e.g., Lippe et al. 1980; Nudo and Masterton 1984; Schwartz et al. 1979), interpretation of the selective contralateral labeling would depend on the locus of the binaural interactions seen in IC neurons. Approximately 70% of cells in IC are binaurally influenced, being either excited or inhibited by ipsilateral stimulation (see Sect. 7.2.3 and Table 4A). If ipsilateral inhibition were produced by inhibitory processes intrinsic to IC, increased activity in a large number of inhibitory and excitatory terminals in ipsilateral IC would be produced by monaural stimulation, and would be expected to result in 2-DG labeling. If, however, ipsilateral inhibition in IC simply reflected that in SOC, there would be a much smaller amount of terminal activity consequent on ipsilateral monaural stimulation, and the differential banding observed by Servière et al. (1984) would be expected. Resolution of these uncertainties requires more detailed knowedge of the nature of the activity changes underlying increased 2-DG labeling, which may be different in different structures, and of the mechanisms underlying binaural interactions in IC neurons.

quency organization of auditory nuclei. Computer graphic procedures of the type applied by Bourk et al. (1981) to electrophysiological data from CN do not appear to have been applied to IC. The application of such procedures to the spatial organization of 2-DG labeling and of fibrodendritic lamination in ICC offers the prospect of detailed accounts of three-dimensional frequency and laminar organization and of the relationships between them.

7.2.2 Monaural Response Characteristics

Almost all IC neurons receive excitatory input from the contralateral ear (see following section for detailed account of binaural properties) and most accounts of monaural response characteristics have been restricted to responses to contralateral stimulation. As noted in the preceding section, there have been two recent studies in which detailed comparisons have been made between the ipsilaterally and contralaterally elicited monaural responses of EE cells (Kuwada et al. 1984; gerbil: Semple and Kitzes 1985).

1. Response Patterns. A variety of temporal patterns of response is seen in ICC. For a given unit, response pattern may alter dramatically with variations in frequency and intensity, such that the variation within a units' response area can be as great as that between units (e.g., Geisler et al. 1969; Kuwada et al. 1984; Rose et al. 1963; monkey: Ryan and Miller 1978; kangaroo rat: Stillman 1971b; mouse: Willott et al. 1977). A broad distinction has commonly been made between units exhibiting onset and sustained responses to tone bursts or noise bursts (e.g., Erulkar 1959; Merzenich and Reid 1974; Rose et al. 1963). Units with onset responses at CF are the most likely to exhibit the same pattern throughout their response area, and Rose et al. (1963) reported that almost 50% of their sample were of this sort. While some neurons respond in a sustained fashion throughout the stimulus, responses broadly classified as sustained can vary considerably in temporal pattern, and responses at stimulus offset are seen in some cases. One commonly observed pattern comprises an onset response, followed by a silent period (or pause) and a resumption of activity for the remainder of the stimulus duration (Geisler et al. 1969; Hind et al. 1963; Kuwada et al. 1984; Leiman and Hafter 1972; Moore and Irvine 1981; Rose et al. 1963). Although such units have sometimes been designated "pausers," by analogy with DCN units whose discharge patterns are also characterized by a pause, there are major differences in the response patterns that suggest the analogy should be resisted. The pause in the discharge of DCN units is characteristically of 5 – 10 ms duration, and all components of the response pattern are clearly apparent in responses to tone bursts as short as 25 – 40 ms (e.g., Godfrey et al. 1975b; Pfeiffer 1966a; see Fig. 19). By contrast, the pause in the discharge of ICC neurons commonly exceeds 50 ms (e.g., Moore and Irvine 1980), and some such units can appear to have simple onset responses when tone-burst durations of less than 100 ms are used. A final type of sustained response is characterized by very long latencies of 50 – 70 ms or more (Hind et al. 1963; mouse: Willott 1983), and thus resembles the sustained component of those responses in which onset and sustained components are separated by a silent period.

Although phase-locking has been reported in some low-frequency ICC neurons (Bock et al. 1972; Geisler et al. 1969; Kuwada et al. 1984; Rose et al. 1966), it apparently occurs relatively infrequently in ICC neurons. The most detailed data have been obtained by Kuwada et al. (1984) in the course of their study of low-frequency IPD-sensitive cells (see Sect. 7.2.5). Only 18% of 82 such cells examined in detail exhibited phase-locking, and it was seldom observed at frequencies above 600 Hz (and never above 1200 Hz). As noted in Sect. 5.1.3, only limited data are available on phase-locking in SOC; it is clear, however, that there is a major reduction in both the number of cells exhibiting phase-locking and the upper frequency limit of its occurrence from CN to IC (cf. de Ribaupierre et al. 1980). The implications of this change for the brainstem processing of spectral information are considered in Sect. 8.1.

Purely inhibitory responses have been reported only infrequently in anesthetized animals (e.g., Semple and Aitkin 1979; gerbil: Semple and Kitzes 1984). That this reflects the absence or low level of spontaneous activity in anesthetized preparations is indicated by the report of Bock et al. (1972) that 35% of their sample of IC units in unanesthetized cats showed substantial suppression of spontaneous activity at some point in their response area. Ryan and Miller (1978) similarly reported sustained suppression in IC of unanesthetized rhesus monkeys, notably at frequencies above and below CF, where it forms inhibitory sidebands.

As noted earlier, neurons with a given form of sustained response pattern to moderate-intensity CF tones commonly exhibit changes in response pattern as frequency and intensity are varied, often exhibiting onset responses at the edges of the response area (e.g., Geisler et al. 1969; Rose et al. 1963; rhesus monkey: Ryan and Miller 1978; kangaroo rat: Stillman 1971b). Many investigators have interpreted the diversity and variability of the temporal patterns of extracellularly recorded discharges in terms of the interplay of excitatory and inhibitory synaptic influences on the cells (e.g., Bock et al. 1972; Hind et al. 1963; Rose et al. 1963; squirrel monkey: Ryan and Miller 1978). This interpretation is supported by the observations made by Nelson and Erulkar (1963) in one of only two intracellular studies of IC neurons. They found that a given neuron could respond to a tone burst with an excitatory postsynaptic potential (EPSP), an inhibitory postsynaptic potential (IPSP), or some combination of the two, and that changes in frequency commonly produced different changes in the excitatory and inhibitory components of the response. Although this variability prohibits classification of neurons in terms of response patterns (cf. Stillman 1971b), it is nevertheless the case that some units exhibit a relatively characteristic pattern throughout much of their response area (e.g., Rose et al. 1963; squirrel monkey: Ryan and Miller 1978). In later sections it will therefore be convenient to refer to onset units and to those with onset and sustained components separated by a pause, but this should not be taken to imply the existence of neural types.

Similar response patterns and variability have been described in a wide range of mammals (rhesus monkey: Ryan and Miller 1978; guinea pig: Popelar and Syka 1982; rabbit: Aitkin et al. 1972b; Syka et al. 1981; kangaroo rat: Stillman 1971; gerbil: Semple and Kitzes 1985; rat: Syka et al. 1983; mouse: Willott 1983; various echolocating bats: e.g., Möller et al. 1978; Pollack et al. 1978; Vater et al. 1979), although there appear to be differences between species in the relative frequency of occurrence of different response patterns. Thus, Syka and his colleagues have reported that onset responses at CF are characteristic of the majority of ICC neurons in the rat (64%; Syka et al. 1983), but of relatively small proportions in the rabbit (7%; Syka et al. 1981) and guinea pig (17%; Popelar and Syka 1982). The fact that the latter differences were observed in studies in a single laboratory, presumably using standard procedures, suggests that they reflect real species differences rather than differences in anesthesia or in some other procedural variable.

2. Rate–Intensity Functions. ICC neurons in the cat and in a variety of other mammals have limited dynamic range, and many cells (approximately 50% in most studies) have nonmonotonic rate–intensity functions at CF or at some frequency within their response area (Erulkar 1959; Rose et al. 1963; rhesus monkey: Ryan and Miller 1978; gerbil: Semple and Kitzes 1985; guinea pig: Popelar and Syka 1982; mouse: Willott et al. 1977). Monotonic rate–intensity functions typically saturate within 10–30 dB of threshold; nonmonotonic functions have a maximum within the same range, the response decreasing with further increases in intensity. Although there is considerable variation between nonmonotonic cells in the extent of the decrease, and hence a degree of arbitrariness in the definition of nonmonotonicity (see Semple and Kitzes 1985 for discussion), the decline is in many cases marked enough to generate inverted-V-shaped functions (e.g., Semple and Aitkin 1980; rhesus monkey: Ryan and Miller 1978). Such cells consequently have a clearly defined "most effective SPL" or "best intensity," as has been described for neurons in auditory cortical fields of the cat (Phillips and Irvine 1981a; Phillips and Orman 1984), monkey (Brugge and Merzenich 1973), and echolocating bat (e.g., Suga 1977).

Bock et al. (1972) reported no nonmonotonic rate–intensity functions in a sample of IC units in unanesthetized cats. Although the sample size and unit locations were not specified, these data raise the possibility that nonmonotonicity is an artifact of anesthesia. That this is probably not the case is indicated by Ryan and Miller's (1978) finding of approximately 50% nonmonotonic functions in a sample of 49 units in ICC of the unanesthetized monkey.

3. Response Latency. The latency to initial discharge of ICC neurons decreases monotonically with increasing intensity, even for those cells with nonmonotonic rate–intensity functions (Hind et al. 1963; Kuwada et al. 1984; Rose et al. 1963; rhesus monkey: Ryan and Miller 1978; gerbil: Semple and Kitzes 1985). For units with onset response components, the range of minimum latencies for responses to CF tone bursts or noise bursts (i.e., the value at which the latency function reaches an asymptotic minimum) is approximately 6–15 ms (Hind et al. 1963; Irvine and Jackson 1983; Kuwada et al. 1984; rat: Syka et al. 1983). Sustained responses consistently have longer latencies, and much wider ranges are obtained when units whose response consists of a delayed response with no onset component are included: up to 120 ms in cat (Kuwada et al. 1984), 4.5–36 ms in rhesus monkey (Ryan and Miller 1978), and 6–40 ms in rabbit (Syka et al. 1981). Kuwada et al. (1984) reported a strong positive correlation between ipsilateral and contralateral latencies of low-frequency IPD-sensitive neurons with onset responses, and a weaker positive relationship for those with sustained responses. Semple and Kitzes (1985) have reported that for EE neurons in ICC of the gerbil, minimum latencies for contralateral stimulation at CF are significantly shorter than those for ipsilateral stimulation.

4. Complex Stimuli. Only limited data are available on the responses of ICC neurons to complex stimuli in mammals other than the echolocating bat. Experiments in which sensitivity to interaural disparities in broad-band or complex stimuli have been examined in the cat will be considered in Sects. 7.2.4 and 7.2.5, but the question of how spectral or temporal complexity per se is coded by IC neurons was not explicitly addressed by these studies. Examination of these issues has been confined to a small number of studies of sensitivity to time-varying (FM and AM) stimuli (Nelson et al. 1966; Watanabe and Ohgushi 1968; chinchilla: Mogus 1972). Some IC neurons are sensitive to the direction and/or rate of modulation of such stimuli, and Yin et al. (1984), in a study of high-frequency ITD sensitivity, demonstrated phase-locking in the responses of high-CF neurons to the modulation frequency of AM high-frequency sounds. As was the case in CN, these results serve to demonstrate the existence of sensitivity to time-varying stimuli, but do not provide an adequate basis for systematic analysis of the processing of such stimuli. More detailed accounts of some of these data can be found in reviews by Erulkar and his colleagues (Erulkar 1975; Erulkar et al. 1968b). An account of the extensive literature on the sensitivity to complex stimuli of brainstem neurons in the specialized echolocating bats is beyond the scope of this review (see Pollack 1980; Suga and Schlegel 1973 for reviews).

5. Other Inferior Colliculus Subdivisions. The preceding account of monaural response properties has been restricted to ICC, except in those cases in which units were not histologically localized to particular IC subdivisions. There is almost no information in the literature on the parametric response properties of neurons in ICX and ICP, other than their broad frequency tuning. Aitkin et al. (1978b) reported that onset and sustained responses occurred in approximately equal proportions in cat ICX, while Ryan and Miller (1978) noted that response latencies of a small sample of units in ICX of unanesthetized rhesus monkeys were longer than those of ICC units. Some ICX units in the cat have been reported to respond well to spectrally complex sounds (Aitkin et al. 1981), but there has been no detailed comparison of responses to pure tones and complex stimuli. In accordance with the anatomical evidence of somatosensory projections to ICX, some neurons in the cat are influenced by electrical stimulation of the dorsal columns (Aitkin et al. 1978b), and a considerably smaller proportion are responsive to tactile stimulation (Aitkin et al. 1978b, 1981). Like their frequency tuning curves, the tactile receptive fields of ICX units are extremely large. Only a small proportion of ICX units in the cat are responsive to both acoustic and tactile stimulation, suggesting that there might be a segregation of auditory and somatosensory afferents akin to the laminar segregation described by Willard and Ryugo (1983) in the mouse.

7.2.3 Binaural Response Characteristics and Segregation

The major forms of binaural input to ICC neurons reflect the major sources of ascending afferents to the nucleus. Although comparison of the data reported in

Table 4. Percentages of neurons in central nucleus of inferior colliculus of the cat showing various forms of binaural input

Study	Sample size	Binaural category					
		EO/mon	EI (IID-sensitive)	EE	EO/F	Delay-sensitive	Other
A							
Aitkin et al. (1975)	87	20	38	29	10	NA	3
Roth et al. (1978)	546	33	32	13	NA	20	2
Semple and Aitkin (1979)	632	17	30	29	3	16	5
Combined samples	1265	24	31	41 (EE, EO/F and Delay-sensitive combined)			4
B							
Semple and Aitkin (1979)	338 (CF>3.2 kHz)	29	40	22	NA	0	9
Irvine and Gago (1985)	294 (CF>3.0 kHz)	39	47	4	2	NA	8
Combined samples	632	34	43	14	1	0	8

A samples comprising cells with CFs covering entire frequency range; **B** samples restricted to cells with CF greater than approximately 3 kHz. Binaural categories defined in text. NA indicates that binaural category is not applicable to that study.

different studies is complicated to some extent by the fact that they have employed different classification systems, most investigators have used relatively limited sets of categories between which correspondences can be established. Data from three studies of ICC in the cat (Aitkin et al. 1975; Roth et al. 1978; Semple and Aitkin 1979) in which data were obtained from large samples encompassing a comprehensive CF range are summarized in Table 4A. Together, these studies provide data on a total sample of 1265 single neurons; where the classification systems allow it, the proportion of the combined sample exhibiting a particular form of binaural input is shown. In each of the three studies, a significant proportion (24% of the combined sample) were excited by stimulation of the contralateral ear, but uninfluenced by ipsilateral stimulation (EO/mon), presumably reflecting the major projections to ICC from contralateral CN and ipsilateral VNLL. A slightly larger proportion of cells (31% of the combined sample) received EI input and were sensitive to IIDs. Semple and Aitkin (1979) distinguished between a major group of EI cells receiving strong ipsilateral inhibitory input (26% of their sample) and a smaller group in which the inhibition was relatively weak and was effective only at large IIDs favoring the ipsilateral ear (4% of their sample, which they included in the category "other"). For the purposes of comparison, these two groups have been combined in Table 4A, but the distinction is an important one and will be returned to in the following section. Although EI binaural properties could be generated by mechanisms intrinsic to IC, a significant proportion of such cells would be expected in ICC as a consequence of the major ascending projections from ipsilateral MSO and

DNLL and contralateral LSO. As discussed in Sect. 5.2.1, the majority of LSO neurons receive IE binaural input (i.e., contralateral inhibition and ipsilateral excitation). When such a neuron projects contralaterally, the binaural input characteristic of its target neuron (or neurons) is EI, since the ears contralateral and ipsilateral to the LSO neuron are ipsilateral and contralateral, respectively, to the target neuron (or neurons).

Comparison of the remaining classes of binaural neurons is complicated to a greater extent by the different classification systems employed. Aitkin et al. (1975) used a purely descriptive system of the type developed in Sect. 5.2.1, and classified most of the remaining cells as EE or EO/F. Roth et al. (1978) and Semple and Aitkin (1979) distinguished a group of cells that were sensitive to IPDs, which they designated "delay-sensitive," and classified as EE only those cells that were excited by monaural stimulation of each ear, but were *not* delay-sensitive. Although delay-sensitive neurons are generally characterized by excitatory and inhibitory input from each ear (Sects. 5.2.4 and 7.2.5), the majority would be descriptively classified as EE or EO/F at zero delay (Kuwada et al. 1984; Semple and Aitkin 1979). Although no detailed comparison across studies is possible, neurons classified as delay-sensitive, EE, or EO/F comprise approximately 40% of the combined samples in Table 4A. The remaining cells, those in the category "other," were described in any detail only by Semple and Aitkin (1979); they comprised a small number of OO/F cells (see Sect. 5.2.1), cells characterized only by suppression of spontaneous activity (both IO and II), and a few cells for which binaural interaction varied in a complex way with intensity.

The data presented in Table 4A are based on samples covering the entire CF range, and therefore obscure the fact that patterns of binaural input are related to CF (e.g., Moore and Irvine 1980; Roth et al. 1978; Semple and Aitkin 1979). The most dramatic form of this association is that delay-sensitivity is a characteristic only of cells with CF below approximately 3 kHz (e.g., Kuwada and Yin 1983; Rose et al. 1966; Roth et al. 1978; Semple and Aitkin 1979; see also Sect. 5.2.4). Conversely, EO/mon and EI patterns occur more frequently in cells with CF above 3 – 5 kHz (Irvine and Gago 1985; Roth et al. 1978; Semple and Aitkin 1979). EE cells (other than those that are delay-sensitive) are distributed rather more homogeneously with frequency, but more commonly have low than high CF (Semple and Aitkin 1979). The association between binaural input pattern and CF is reflected in the distribution across binaural classes obtained in samples of ICC neurons with high CFs (above approximately 3 kHz), as shown in Table 4B. EO/mon and EI cells comprise 34% and 43%, respectively, of the combined high-frequency sample, and delay-sensitive, EE, and EO/F cells together comprise only about 14%. Table 4B is based on the data of Semple and Aitkin (1979) and Irvine and Gago (1985); the data of Roth et al. (1978) show similar trends, but could not be included because exact cell counts for units in particular CF ranges could not be derived from their histograms. The association between CF and binaural input pattern is also reflected in the fact that although the same forms of binaural input are observed in ICC of a wide range of mammals (rat: Silverman and Clopton 1977; Syka et al. 1983; guinea pig: Popelar and Syka 1982; rabbit: Aitkin et al. 1972; Syka et al. 1981; echolocating bats: Schlegel 1977; Wenstrup et al. 1985; brush-tailed possum: Aitkin et al. 1978a), the proportions can differ significantly from those in the cat. The difference is most marked in the rat, in which 70% – 85% of ICC neurons are reported to receive EI binaural input. This characteristic is presumably related to the high-frequency hearing and limited development of MSO in this species (Sect. 5.1.1); as noted previously, MSO in the rat has also been reported to contain an unusually high proportion of EI cells (Inbody and Feng 1981; Sect. 5.2.2).

Some degree of segregation of cells with different binaural input characteristics would be expected within the laminated central nucleus as a consequence of its tonotopic organization and the association between binaural input pattern and CF. Thus, Semple and Aitkin (1979) reported that delay-sensitive units are most frequently found dorsally, laterally, and rostrally in ICC, in the region in which low CFs are most strongly represented. Similarly, EI cells are most commonly found ventrally in ICC, as would be expected on the basis of IC's frequency organization. Recent evidence suggests, however, that there is also a partial segregation of binaural classes in ICC beyond that consequent on frequency organization. Thus, within ventral ICC, EI (IID-sensitive) cells are encountered in greatest proportion rostrally and EO/mon cells are encountered in greatest proportion caudally and laterally (Roth et al. 1978; Semple and Aitkin 1979). The latter observation is in good accord with the caudal concentration of terminations of projections from contralateral CN and ipsilateral VNLL (Brunso-Bechtold et al. 1981). As this correspondence indicates, the partial segregation of cells with different binaural input characteristics is undoubtedly related to the partial segregation of the terminations in ICC of the afferents from different brainstem structures. As noted previously in connection with the latter segregation, it will be of great interest to determine whether regions of ICC characterized by particular patterns of binaural input and afferent terminations correspond to the subdivisions defined by Morest and Oliver (1984).

Since the laminated central nucleus is characterized by a single tonotopic representation, partial segregation of neurons with different binaural input characteristics, but with similar CFs, necessarily implies a partial segregation within frequency-band sheets. The fact that a point on the cochlea is represented as a sheet of neurons in central auditory nuclei has prompted the suggestion that some other feature or features might be topographically organized across the two dimensions of the sheet (e.g., Merzenich and Reid 1974). In primary auditory cortex (area AI), in which a given frequency band is represented along a strip of the two-dimensional cortical sheet, there is evidence to suggest that binaural properties might be topographically represented across the frequency-band strips (Imig and Adrian 1977; Middlebrooks et al. 1980), although the nature of this organization is still in dispute (Phillips and Irvine 1979, 1983; see Aitkin et al. 1984b for review). In the ICC of most mammals it is difficult to test this hypothesis because of the difficulty of recording from a significant sample of neurons within a single frequency-band sheet in a single animal. The best evidence is that of Roth et al. (1978) and Semple and Aitkin (1979), who in some horizontal penetrations obtained sequences of neurons with similar CF, indicating that all recordings had been made within a restricted set of adjacent frequency-band sheets (see Fig. 41A). Although there was significant intermingling of cells with different binaural characteristics in these penetrations, both studies also reported sequences of cells with the same binaural input (e.g., EI cells in track 3 in Fig. 41A and the caudal sequence of EO/mon cells in Fig. 41B), suggesting the occurrence of such cells in partially segregated clusters. Recently, Wenstrup et al. (1985) have determined the binaural response characteristics of neurons and neuron clusters in the large division of the central nucleus of IC in the mustache bat in which the dominant (60 – 64 kHz) component of its constant-frequency echolocation cry is represented. They have obtained evidence for segregation of cells of three binaural classes (viz., EO/mon, EI, and EE) in what is essentially a massively enlarged frequency-band sheet.

The "inhibitory bands" demonstrated by Webster et al. (1984a, 1985) in 2-DG studies of cat IC could also be interpreted as indicating segregation with respect to binaural characteristics, and Webster et al. discuss their results in terms of separate EE and EI bands. It is not clear, however, that the data require interpretation in these terms. The majority (approximately 78%; see Table 4B) of ICC cells with CFs at the high frequencies used by Webster et al. (1984a, 1985) to demonstrate inhibitory bands in ICC ipsilateral to the stimulated ear receive EI or EO/mon input (Irvine and Gago 1985; Moore and Irvine 1980; Roth et al. 1978; Semple and Aitkin 1979). That is to say, the frequency-band sheets in which these frequencies are represented are composed predominantly of cells receiving such input. Stimulation of the contralateral ear would evoke increased activity in excitatory terminals and predominant excitation of the neurons in such a sheet, and would therefore result in a band of increased 2-DG uptake. Conversely, stimulation of the ipsilateral ear would produce predominant suppression of neuronal activity and – if EI properties in IC reflect binaural interaction in SOC – a reduction in terminal activity, and would therefore result in a band of reduced 2-DG uptake. Thus, the *same* frequency-band sheet of neurons and axon terminals would be differentially labeled by contralateral and ipsilateral stimulation. On this interpretation, the existence of inhibitory bands carries no implication of segregation of binaural classes beyond that consequent on the frequency organization of ICC. An interesting feature of the "inhibitory bands" described by Webster et al. (1984a, 1985) was that they tended to be wider than the excitatory bands produced in the IC contralateral to the ear to which the tonal stimulus was delivered. One interpretation of this difference is that a given frequency has an ipsilateral inhibitory influence over IC neurons with a wider CF range than that of the neurons on which it has a contralateral excitatory influence.

In contrast to the physiological and anatomical evidence for partial segregation of the terminations in ICC of projections from different sources is evidence indicating some degree of convergence of different projections. Semple and Aitkin (1980) examined the response characteristics of a sample of ICC neurons ($N = 33$) that were activated at short latency (<2.0 ms) by electrical stimulation of DAS. Only 9 of the units were EO/mon, while 18 were EI, 11 exhibiting strong ipsilateral inhibition, and 7 relatively weak inhibition. The inhibition in the strongly inhibited EI units was stronger than that seen in DCN neurons receiving binaural input (see Sect. 4.2.4). Thus, if these cells were orthodromically activated by DAS stimulation,[19] their EI characteristics would have to reflect additional input from some other source. Semple and Aitkin (1980) suggested that such input might be derived from contralateral LSO, a proposal that is supported by the observation that EI and EO/mon cells are intermingled in caudal and

[19] Footnote see p. 164.

ventral ICC (Irvine and Gago 1985; Roth et al. 1978; Semple and Aitkin 1979), and by Semple and Aitkin's (1981) report of a case in which a small injection of HRP into this region resulted in labeled cells mainly in contralateral DCN and LSO. Weak ipsilateral inhibition (as seen in some cells activated by DAS stimulation) could reflect a weak LSO input to cells receiving their major input from DCN, but strong ipsilateral inhibition would require a relatively strong LSO input compared with that from DCN. Convergence of this sort might provide an explanation of the puzzling report by Bengry et al. (1977) that lesions of DAS and IAS in the rat resulted in a potentiation of the effect of ipsilateral stimulation on EI cells in IC contralateral to the lesion. Such lesions would remove a source of contralateral excitatory drive from cells receiving convergent input from DCN and LSO and could thus result in an apparent potentiation of ipsilateral inhibitory input.

Semple and Aitkin (1979) also noted that convergence is indicated by the almost complete absence of neurons in ICC with the binaural input characteristics that would be expected on the basis of the projection from ipsilateral LSO. They originally made this observation with reference to the absence of IE cells in ICC, but subsequent data have indicated that the ipsilateral projection from LSO to ICC is derived from the low-frequency lateral limb (Glendenning and Masterton 1983), which receives only a weak projection from MNTB (Glendenning et al. 1985) and in which most neurons consequently receive OE/mon rather than IE input (Boudreau and Tsuchitani 1968; Guinan et al. 1972a, b; see Sects. 5.1.4 and 5.2.2). Cells with OE/mon input are almost never encountered in ICC, however, and Semple and Aitkin's (1979) suggestion that convergence of ipsilateral LSO input with EO/mon input from contralateral CN would generate EE response characteristics provides a plausible account of their absence.

The binaural input characteristics of cells in ICX and ICP have received little attention. Aitkin et al. (1975, 1978b) reported that neurons in ICX exhibit the same patterns of binaural input, and in similar proportions, as neurons in ICC. In a small sample ($N = 22$) of ICP neurons described by Aitkin et al. (1975), EO/mon cells were by far the dominant category, but the difference from the other IC divisions suggested by these data is qualified by the small size of the sample.

[19] Although these data suggest convergence onto ICC neurons, it should be noted that ICC is the source of descending projections to DCN (Kane 1977c; Kane and Conlee 1979; Kane and Finn 1977), and that some of the cells studied by Semple and Aitkin (1981) might have been antidromically activated by DAS stimulation. They discounted this possibility on the grounds that the activation latencies were longer than the latency of the initial component of the field response to DAS shock, which they identified as the "tract" response (i.e., the response of presynaptic axons). However, the shortest latencies in their sample were in the range 1.26 – 1.74 ms (reported as 1.5 ms after rounding) and the latency of the negative peak of the tract response in most of their figures is approximately 1.0 ms (their Figs. 2 and 3). If the tract response mainly reflects activity in the large-diameter axons of DCN giant and fusiform cells, it is not inconceivable that ICC cells with smaller-diameter axons descending to DCN might be antidromically activated at latencies somewhat longer than the tract response peak. The experimental paradigm employed by Semple and Aitkin (1980) offers a valuable method for investigating convergence of afferents from different sources onto single ICC neurons, and this issue merits investigation with use of the collision test for antidromic activation.

7.2.4 Binaural Interaction: Coding of Interaural Intensity Differences

Studies of the sensitivity to IIDs of ICC neurons in the cat have consistently confirmed that the major class of neurons exhibiting such sensitivity is that receiving EI binaural input (Geisler et al. 1969; Hind et al. 1963; Moore and Irvine 1981; Rose et al. 1963; Roth et al. 1978; Semple and Aitkin 1979). In agreement with Goldberg and Brown's (1969) observations in SOC, the majority of EE cells are reported to be insensitive to IIDs (Geisler et al. 1969; Moore and Irvine 1981), although some exceptions to this generalization will be considered in Sect. 7.5.

The form of IID sensitivity characteristic of EI cells in IC is illustrated by the functions in Fig. 43 (full lines). For each of the cells shown in these figures, response is maximal over a range of positive IIDs (corresponding to contralateral azimuths) and, for all but one response, is totally suppressed over a range of negative IIDs (corresponding to ipsilateral azimuths). Although the functions for all these cells thus have the same basic form, the steepness of the function, and the IID range over which the decline in response occurs, differ between individual neurons. The former feature can be quantified by measuring the 80% dynamic range of the function, i.e., the IID range over which the response declines from 90% to 10% of maximum. The second feature can be quantified in terms of the IID value at which the response is 50% of the maximum response (i.e., the half-maximal IID). The significance of variation in these parameters will be considered in Sects. 7.5 and 8.2.4.

If the azimuthal sensitivity of these cells were determined primarily by their IID sensitivity, they would be expected to respond over a range of contralateral azimuths and to be inhibited over some or all of the range of ipsilateral azimuths. Although the maximum response would be expected to occur at some contralateral azimuth, its specific location cannot readily be predicted as it would be jointly determined by the shapes of the IID sensitivity function and the IID–azimuth function at the stimulus frequency.

The data presented in Fig. 43 were obtained in experiments in which IIDs were generated by varying the ipsilateral and contralateral intensities symmetrically about a specified intensity that has been termed the average binaural intensity (ABI).[20] For example, at the ABI of 50 dB used to obtain the data in Fig. 43A, an IID of +20 dB is generated by presenting 60 dB contralaterally and 40 dB ipsilaterally, and an IID of −20 dB is generated by presenting 40 dB contralaterally and 60 dB ipsilaterally. The advantage of this (ABI-constant) method of IID generation is that it approximates the combinations of near-ear amplification and far-ear attenuation by which IIDs are produced when a sound source is located at different azimuthal positions on an arc in the horizontal plane (Sect. 2.3), and thus facilitates extrapolation to the free-field situation. The method of generating IIDs used in most earlier studies was to hold the intensity of the stimulus to one ear (usually that providing the excitatory input to EI or IE cells) constant and to vary the intensity at the other ear above and below the constant

---

[20] The value about which the monaural intensities are varied is not in fact their arithmetic mean, and the designation "average binaural intensity" is therefore inaccurate. The term will nevertheless be retained here and will be defined as the base intensity about which the monaural intensities are symmetrically varied.

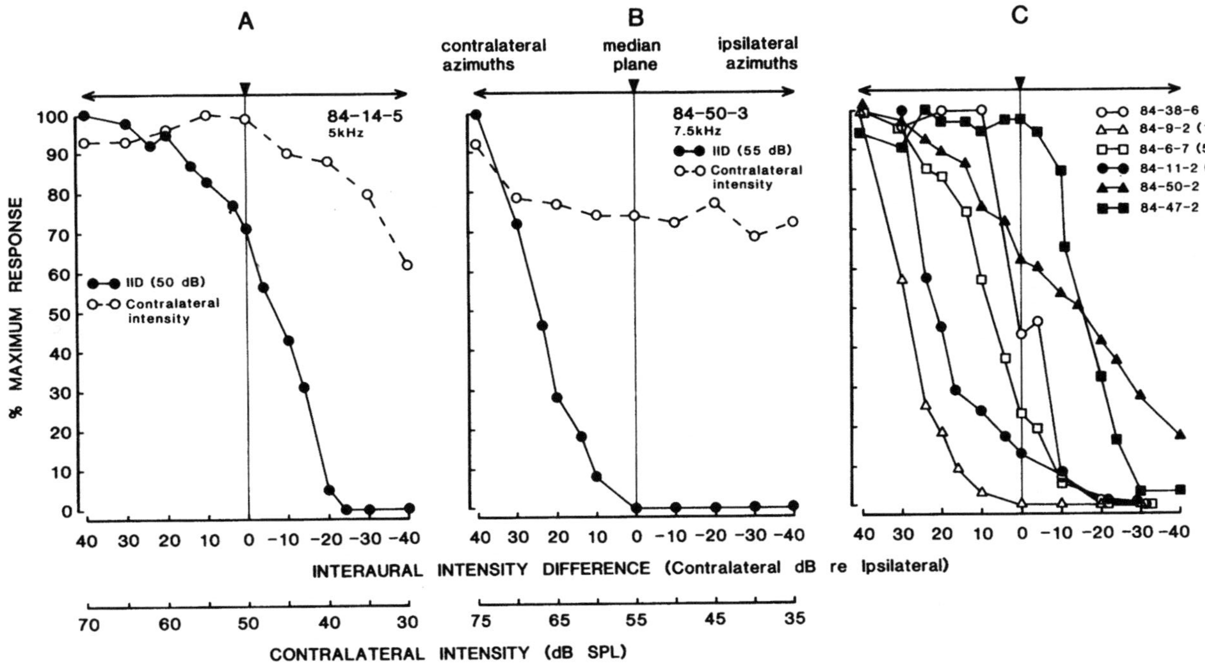

Fig. 43 A – C. Representative interaural-intensity-difference (IID)-sensitivity functions for EI neurons in central nucleus of inferior colliculus of cat. All functions have been normalized to percentage of maximum response for ease of comparison. **A, B** *Full circles* (*full lines*) show variation in response as a function of IID at average binaural intensity and frequency indicated, while *open circles* (*broken lines*) show variation in response to monaural contralateral stimulation over intensities used to generate IIDs. Contralateral intensities associated with particular IIDs on upper abscissa are shown on lower abscissa, **C** IID functions for a number of EI neurons, each at characteristic frequency of cell (indicated in parentheses after unit identification numbers), are superimposed to illustrate variation between cells in position and slope of cutoff in function. (D. R. F. Irvine and G. Gago 1984, unpublished work)

level. This procedure – which might be termed the "excitatory-monaural-intensity-constant" (EMI-constant) method – involves combinations of intensities at the two ears that bear little relationship to those involved in the generation of IIDs by azimuthal displacements in the free field.

When the ABI-constant method is used to examine the IID sensitivity of an EI cell, the resultant function could reflect sensitivity to IID per se or to the changes in contralateral intensity that are associated with changes in IID. The effects of these factors can be separated by comparing the binaural function with the contralateral rate–intensity function over the range of contralateral intensities involved in generating the IIDs. This procedure is illustrated in Figs. 43A and B, in which the contralateral rate–intensity function is presented as a broken line. In these cases, the comparison indicates that the total suppression of the response at negative IIDs is attributable to ipsilateral inhibition and reflects IID sensitivity. Although this is generally the case for EI cells with steeply sloping IID-sensitivity functions, the detailed form of the function is in many cases a joint product of sensitivity to IID and to the changes in contralateral intensity. This is particularly the case for cells with nonmonotonic monaural rate–intensity functions (Irvine and Gago 1985).

As noted previously in discussion of binaural categories in ICC, the inhibitory input to some EI cells is weak and/or influences the cell's response only at large negative IIDs. An example of the shallow IID-sensitivity functions characteristic of some such cells is provided by unit 84-50-2 in Fig. 43C. Because it seems unlikely that such cells can play an important role in the neural coding of IIDs in the physiological range, Semple and Aitkin (1979) placed them in their category "other" rather than with the more usual IID-sensitive EI cells. It is not clear, however, that this distinction can be made without recourse to a relatively arbitrary criterion, since the distribution of IID dynamic ranges appears to be continuous, with no clear indication of groups with steep and shallow slopes (D. R. F. Irvine and G. Gago 1984, unpublished work).

Differential sensitivity of onset and sustained components to variation in IID is commonly observed in neurons with the common ICC response pattern in which these two components are separated by a pause. Most frequently, the sustained component is much more sensitive than the onset component, although in some cases both components show similar sensitivity (Geisler et al. 1969; Hind et al. 1963; Moore and Irvine 1981; D.R.F. Irvine and G. Gago 1984, unpublished work). As noted in Sect. 6.2, differential sensitivity of onset and sustained response components was also reported by Brugge et al. (1970) in DNLL. It appears to be an emergent feature at these higher levels, since this response pattern is rarely seen in LSO (Tsuchitani 1982), and differential sensitivity of onset and late response components does not seem to have been reported in SOC. Indeed, Boudreau and Tsuchitani (1968) emphasized that the first spikes in the responses of IE cells in LSO could be inhibited by ipsilateral stimulation. Neurons in IC with *pure* onset responses, however, can have IID-sensitivity functions with slopes as steep as those of the sustained components of cells with two-component response patterns, and such functions are commonly exhibited by "on" units in SCD (see Sect. 7.5 and Fig. 52).

Most studies of IID sensitivity have been restricted to tones at the unit's CF and at a single intensity, although Rose et al. (1966) noted that the inhibitory effect in

EI cells could vary with changes in the frequency, intensity, and duration of stimulation. Geisler et al. (1969), using the EMI-constant method, found that a unit's IID function for broad-band noise stimulation generally resembled that for tonal stimulation at CF. For stimuli of both sorts, the responses of most neurons were also sensitive to absolute intensity levels (i.e., to ABI), such that a given discharge rate could not unequivocally indicate IID. Benevento et al. (1970) examined the effects of changes in absolute intensity on the sensitivity to IIDs in click stimuli of ICC units in unanesthetized, paralyzed cats, using the ABI-constant method. Although they did not classify units in terms of their binaural input patterns, the largest group of IID-sensitive cells was apparently EI and had monotonic IID-sensitivity functions of the same sort as described in anesthetized preparations. This is the only account of IID sensitivity in unanesthetized preparations, and confirmation of the conventional form of IID sensitivity under these conditions provides an important validation of the data obtained from anesthetized preparations. Benevento et al. (1970) found that for some neurons, IID-sensitivity functions were invariant with changes in intensity, but for others spike counts and/or the function shape varied with changes in ABI. Irvine and Gago (1985) have described similar effects in a preliminary report of the effects of variation in ABI on the sensitivity to IIDs in CF tone bursts of EI cells in the central nucleus of the cat. In these studies, marked changes in the form of the functions were found in units with nonmonotonic monaural rate–intensity functions.

Both the status of EI cells as the major class of IID-sensitive neurons in ICC, and the characteristic form of this sensitivity, have been confirmed in a number of other mammals (e.g., rat: Silverman and Clopton 1977; Syka et al. 1983; kangaroo rat: Stillman 1972; rabbit: Aitkin et al. 1972b; guinea pig: Popelar and Syka 1982; brush-tailed possum: Aitkin et al. 1978a; gerbil: Semple and Kitzes 1985; various echolocating bats: Harnischfeger 1980; Harnischfeger et al. 1985; Wenstrup et al. 1985). The effects of variation in absolute intensity on the IID sensitivity of ICC neurons in the kangaroo rat were examined by Stillman (1972), who reported that IID-sensitivity functions either remained invariant, or were systematically shifted, but remained parallel, when absolute intensity was varied. A preliminary report by Semple and Kitzes (1984) has described similar observations on ICC neurons in the gerbil.

Wenstrup et al. (1985) have recently reported a topographic organization of the IID sensitivity of EI cells in the enlarged 60-kHz representation in ICC of the mustache bat. The value of the half-maximal IID shifted systematically as a function of neuronal position within this isofrequency region. A similar topographic organization has been described in SCD of the cat (Irvine and Wise 1983a; Wise and Irvine 1985), and this issue will be examined in more detail after the SC data have been described (see Sects. 7.5 and 8.2.4, and Fig. 55).

Although EI cells undoubtedly constitute the major class of IC neurons exhibiting sensitivity to IIDs, there has been a small number of reports of such sensitivity based on facilitatory binaural interactions, both in the cat (Benevento et al. 1970) and other mammals (kangaroo rat: Stillman 1972; gerbil: Semple and Kitzes 1984). Such sensitivity is seen in OO/F, EO/F, and EE/F cells, and is manifested in nonmonotonic functions with peaks in the region of zero IID. Sensitivity of this sort has been described in greater detail in SCD of the cat (Hirsch et al. 1985; Irvine and Wise 1983a; Wise and Irvine 1983, 1984, 1985) and in primary auditory cortex (Kitzes et al. 1980; Phillips and Irvine 1981b), and will be discussed in more detail in Sect. 7.5. Intermediate forms of sensitivity involving mixed facilitatory and inhibitory binaural interactions have also been described in IC (kangaroo rat: Stillman 1972; gerbil: Semple and Kitzes 1984) and in SCD (Wise and Irvine 1984, 1985), and are also discussed in Sect. 7.5.

This account of IID coding in ICC has been restricted to a description of the major features of the IID sensitivity of single neurons. No attention has been given here to the implications of these data for the manner in which IID (and auditory azimuthal location) is represented by the activity of such neurons, or to the consequences for such representations of variations in the form of IID sensitivity with changes in other stimulus parameters. These issues will be considered in Sect. 8.2.4, in the light of other relevant information to be examined in the following four sections (Sects. 7.2.5 and 7.3 – 7.5).

7.2.5 Binaural Interaction: Coding of Interaural Time Differences

Rose et al. (1966), in their classical paper on neural mechanisms in IC relevant to sound localization, were the first to report periodic delay functions of the type described in Sect. 5.2.4. Periodic IPD-sensitivity functions have subsequently been described in a number of studies of IC in the cat (Geisler et al. 1969; Kuwada and Yin 1983; Yin and Kuwada 1983a, b) and other species (rabbit: Aitkin et al. 1972b; kangaroo rat: Stillman 1971). In a small number of cells, Rose et al. (1966) were able to obtain delay functions at a number of different frequencies. The functions were in all cases periodic (e.g., Fig. 44), but for some cells the curves for different frequencies reached the same relative amplitude at a particular delay (e.g., all three curves in Fig. 44 reach their maximum response at a delay of approximately 140 μs). Rose et al. (1966) termed the delay at which the different curves for a given neuron reached the same relative amplitude the "characteristic delay" (CD) of the neuron, and suggested that neurons with CDs could be the basis for a place coding of ITD of the type proposed by Jeffress (1948). The fact that many IPD-sensitive neurons have a CD has been confirmed in subsequent investigations (Geisler et al. 1969; Yin and Kuwada 1983b; kangaroo rat: Stillman 1971), and the concept of CD has become a major, but controversial (e.g., Stillman 1971), element in our understanding of spatial coding (see Yin and Kuwada 1983b, 1984 for reviews).

In most of the studies cited here, delay sensitivity functions were obtained for only a small number of neurons, and CDs were investigated using only a small number (usually two or three) frequencies. These small samples reflect the slowness of data collection when delay stimuli are generated using passive delay lines. Recently, Yin and Kuwada and their colleagues (Kuwada and Yin 1983; Kuwada et al. 1979, 1980, 1984; Yin and Kuwada 1983a, b, 1984; Yin et al. 1983, 1984, 1986) have presented detailed, quantitative data on IPD sensitivity and CD in large samples of low-frequency neurons in ICC of the cat. Collection of extensive quantitative data in these studies was made possible by the fact that delay stimuli were generated using a digital stimulus system (Rhode 1976) and by the discovery that evidence on IPD sensitivity could be obtained with great rapidity by the use

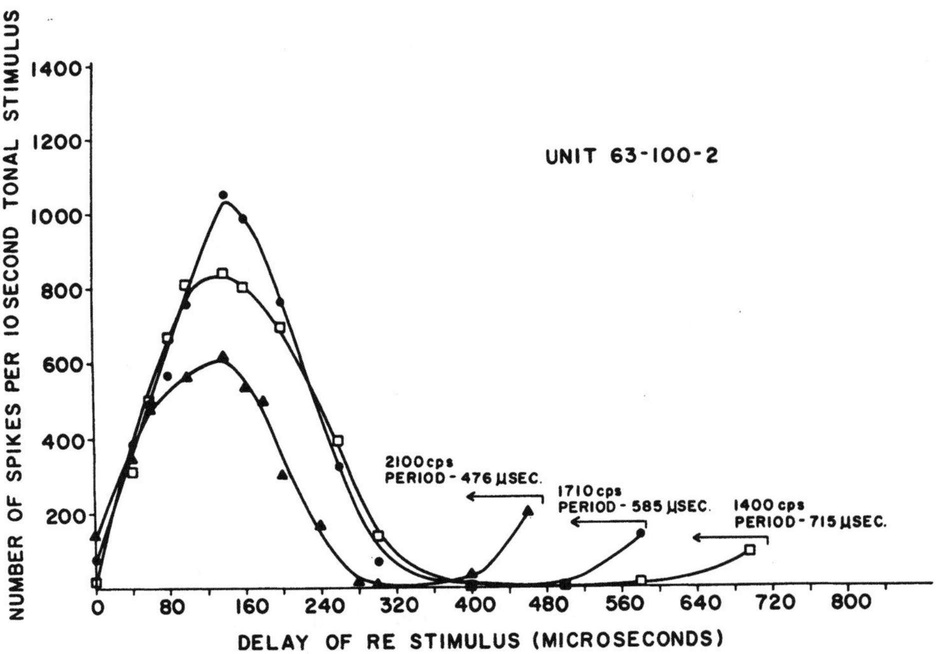

Fig. 44. Interaural delay curves generated for a neuron (unit 63-100-2) in ICC of the cat at three frequencies indicated in cps (cycles per second). Intensity of the stimuli: 48 dB SPL for 1400 and 1710 Hz (cps); 50 dB SPL for 2100 Hz. The period of each frequency is indicated by the *vertical stroke* and attached *horizontal arrow*. Note that the period of each curve equals that of the stimulating frequency, but that each curve has a maximum at the same "characteristic" delay of 140 µs. (Rose et al. 1966)

of a "binaural beat stimulus." When a sine wave is presented to one ear at a frequency slightly different from that presented to the other, the consequence is a continuous, cyclic change in interaural phase with a repetition frequency equal to the frequency difference between the two tones (the "beat frequency"). Yin and Kuwada (1983a) found that almost all low-frequency ICC neurons with periodic delay sensitivity functions were also sensitive to binaural beats, and that for almost all such cells the IPD-sensitivity function obtained at a low beat frequency closely approximated that obtained with conventional delays. The great advantage of the binaural beat stimulus is that it allows data on a given cell's IPD sensitivity to be obtained almost ten times more rapidly than is possible even with digitally synthesized delay stimuli. The following account of delay sensitivity will be based largely on the extensive data provided by these recent studies.

Kuwada and Yin (1983) found that most (approximately 80% of a sample of 197) low-frequency ICC neurons (CF < 2.5 – 3.0 kHz) had periodic interaural delay functions to CF tones. Most (93%) of these IPD-sensitive cells were insensitive to onset disparity (TTD), confirming the fact that these forms of ITD are processed by separate mechanisms. In contrast to MSO, in which almost all of the small sample of IPD-sensitive cells studied by Goldberg and Brown (1969) were excited by, and exhibited phase-locked responses to, monaural stimulation of either ear

(Sect. 5.2.4), IPD-sensitive cells in ICC exhibited diverse response patterns and patterns of binaural input. Thus, although the majority of IPD-sensitive cells would be classified as EE on the basis of monaural stimulation, a large proportion are EO and exhibit binaural facilitation and inhibition at various ITDs (Aitkin et al. 1975; Geisler et al. 1969; Kuwada et al. 1984; Rose et al. 1966). Only a small proportion of IPD-sensitive cells exhibit monaural phase-locking (18%; Kuwada et al. 1984; see also Geisler et al. 1969) and sensitivity is also exhibited by cells with onset responses to monaural stimulation of one or both ears. These observations strongly suggest that IPD sensitivity in ICC reflects earlier processing of this cue in MSO, a conclusion supported by the fact that the bushy/large spherical cells in anterior AVCN that preserve monaural phase information do not project to IC (see Sect. 7.1.3).

In a small number of neurons that exhibited phase-locked responses to monaural stimulation of either ear, Kuwada et al. (1984) obtained data in support of the coincidence detector model (Sect. 5.2.4). In accordance with Goldberg and Brown's (1969) observations in MSO of the dog, the optimal delay for each of these neurons was almost identical to the difference between the monaural phase-locking points. The coincidence detector model will be considered in more detail in the context of a general discussion of models of binaural interaction in Sect. 8.2.2.

Within the physiological ITD range, the great majority (80% or more) of IPD-sensitive cells in the cat (Kuwada and Yin 1983a; Roth et al. 1978) and kangaroo rat (Stillman 1971) respond maximally when the contralateral ear is leading. They would therefore be expected to respond maximally to sounds in the contralateral half of space (see more detailed discussions in Sects. 7.5 and 8.2.5).

Yin and Kuwada (1983a, b) examined the effects of variation in absolute intensity and of IIDs on delay sensitivity. While some cells were insensitive to these variations, others showed marked changes in delay functions with changes in absolute intensity and/or IID, and there was a continuous distribution of sensitivity between these extremes. Given the dependence of IPD sensitivity on the timing of phase-locked input to MSO neurons, variation in the form of such sensitivity with intensity would be expected if the phase angle of discharge of monaural responses at lower levels of the auditory pathway varied with intensity. Although Anderson et al. (1971) reported that the mean phase angle of discharge of AN fibers in squirrel monkey was relatively invariant with the intensity of tones at fiber CF (but not at other frequencies – see Sect. 3.4), variation in mean phase angle of discharge with intensity has been reported for cells in CN of cat (Lavine 1971) and SOC of kangaroo rat (Crow et al. 1978).

An advantage of the binaural beat stimulus is that it allows investigation of responses to dynamic phase changes. The direction of the frequency difference between the ears determines the direction, and its magnitude the rate, of phase change. The responses of most neurons examined were uninfluenced by changes in either the direction or rate of phase change, but some cells were highly sensitive to one parameter or the other. Such cells would be expected to show selective sensitivity to moving sound sources in the free field.

As Yin and Kuwada (1983b) have pointed out, most previous studies of CD used a small range of frequencies and identified CD by visual examination and without formulating explicit criteria. Although Rose et al. (1966) defined CD in terms of discharge at the same *relative amplitude*, subsequent studies identified CDs only at the peaks or the troughs of the functions. The difficulty of identifying CD by purely visual criteria is illustrated by the delay functions obtained for one cell at eight frequencies in Fig. 45 A. Even when each of the curves is normalized as a percentage of maximum response (Fig. 45 B), it is not clear whether the cell has a CD and, if it has, whether it is located at the peak, trough, or some other point. Yin and Kuwada (1983 b) developed a quantitative analytic procedure which provides an objective criterion for the presence of CD and returns values for CD and for what they termed "characteristic phase" (the relative amplitude of the periodic functions at which the CD occurs). In a linear system, a time delay produces a linear phase-frequency plot, and Yin and Kuwada (1983 b) therefore took linearity of the relationship between the mean interaural phase (calculated by Goldberg and Brown's 1969 vector averaging method) and frequency as evidence of the existence of a CD. The slope of the line is equal to the value of CD and the phase intercept is equal to the characteristic phase. Using this analytic method, Yin and Kuwada (1983 b) found that a large proportion of a sample of 82 cells studied in detail satisfied the criteria for CD at a high level of statistical significance, and that for the majority of these cells the CD did not fall at the peak or trough of the response. Thus, for the cell for which data are presented in Fig. 45, the CD was +5 μs and the characteristic phase value indicated that the CD occurred midway between the peak and the trough (Yin and Kuwada 1984).

Although the reality of CDs as originally defined by Rose et al. (1966) is unequivocally established by these data, their significance for the neural representation of space is unclear (see Yin and Kuwada 1983 b, 1984 for discussion). As Yin and Kuwada (1983 b) pointed out, it is difficult to understand how the CNS could use the occurrence of some percentage of maximum discharge in single neurons to code an ITD. Furthermore, the fact that a given discharge rate can be elicited at different delays at different frequencies (see Figs. 44 and 45 A) and intensities, indicates that a single neuron cannot unequivocally signal a particular ITD. It seems clear that ITD must be coded in terms of the pattern of activity across a population of neurons. Whether or not CD is the critical feature of a given cell's response that determines its contribution to various patterns of activity, or whether it is some other feature of the functions (e.g., the positions of the peaks, troughs, and/or slopes) that is important is not known (cf. Yin and Kuwada 1983 b). A similar point was made by Stillman (1971) when he noted that the CD of most of a sample of IC neurons in the kangaroo rat occurred at delays outside the naturally occurring range of ITDs in this species. Although the CDs in this study were identified at either peaks or troughs by visual inspection, and are therefore open to question, Stillman's observation that the steeply sloping portion of most delay functions occurred across the physiological range of ITDs suggests that the slopes of the functions might be their important feature. These issues are discussed in more detail with reference to both IID and ITD coding in Sect. 8.2.4.

Yin et al. (1983) have investigated the possibility that delay sensitivity might be topographically organized within frequency-band sheets in ICC. Although preliminary analysis indicates that, in some experiments, delay sensitivity is not randomly distributed across a frequency-band sheet, a consistent topography has not yet been revealed. Preliminary data have also been reported from experi-

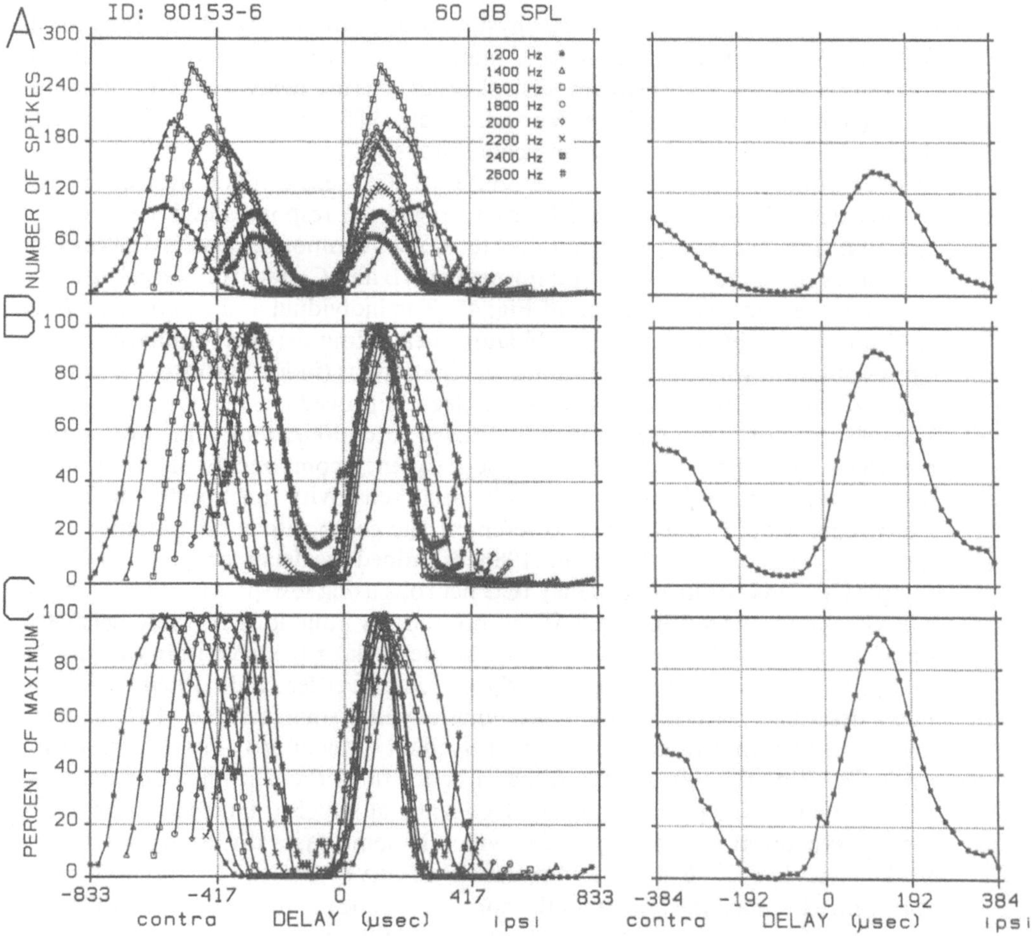

Fig. 45 A – C. Superimposed interaural delay curves (*left*) and composite curves (*right*) for an ICC neuron (80153-6). In all cases, stimulating frequencies (identified by symbols in *key*) ranged from 1200 to 2600 Hz in 200-Hz increments. Stimulus intensity in all cases was 60 dB SPL. Each composite curve was calculated from the corresponding superimposed delay curves by averaging all the individual delay curves at each delay up to that equal to the period of the highest stimulating frequency. By convention, delays of the ipsilateral stimulus are positive. **A** Superimposed and composite curves plotted from spike count data obtained using interaural delay stimulus (stimulus duration 500 ms repeated four times), **B** data from **A** are replotted with each interaural delay curve normalized to its maximum discharge rate, and composite curve is calculated from normalized delay curves, **C** superimposed normalized interaural delay and corresponding composite curve obtained for same cell using binaural beat stimulus (stimulus duration 3 s repeated five times with beat frequency −1 kHz). From Yin and Kuwada (1984) Neuronal mechanisms of binaural interaction. In: Edelman, Gall, Cowan (eds) *Dynamic Aspects of Neocortical Function*. Copyright © 1984, reprinted by permission of John Wiley & Sons, Inc.

ments in which ICC neurons were physiologically characterized and then intracellularly labeled with iontophoretic injections of HRP (Kuwada et al. 1980). These data suggest that fusiform (disk-shaped) cells are delay sensitive and multipolar (stellate) cells insensitive (Kuwada et al. 1980, and personal communication), but do not yet allow definite conclusions.

In attempting to characterize the responses of ICC cells to ITDs over a range of frequencies, Yin and Kuwada (1983b) averaged the responses of a given cell at different frequencies to yield what they termed a "composite curve". The composite curves for spike count (A) and normalized (B and C) data for one cell are presented in the right-hand column of Fig. 45. The individual and composite curves in C are those obtained with the binaural beat stimulus; their similarity to the curves obtained with the conventional delay stimulus (in B) illustrates the equivalence for most cells of these two data collection procedures.

If a cell's response to an interaurally delayed, broad-band stimulus were simply the linear sum of its responses to those frequency components of the stimulus that fell within its response area, then its delay sensitivity function for the broadband signal would be expected to resemble the composite curve obtained from the nonnormalized data. Yin et al. (1986) examined the delay sensitivity of a large sample ($N = 144$) of low-frequency ICC neurons using low-pass (0 – 4 kHz) noise stimuli. As illustrated in Fig. 46, the noise delay functions of these cells (full lines) were diverse in form; some had a single peak or trough within the physiological range, while others were clearly periodic. Geisler et al. (1969) had previously described periodic noise delay functions in a small sample of IC neurons and had noted that in most cases the periodicity was approximately that of the cell's CF. Yin et al. (1986) found that a better fit to the periodicity of the noise delay function was provided by the frequency at which the neuron showed the highest synchronized discharge rate, which in some cases differs from CF. The broken lines in Fig. 46 give the composite tone curve for each cell; the close correspondence between the noise and the composite tone curve in each case is illustrative of the fact, supported by quantitative analysis, that to a first approximation the noise delay curves are accurately predicted on the assumption of linear summation. An interesting feature of noise delay curves (Yin et al. 1986) and composite tone curves (Yin and Kuwada 1983b) is that their shape is relatively invariant with changes in intensity, although response rate changes. It therefore appears that the problems posed by variations in the form of pure-tone delay sensitivity functions with intensity may not arise with more natural, spectrally complex sounds.

Yin et al. (1986) found that there was no evidence of delay sensitivity when different (i.e., uncorrelated) noise signals with the same spectral composition were presented to the two ears. This observation indicates that sensitivity is based on delays in the fine structure of the signal and suggests that the CNS performs a cross-correlation on the input signals. Chan et al. (1985) have presented strong evidence for the cross-correlation model by demonstrating a monotonic, positive relation between the degree of modulation of noise delay responses and the correlation between the input noises (Fig. 47). The cross-correlation model will be discussed further in Sect. 8.2.2.

It was noted in Sect. 2.3 that, although OTDs provide a cue to the azimuthal location of pure tones only at low frequencies, such disparities in the envelopes of

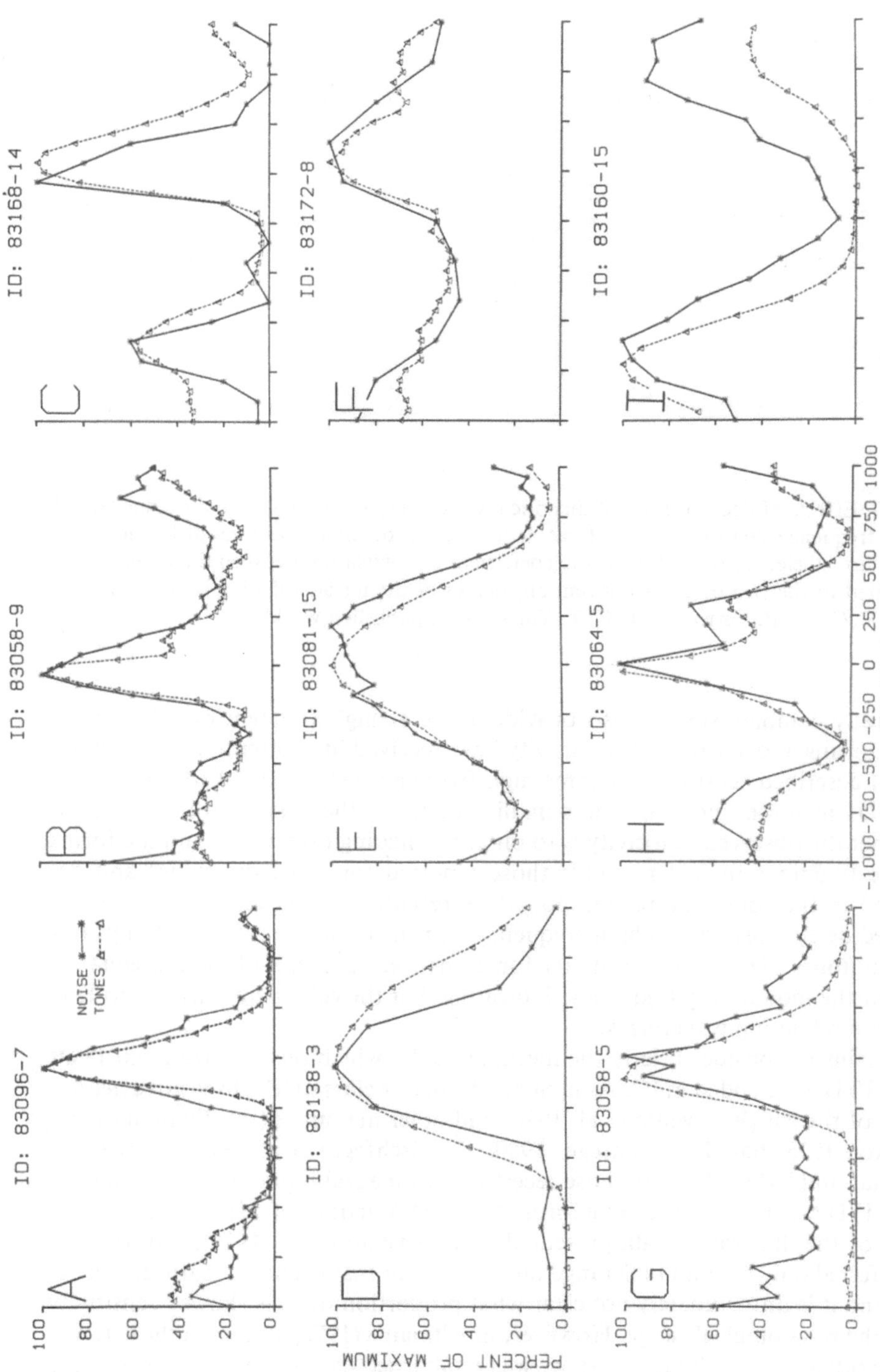

Fig. 46 A – I. Comparisons of noise delay curves (*stars*) and tone composite curves (*triangles*) for nine representative low-frequency neurons in ICC of the cat. Each curve has been normalized to its maximal spike count for ease of comparison. Unit identification number (*ID*) is shown at top of each panel. By convention, delays of the ipsilateral stimulus are positive. (Yin et al. 1986)

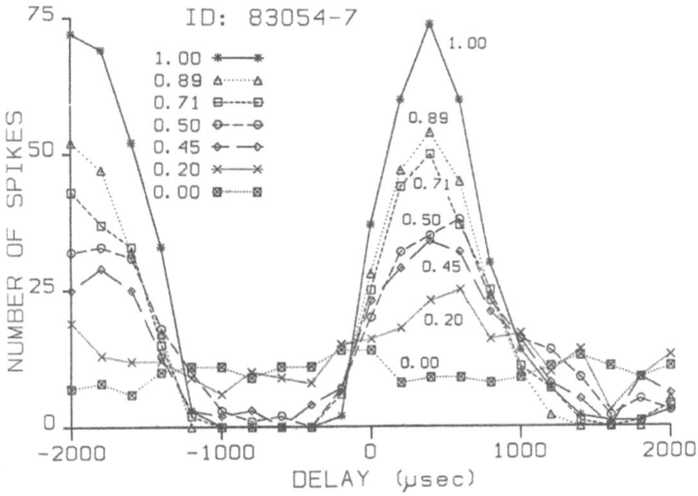

Fig. 47. Effects of decorrelation of dichotically presented noise signals on the responses of a low-frequency neuron in ICC of cat as a function of interaural time difference. The curves are labeled by the values of the coefficient of correlation between the noise signals presented to the two ears. By convention, delays of the ipsilateral stimulus are positive. (J. C. K. Chan, P. Verro and T. C. T. Yin 1984, unpublished work)

spectrally complex sounds also provide a cue at high frequencies. The neural mechanisms underlying such sensitivity have received little attention. Crow et al. (1980) described sensitivity of some high-frequency cells in IC of the guinea pig to ITDs in noise and two-tone stimuli. However, their data do not establish whether the observed sensitivity is to onset or ongoing disparities, and the forms of the functions in fact resemble those reported for onset disparities and described in the following paragraph. More recently, Yin et al. (1984) have described delay sensitivity in high-frequency IC neurons in the cat to AM high-frequency tones. The delay sensitivity functions were cyclic, with period equal to that of the modulating frequency, indicating that the cells were sensitive to IPD of the modulating waveforms.

It remains to consider briefly the mechanisms by which onset or transient ITDs (i.e., TTDs) are coded in IC. As in SOC, the major class of TTD-sensitive neuron in IC of the cat (Benevento et al. 1970) and other mammals (rat: Flammino and Clopton 1975; bat: Harnischfeger 1980; Harnischfeger et al. 1985; guinea pig: Popelar and Syka 1982) are those receiving EI binaural input and having monotonic ITD functions of the same form as the IID functions of cells of this class. In most of the illustrative data presented by Benevento et al. (1970), sensitivity is manifested only over an ITD range far in excess of that occurring naturally in the cat, and it is unfortunately not clear what proportion of cells showed sensitivity over the physiological range. However, in Altman's (1978) studies in the cat, and in the other studies cited, sensitivity was established over the naturally occurring range of ITDs in the species used. Benevento et al. (1970) reported that while some cells in IC of paralyzed, unanesthetized cats were sensitive to both IIDs and ITDs in click stimuli, a larger number were sensitive to one, but not the other. A similar dissociation between IID and ITD sensitivity was reported by Altman (1978) in anesthetized cats and by Harnischfeger (1980) in the bat. These observations bear on models of IID sensitivity (specifically on the latency hypothesis) and will be considered further in Sect. 8.2.2. The fact that EI cells are the major class of cells in IC that are sensitive to TTD, together with Kuwada and Yin's (1983) report that most IPD-sensitive cells were insensitive to TTD, supports the claim that these forms of ITD are processed by different mechanisms. Benevento et al. (1970) also reported that approximately 20% of TTD-sensitive cells had peaked functions, based on binaural facilitation. These functions thus resemble the peaked IID-sensitivity functions that also reflect facilitatory binaural interaction.

Altman (1968) presented binaural click trains in which there was a progressive change (a diminution followed by an increase) in ITD between successive click pairs. A small proportion of IC cells were found to respond differentially as a function of the direction of TTD variation. These data are therefore analogous to those obtained by Yin and Kuwada (1983b) with the dynamic IPD change produced by the binaural beat stimulus. Both sets of data suggest that only a small proportion of IC neurons would be sensitive to the direction of sound source movement.

7.3 Auditory Spatial Representation in Midbrain of Barn Owl

As noted in Sect. 2.3, the barn owl is a nocturnal predator that can locate prey acoustically in total darkness (e.g., Konishi 1973a, b; Payne 1971). Sound localization by the barn owl is most accurate for spectrally complex sounds with energy in the range 5–9 kHz (Konishi 1973a), and such sounds can be localized with errors of less than 2° in elevation and azimuth (Knudsen et al. 1979; Payne 1971). Studies by Knudsen and Konishi and their colleagues (Knudsen 1982, 1983b, c, 1984a, b; Knudsen and Konishi 1978a, b, c, 1979, 1980; Knudsen et al. 1977; Moiseff and Konishi 1981, 1983a, b; Takahashi et al. 1984) of the manner in which auditory spatial location is represented in the midbrain of the barn owl have led to a recent resurgence of interest in auditory spatial representation in the mammalian auditory system. The impact of these studies has derived both from the development of a method for the controlled presentation of acoustic stimuli in the free field, and from the form of the spatial representation discovered in the owl midbrain.

The apparatus used to provide precise control over the spatial location of an acoustic stimulus in these experiments is illustrated in Fig. 48A (see Knudsen and Konishi 1978b for more detailed description). A small speaker is mounted on a semicircular track 2 m in diameter, along which it can be moved under stepping motor control to vary sound source azimuth. The track or "hoop" can be rotated around its horizontal axis (which is aligned with the owl's interaural axis) to vary the elevation of the sound source. The speaker can thus be moved to any point on the surface of a sphere of 1 m radius centered on the owl's head. The entire system is located within an anechoic chamber.

Using this system, it was found that some neurons in the telencephalon (Knudsen et al. 1977) and the auditory midbrain area (mesencephalicus lateralis dorsalis,

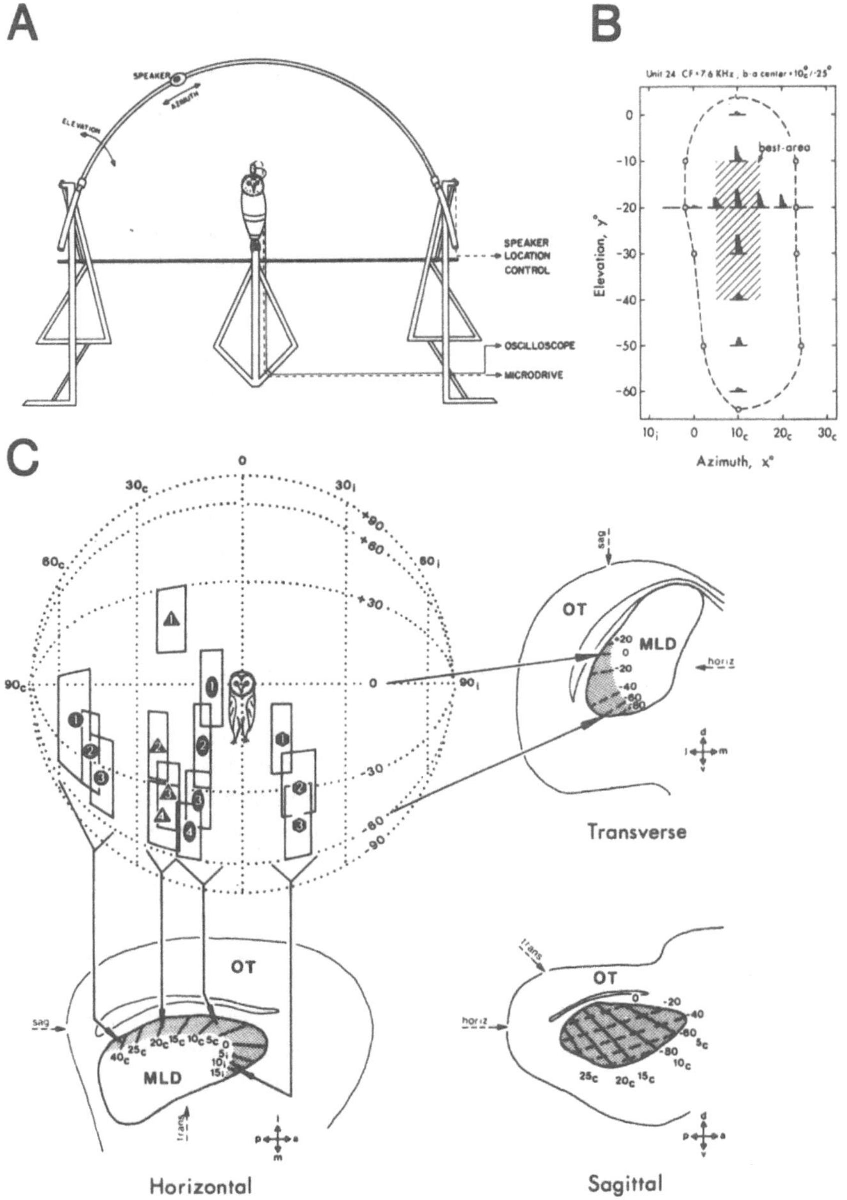

Fig. 48. A Simplified diagram of apparatus ("hoop") used for controlled presentation of free-field stimuli to barn owl (see description in text). Recent studies in cat and other mammals have used similar devices. (Knudsen and Konishi 1978b). **B** Receptive field (RF) plot for a limited field unit in midbrain nucleus (*MLD*) of barn owl. *Broken lines* mark the RF boundaries as projected from actual measurement sites (*open circles*). Azimuths are designated in degrees from median plane contralateral (*c*) and ipsilateral (*i*) to recording side, and elevation is designated in degrees below (−) visual plane. RFs were plotted, and peri-stimulus time histograms were generated, using broad-band noise bursts at 10 dB above the unit's threshold; the position of each histogram corresponds to the speaker location for which it was generated. The approximate extent of the best area is indicated by the

MLD; Knudsen and Konishi 1978a, b) of the barn owl had restricted auditory spatial receptive fields (RFs), in that they responded to a given sound only when it was located in a particular region of frontal space. A representative RF for a "limited field" (LF) MLD unit is shown by the broken lines in Fig. 48B. Although the RFs of these units were relatively large – the dimensions of the RF illustrated in Fig. 48B (viz., approximately 25° azimuthal extent × 70° elevation extent) are almost identical to the mean values reported by Knudsen and Konishi (1978b) – they were characterized by a more restricted central region within which the unit responded most vigorously. The boundaries of this region, the unit's "best area" (hatched region in Fig. 48B) were defined by the speaker positions at which the unit first gave a "submaximal" response. The RFs of MLD units were routinely determined using noise bursts 10 dB above threshold, but for most LF units the borders were invariant with changes in intensity. For some LF units (35%), the RF was also invariant with different types of stimuli, but for the majority the RF boundaries for noise, CF tone, and click were different, in some cases to a marked extent (Knudsen and Konishi 1978b). Under appropriate stimulation conditions, the excitatory RFs of LF units could be shown to have large inhibitory surrounds (Knudsen and Konishi 1978c).

LF units were located only along the lateral and anterior borders of MLD, had a restricted range of CFs (5 – 8 kHz), and most commonly exhibited onset responses. The most striking feature of this region of MLD was that the location of the best areas of LF units varied systematically as a function of neuronal position, in such a way as to create a physiological map of auditory space (Knudsen and Konishi 1978a, b). Data from a single subject illustrating this spatial representation are presented in Fig. 48C. It is apparent that sound azimuth is represented approximately rostrocaudally along the horizontal plane, and sound elevation is represented dorsoventrally in the transverse plane, within this "space-mapped" region of MLD. Most of the space-mapped region is devoted to the representation of the contralateral half of frontal space and to elevations below the interaural axis.

Fig. 48 (*continued*) *diagonal lines.* Details of unit's characteristic frequency (CF = 7.6 kHz) and best area (*b.a.*) center (10°_{c}; −25°) are shown above plot. (Knudsen and Konishi 1978b) **C** Representation of auditory space in the midbrain nucleus (*MLD*), as defined by centers of unit best areas. *In the upper left,* coordinates of auditory space are depicted as a *broken-line globe* surrounding the owl. Projected onto the globe are the best areas (*full-line rectangles*) of 14 units that were recorded in four separate penetrations. The large numbers backed by similar symbols represent units from the same penetration; the numbers themselves signify the order in which the units were encountered, and are placed at the centers of their best areas. Penetrations were made with the electrode oriented parallel to the transverse plane at positions indicated in the horizontal section by the *full arrows.* Below and to the right of the globe are illustrated three histological sections through MLD in the horizontal, transverse, and sagittal planes. The *stippled* portion of MLD corresponds to the space-mapped region; the remaining portion is the tonotopic region. Isoazimuth contours, based on best area centers, are shown as *full lines* in the horizontal and sagittal sections; isoelevation contours are represented by *broken lines* in the transverse and sagittal sections. On each section, *broken arrows* indicate planes of the other two sections. *Full, crossed arrows* to the lower right of each section define the orientation of the section: *a* anterior; *d* dorsal; *l* lateral; *m* medial; *p* posterior; *v* ventral; *OT* optic tectum. (Knudsen and Konishi 1978a; copyright 1978 by the AAAS)

Within the larger medial portion of MLD, the spatial RFs of units were more diverse and generally larger than those in the space-mapped region. Some units responded regardless of the spatial location of the stimulus ("space-independent" units), some had multiple excitatory areas ("complex-field" units), and some had RFs with indistinct borders that expanded considerably with increases in intensity ("space-preferring" units). Neurons in this medial region generally had sustained responses to tone- and noise-burst stimuli, and their CFs extended over most of the barn owl's audibility range of 0.1 – 10 kHz (Konishi 1973a). As illustrated in Fig. 49, the medial region, unlike the space-mapped region, was characterized by precise tonotopic organizaton (Knudsen and Konishi 1978b).

The barn owl's MLD thus contains two functionally distinct regions, one of which contains a full frequency representation and is tonotopically, but not spatially organized, while the other contains a topographic map of auditory space. The avian MLD is generally regarded as homologous to the mammalian IC (e.g., Karten 1967). Knudsen (1983a) has recently argued that the neuronal architecture, connections, and physiology of the tonotopic and space-mapped regions of MLD suggest that they are homologous to the mammalian ICC and ICX, respectively, and has proposed that the latter terms be applied to the owl. Although the similarities might be taken to imply homologies, the adoption of this terminological proposal would appear to prejudge the issue, at a time when the parcellation of IC is itself under review and the existence of a unitary ICX is open to question (see Sects. 7.1.1 and 7.1.2). For this reason, the divisions of MLD will be referred to here by the functionally derived terms "tonotopic" and "space-mapped."

The "best area" map of auditory space described by Knudsen and Konishi (1978a, b) in MLD is of the same general form as the RF-based maps of visual and somatosensory space that are characteristic of retinotopically and somatotopically organized CNS structures across a wide range of vertebrates. In the visual and somatosensory systems, such maps are a direct consequence of the joint facts that the spatial location of a stimulus determines the point on the receptor epithelium at which it produces activation and that projections from the receptor to central structures are topographically organized. In the auditory system, however, it is the frequency of an acoustic stimulus that determines the point on the receptor epithelium at which it produces activation, and the consequence of topographically organized central projections is frequency mapping (i.e., tonotopicity) in central structures. As Knudsen and Konishi (1978b) emphasized, the spatial selectivity of MLD neurons reflects neural comparisons of binaural inputs, and the resultant map is therefore the consequence of binaural integration. Subsequent studies have clarified both the nature of the binaural disparities on which the owl's localization ability is based and the brainstem mechanisms by which these disparities are processed.

The importance of interaural disparities for auditory localization by the barn owl and for the generation of restricted auditory spatial fields in MLD is confirmed by the effects of monaural occlusion on both behavioral localization (Knudsen and Konishi 1979) and single unit RFs (Knudsen and Konishi 1980). Such occlusion causes a shift of RF location in *both* elevation and azimuth (and corresponding errors in both dimensions in behavioral localization), indicating that interaural disparities provide the owl with cues for both dimensions of space (Knudsen

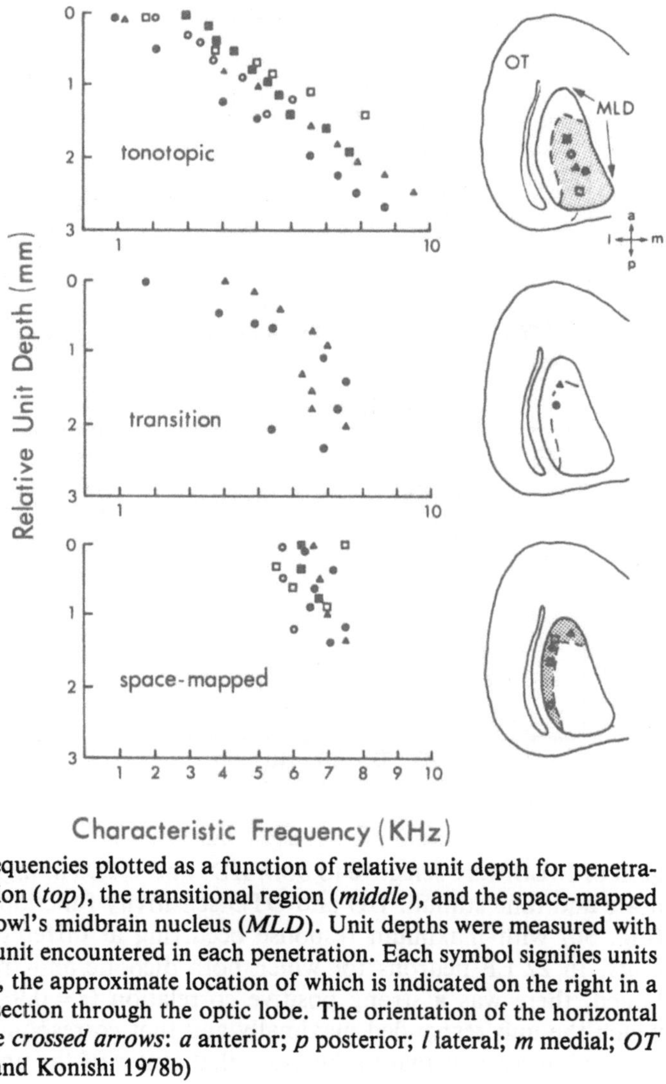

Fig. 49. Characteristic frequencies plotted as a function of relative unit depth for penetrations in the tonotopic region (*top*), the transitional region (*middle*), and the space-mapped region (*bottom*) of barn owl's midbrain nucleus (*MLD*). Unit depths were measured with respect to the first MLD unit encountered in each penetration. Each symbol signifies units from a single penetration, the approximate location of which is indicated on the right in a diagram of a horizontal section through the optic lobe. The orientation of the horizontal section is indicated by the *crossed arrows*: *a* anterior; *p* posterior; *l* lateral; *m* medial; *OT* optic tectum. (Knudsen and Konishi 1978b)

and Konishi 1979, 1980). Knudsen and Konishi (1979) identified the major cue for elevation as IIDs in high-frequency components of the sound, which are produced as a consequence of the vertical asymmetry in the location of the openings of the ear canals (Knudsen and Konishi 1979, 1980; Konishi 1973b; Payne 1971). Although Knudsen and Konishi (1979) proposed that *onset* ITDs (TTDs) were used by the owl to determine azimuthal location, subsequent behavioral and electrophysiological studies by Moiseff and Konishi (1981, 1983a) have established that the major cue for azimuthal location is provided by OTDs. Moiseff and Konishi (1981) developed a technique that allowed them to shift from free-field to dichotic stimulation conditions during recording from a single neuron. Under dichotic stimulus conditions, individual neurons in the space-mapped re-

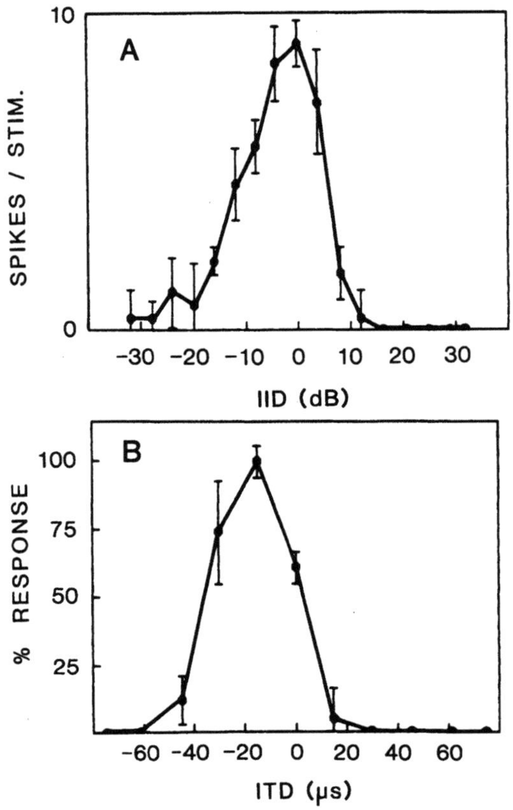

Fig. 50A,B. Representative **A** IID- and **B** ITD-sensitivity functions for a spatially selective "limited field" neuron in MLD of the barn owl. *Points* and *error bars* in each plot are means and standard deviations of responses to five presentations of a band-pass noise stimulus. In **A**, IID was varied (ABI = 51 dB) while ITD was held constant at −15 µs (optimal value shown in **B**). Negative IID values indicate that the contralateral stimulus was more intense. In **B**, ITD was varied while IID was held constant at 0 dB (optimal value shown in **A**), with ABI = 46 dB. Negative ITD values indicate that contralateral stimulus was leading. (Redrawn from Moiseff and Konishi 1983b)

gion of MLD were sensitive to *both* IIDs and OTDs in high-frequency band-pass noise and tone stimuli. IID and OTD sensitivity functions were sharply peaked (Fig. 50) with maximum response occurring at different values for different units. In 22 LF neurons for which both dichotic and free-field data were obtained, there was a strong positive correlation ($r = 0.93$) between the ITD at which the unit responded maximally and that corresponding (as established by physical measurements) to the azimuthal center of the unit's RF determined using free-field stimulation. None of the LF units were sensitive to TTDs. These data establish clearly that the restriction in azimuth and elevation of the spatial RFs of LF units reflects sensitivity to restricted ranges of OTD and IID, respectively. The behavioral importance of these interaural disparities was further established by experiments in which dichotic stimuli with either OTDs or IIDs were presented through custom-made "stereo headphones" to awake owls (Moiseff and Konishi 1981, 1983a). Stimuli containing OTDs elicited head orientation responses to an azimuth appropriate to the stimulus disparity, while stimuli containing IIDs elicited orientation responses to appropriate elevations.

All neurons in the space-mapped region of MLD have sharply peaked sensitivity functions for *both* IID and OTD, and respond *only* to binaural stimuli with the appropriate combination of disparities (Fig. 50; Moiseff and Konishi 1981, 1983a, b). In terms of the classification system used previously, all of these units would be classified as OO/F, the binaural facilitation depending on the appropriate disparity values[21]. Moiseff and Konishi (1983 b) have also reported an extensive study of the patterns of binaural input and of sensitivity to IID and OTD exhibited by neurons in lower brainstem nuclei of the barn owl. Their results indicate that the binaural properties and disparity sensitivity of MLD units differ qualitatively from those of units in lower brainstem nuclei, and that there are separate brainstem pathways for IID and OTD processing. The only lower brainstem neurons sensitive to OTD were cells receiving EE binaural input (although not all such cells were sensitive), but the data presented do not indicate whether the delay sensitivity functions of the cells were periodic. EI cells in lower brainstem nuclei were sensitive to IIDs and exhibited monotonic sensitivity functions of the type that have been described in previous sections for mammalian brainstem neurons. No neurons in lower brainstem nuclei were sensitive to *both* IID and OTD, and the nuclei containing neurons sensitive to each disparity comprised separate pathways. Unfortunately, no data were presented on the properties of neurons in the tonotopic medial region of MLD. Knudsen (1983 a) has shown that the space-mapped region of MLD receives almost all of its extrinsic afferents from the medial tonotopic region, and receives no direct input from lower brainstem auditory nuclei. It is consequently unclear whether the emergence of OO/F cells with sharply peaked sensitivity functions for both IID and ITD takes place in the space-mapped region itself or in the medial region of MLD. The spatial RF characteristics of neurons in the tonotopic region suggest the former is the case, but direct evidence is lacking.

The sensitivity of LF neurons in MLD to OTDs in high-frequency (7 – 9 kHz) tones and band-pass noise implies the existence of phase-locking at much higher frequencies in the owl than in mammals. Sullivan (1985) and Sullivan and Konishi (1984) have recently presented evidence for phase-locking at frequencies up to 9 kHz by neurons in the magnocellular nucleus, one of the two separate nuclei comprising the region in the barn owl that is homologous to the mammalian CN. The mechanisms by which phase-locking at such high frequencies is achieved in AN fibers and magnocellular neurons is unknown, but the control procedures reported by Sullivan and Konishi (1984) leave little doubt as to the reality of the phenomenon. The occurrence of phase-locking at these frequencies suggests that high-frequency OTD sensitivity in the owl is based on the coincidence detection mechanism described in the cat and dog (see Sects. 5.2.4 and 7.2.5). That this is so, and that the sharp OTD tuning of MLD neurons is a consequence of the high CF of the cells, is indicated by observations reported by Yin et al. (1986). For ICC units in the cat, they examined the relationship between the sharpness of the peaks of the noise delay functions and a measure of the frequency at which synchronized discharge rate was maximal. They found that the sharpness increased with increasing frequency, and that if the regression line for the cat data was extended to high frequencies, data for the owl fell very close to it.

Sullivan (1985) and Sullivan and Konishi (1984) also obtained evidence of a segregation of phase and intensity coding in the barn owl homolog to CN. Magno-

---

[21] In fact, an unspecified number of units exhibited monaural inhibition of spontaneous activity, and some MLD units were therefore described by Moiseff and Konishi (1983 b) as receiving IO, OI, or II rather than OO input. All of these units showed binaural facilitation.

cellular neurons showed phase-locking up to high frequencies, but had very restricted intensity dynamic ranges. Conversely, neurons in the other major subnucleus, the angular nucleus, exhibited little or no phase-locking, but had significantly greater dynamic ranges. The proposal that the separate brainstem pathways for ITD and IID might derive their input selectively from these phase- and intensity-specific subnuclei is supported by the results of experiments in which lidocaine was injected into one or the other subnucleus during recordings from LF neurons in MLD (Takahashi et al. 1984). Injections into the magnocellular nucleus produced a reversible change in the OTD sensitivity of MLD cells without any change in IID sensitivity, while injections into the angular nucleus had the opposite pattern of effects.

The response properties of neurons in the barn owl's magnocellular nucleus – primarylike PSTHs, phase-locking, and irregular firing patterns (Sullivan 1985) – resemble those of the bushy/spherical cells in mammalian AVCN. As discussed previously, bushy cells in AA project to MSO and are considered to be the origin of an IPD pathway in the brainstem of mammals with low-frequency hearing (see Sects. 4.3 and 5.4). The properties of neurons in the angular nucleus (little or no phase-locking, chopper PSTHs, and regular discharge) resemble those of "chopper" units in mammalian VCN, which have been identified with stellate cells (see Sect. 4.2.5). However, the projections of stellate cells are primarily to IC and to the periolivary nuclei rather than to the major SOC nuclei thought to be involved in IID coding. Sullivan's (1985) suggestion that the magnocellular and angular nuclei in the owl are analogous[22] to different populations of mammalian VCN neurons will be considered further in Sect. 8.2.1.

7.4 Auditory Spatial Receptive Field Characteristics of Mammalian Inferior Colliculus Neurons

The demonstration of an RF-based map of auditory space in the midbrain of the barn owl has naturally raised the question of whether a similar representation might exist in the mammalian auditory system, either at IC or at some other level. This question has prompted a number of recent studies of the responses to free-field stimuli of IC neurons, using hoop systems of the type described by Knudsen and Konishi (1978a, b). Prior to this recent upsurge of interest, the *azimuthal* sensitivity of IC neurons had been studied in the free field by investigators using either a single, movable speaker (Erulkar 1959) or a horizontal array of fixed speakers (Bock and Webster 1974b; Leiman and Hafter 1972). Erulkar's (1959) observations were restricted to variations in click response latency, but Leiman and Hafter (1972) and Bock and Webster (1974b) described variations in response strength as a function of speaker azimuth. Both the latter studies reported that the most common form of azimuthal sensitivity exhibited by IC neurons was characterized by a maximal response at contralateral azimuths and a progressive decline as speaker position was shifted to ipsilateral azimuths. Since

22 The term "analogous" is used here in the sense proposed by Campbell and Hodos (1970, p. 359) to refer to "correspondence between structures or entities due to similarity in function whether or not they can be traced to a stipulated common precursor".

the majority of IPD- and IID-sensitive cells have maximum response at disparities corresponding to contralateral azimuths, many binaural cells would be expected to exhibit sensitivity of this form. High-frequency EO/mon cells with nonsaturating rate–intensity functions would also be expected to show a progressive decline with speaker movement to the ipsilateral side, as a consequence of decreasing intensity at the contralateral (excitatory) ear. Leiman and Hafter (1972) made two important observations that have unfortunately not been pursued in more recent studies. They examined the azimuthal sensitivity of single neurons over a range of frequencies other than CF, and found marked variations in a given neuron's sensitivity as a function of frequency. They also noted that in cells whose response pattern comprised onset and sustained components separated by a pause, sensitivity to azimuthal position was restricted to the sustained component, the onset component being invariant with speaker direction. This differential sensitivity of the two components parallels their differential sensitivity to IIDs in most units with this response pattern (Sect. 7.2.4). Although Bock and Webster's (1974b) data were not obtained under anechoic conditions, they remain the only demonstration of the azimuthal sensitivity of IC units in unanesthetized cats; as the authors point out, the similarity of their data to those reported by Leiman and Hafter (1972) suggests that these properties of IC units are relatively uninfluenced by barbiturate anesthesia.

There have been only three reports describing spatial RFs of IC neurons in the cat, measured under free-field conditions using a hoop system. Interpretation of the data presented in two of these studies is qualified by the fact that observations were made only at near-threshold intensities. Semple et al. (1983) and Moore et al. (1984a) followed the procedure of determining the CF threshold for each unit at the speaker location at which the unit was most sensitive, and mapping the RF using a CF stimulus 10 dB above that threshold. As noted previously (Sect. 2.3), the pinna in the cat has a directional amplifying effect, the direction in which the greatest amplification occurs defining the *acoustic axis* of the pinna (see Fig. 5). Since most IC neurons receive their sole or dominant excitatory input from the contralateral ear (see Table 4), the point of lowest excitatory threshold for most neurons lies on the acoustic axis of the contralateral ear[23]. The procedure of setting the stimulus level at 10 dB above the on-axis threshold therefore has the consequence that, at all but the lowest frequencies at which the pinna provides little amplification, the stimulus falls below threshold when it is moved off

[23] Suga (1964) similarly showed that lowest CF thresholds for most IC units in echolocating bats were obtained around 40° contralateral azimuth, whereas those of CN units were around 40° ipsilateral azimuth. As noted in Sect. 2.3, directionality reflecting that of the pinna has been described in unit and evoked potential thresholds at a number of levels of the bat auditory system (e.g., Grinnell 1963b; Grinnell and Grinnell 1965; Jen 1980; Schlegel 1977; Shimozawa et al. 1974). Fuzessery and Pollack (1984) recently interpreted differences in the spatial locations at which units of different CF in IC of the mustache bat had lowest thresholds as evidence for binaural processing of interaural spectral differences. But the locations at which the lowest thresholds of their units were obtained corresponded closely to the pinna axis at the CF frequencies, and these thresholds almost certainly reflect monaural effects of the pinna's directionality. A test of their hypothesis requires evidence of unit directionality at intensities well above threshold, i.e., at intensities at which binaural interaction is unequivocally occurring.

the axis by the angular displacement at which the amplification is 10 dB lower. Since the amplifying effect of the pinna becomes more directional (i.e., the 10 dB contour is closer to the axis) as frequency increases (see Fig. 5; Calford and Pettigrew 1984; Phillips et al. 1982), it is not surprising that both Semple et al. (1983) and Moore et al. (1984a) found a monotonic decrease in RF size with increasing frequency. Thus, low-frequency cells were "omnidirectional", being activated by stimulation from all positions in the frontal field (equivalent to Knudsen and Konishi's 1978a "space-independent" units). Units with CFs at intermediate frequencies generally had RFs restricted to the contralateral half of frontal space ("hemifield" RFs), while high-frequency units had circumscribed RFs lying on the acoustic axis of the pinna ("axial" RFs). Semple et al. (1983) obtained data for neurons in both ICC and ICX; there was no apparent difference between the spatial properties of neurons in the two areas, except that ICX cells tended to have higher CFs and, consequently, more restricted RFs. Similar RFs had previously been obtained with similar stimulus procedures in primary auditory cortex (area AI) of the cat (Middlebrooks and Pettigrew 1981). In the study by Moore et al. (1984a), information on variation in discharge rate within the RF was obtained; for the majority of cells with large (i.e., hemifield or omnidirectional) RFs, there was a region of maximal discharge within the field, the position of which corresponded with that of the pinna axis. Similar data have been presented for high-frequency neurons in IC of the big brown bat by Jen and Sun (1984): within 10 dB of the on-axis threshold, all of a large sample ($N = 192$) of units had contralateral RFs with best areas on the pinna axis.

As Semple et al. (1983a) and Moore et al. (1984a) themselves concluded, it seems clear that the RFs obtained at near-threshold levels in their studies reflected purely *monaural* effects attributable to the pinna. Although these data therefore provide compelling demonstrations of the directional amplification produced by the pinna, they throw little light on the existence of a map of auditory space in the cat's IC. In a second phase of their study, however, Moore et al. (1984b) examined the effects of increasing intensity on the RFs of ICC neurons. Of 66 neurons studied in detail, 12 were omnidirectional at all intensities tested, while a further 41 became omnidirectional at higher intensities. Only 13 cells (20%) had RFs that did not expand with intensity; most of these had high CF and contralateral hemifield RFs. These RFs were of the type that would be predicted on the basis of the IID sensitivity of EI cells with sharp cutoffs (see Sect. 7.2.4), and Moore et al. (1984b) suggested that they were generated by cells of this type. It was noted in Sect. 7.2.4, however, that although the IID-sensitivity functions of some EI cells are invariant with intensity, the cutoff in others shifts with increasing intensity, generally in the direction of IIDs corresponding to ipsilateral azimuths. Hemifield RFs with invariant boundaries would presumably be produced only by EI cells with invariant IID-sensitivity functions.

Moore et al. (1984b) distinguished different groups of cells with expanding RFs, whose properties could also be related to those of cells defined in dichotic studies. At intensities more than 10 dB above threshold, the cells in one of these groups ($N = 17$; 26% of whole sample), all of which had low CF (median = 2.1 kHz), had RFs with two regions of peak discharge. Neurons with cyclic delay functions would be expected to have two peak discharge regions if the CF period were less than the maximum possible ITD, and Moore et al. (1984b) therefore suggested that the cells with RFs of this sort might be IPD-sensitive cells. The angular distances between the peaks of these RFs, when converted to ITDs, corresponded to the periods of frequencies of 1.2 – 2.8 kHz, a range encompassing the CFs of most units with two-peak RFs, but no attempt was made to correlate CF with peak separation in individual units. Most of the remaining cells had RFs that expanded with intensity and had either monotonic ($N = 16$; 24%) or nonmonotonic ($N = 14$; 21%) rate–intensity functions at different speaker positions. Although these cells showed no evidence of binaural interaction, Moore et al. (1984b) noted that the proportion of such cells (viz., 45%) exceeded the proportion of EO/mon cells in ICC (see Table 4; Sect. 7.2.3). As noted previously, RFs of this sort might also be produced by cells with other patterns of binaural input, e.g., EI cells with weak ipsilateral inhibition (such as that shown by unit 84-50-2 in Fig. 43C) or EE cells with high ipsilateral threshold and occlusive binaural interaction. A final important point noted by Moore et al. (1984b) was that, for all but high-CF units, there was little sensitivity to variations in elevation, the isorate contours being approximately vertical.

The study by Moore et al. (1984b) has provided the only data on the RFs of IC neurons in the cat at levels well above threshold. However, Aitkin and his colleagues (Aitkin et al. 1984a, 1985) have used a hoop system in anechoic space to measure the azimuthal sensitivity of IC units at a fixed elevation (10° or 20° above the horizontal interaural plane) at different intensities. The results of these studies are generally in agreement with those of Moore et al. (1984b). Aitkin et al. (1984a) studied units in ICC of the cat and in ICC and ICX of the brush-tailed possum; there were no marked differences between species or IC subdivisions, and the data were therefore pooled. Approximately half of the units in their sample (total $N = 131$) were omnidirectional; the remainder were directionally sensitive, either responding in the contralateral hemifield and having a sharp medial border, or having a peak response at a contralateral azimuth. Azimuthal selectivity of the former type is that expected from cells with EI binaural input, and some of these cells with spontaneous activity showed inhibition of ongoing activity for ipsilateral speaker positions. For some of these cells the position of the border was invariant with intensity, while for others it moved, presumably reflecting the different effects of intensity variation on the IID sensitivity of different EI cells (see Sect. 7.2.4). Physical displacement of the contralateral pinna shifted the position of the borders of these RFs in a small number of units so tested, as would be expected from the dependence of the IID–azimuth relationship on pinna position (Sect. 2.3).

Neurons with peaked azimuth functions almost all had low CF and were assumed to be sensitive to IPDs. The properties of cells of this type were examined in greater detail by Aitkin et al. (1985) in a subsequent study of the azimuthal sensitivity of neurons with CF below 3 kHz in ICC of the cat. Approximately half the units in their sample (total $N = 204$) had peaked functions, while 39% were classified as omnidirectional and 9% had two peaks. Aitkin (1983) had earlier reported that HRP injections into the area containing low-frequency cells with peaked functions resulted in labeling predominantly in MSO, and had suggested that the responses of such cells reflected sensitivity to IPDs. A number of features of the cell classes described by Aitkin et al. (1985) are concordant with this hypothesis. Cells with very low CF (range 0.13 – 0.8 kHz) were most likely to be omnidirectional; since the period of the delay sensitivity functions of these cells is large (1250 – 7700 μs) relative to the maximum ITD in the cat (350 – 400 μs), little modulation of their response rate would be expected with change in azimuth. For the azimuth-selective units, the sharpness of the azimuthal tuning generally increased with increasing CF, as would be expected on the basis of IPD sensitivity, although there was considerable scatter for units with any given CF. This scatter may reflect in part the fact that the index of tuning sharpness used was a half-width measure based on the steeper medial slope of the functions rather than the true azimuthal response range at 50% of maximum response. As noted previously, higher-frequency delay-sensitive cells would be expected to have azimuthal sensitivity functions with two peaks, but Aitkin et al. (1984b) did not specify the CF range of cells of this type. Further support for the view that the azimuthal sensitivity exhibited by cells in this study was based on IPD sensitivity was obtained in cats in which the pinnae were removed and one or the other meatus could be reversibly occluded. In units examined under these conditions, azimuthal sensitivity was found to reflect azimuth-dependent facilitatory and suppressive interactions in cells with diverse binaural input patterns (mainly EE and EO), characteristics similar to those of IPD-sensitive cells (Sect. 7.2.5). Perhaps the strongest prediction from the IPD-sensitivity hypothesis is that the sharpness of azimuthal tuning of a given unit should increase directly with stimulus frequency. Aitkin et al. (1985) varied frequency for a small number of units, but they used variations over only 100 – 200 Hz and did not directly test this prediction, although their published data are in accordance with it (see their Fig. 5). In three cats in which recordings were made in multiple penetrations, the data suggested a possible topographic organization of azimuthal sensitivity. Units with more peripheral "best azimuths" (defined as the azimuth eliciting maximum discharge) were located most rostrally and units with medial best azimuths were located caudally.

In all of the studies reviewed in this section, the spatial selectivity of IC units has proved to be broadly in accordance with what would be expected on the basis of the properties of binaural cells revealed in dichotic studies. As discussed in Sects. 7.2.4 and 7.2.5, the vast majority of IID- and IPD-sensitive cells in IC respond maximally over disparity ranges corresponding to contralateral azimuths. In accordance with this observation, free-field studies have consistently shown that the overwhelming majority of directionally selective IC units have contralateral RFs. Further evidence for the contralateral representation of auditory spatial location in the midbrain is examined in Sects. 7.5 and 7.6, and an integrated account of this issue is presented in Sect. 8.2.5. Beyond this broad correspondence, many of the detailed features of the azimuthal or spatial selectivity of IC neurons can be related to the detailed features of IID and IPD sensitivity. However, although many of these correlations are compelling, it should be emphasized that they are indirect. There is a need for techniques of the type developed by Moiseff and Konishi (1981) for the barn owl, allowing change from free-field to dichotic stimulation during recording from individual neurons, to be applied to the cat. Only in this way can hypotheses as to the binaural interaction mechanisms underlying azimuthal or spatial selectivity be directly tested.

The major impetus to the recent interest in the spatial RFs of IC neurons has come from the discovery of a RF-based map of auditory space in the barn owl.

The data reviewed in this section provide little evidence for the existence in the mammalian IC of restricted RFs comparable to those in the space-mapped region of the owl's MLD. At high frequencies, restricted fields and the best areas of large hemifield RFs are all located on the acoustic axis of the pinna, and thus do not provide the necessary substrate for a map of auditory space based on RFs or best areas. The azimuthal RFs of low-frequency units are also extremely broad: if the widths of azimuthal tuning curves at 50% of maximum response are estimated by doubling the half-widths reported by Aitkin et al. (1985), the resultant values are distributed homogeneously over a range of 30° – 130°. As noted previously, these values underestimate the true breadth of tuning, because their half-width measurements were based on the steeper medial slopes of the functions.

The conclusion that IC in the cat, at least, does not contain a map of auditory space based on restricted RFs must be qualified, however. In the barn owl, the space-mapped region of MLD is located laterally and anteriorly, in an area that is possibly homologous to part of mammalian ICX. Neurons in the medial region that is possibly homologous to ICC do not have restricted spatial fields or comprise a map. Most of the free-field data for the cat (and other mammals) have been obtained from ICC, and only limited data are available for ICX. Semple et al. (1983) described the RFs at 10 dB above threshold of 32 units in ICX (mainly the lateral portion) in the cat, and Aitkin et al. (1984a) described the azimuthal sensitivity of 18 units in lateral ICX and 10 units in the rostroventral process of IC in the brush-tailed possum. Although their data suggest that the spatial selectivity of neurons in these areas does not differ from that of ICC neurons, more detailed observations on the responses to free-field stimuli of neurons in the various subdivisions of IC that comprise the traditionally defined ICX are needed to establish with certainty whether or not the mammalian IC contains a region comparable to the space-mapped region of the owl's MLD. The implication of RF-based maps is that spatial location is coded in terms of the locus of activation (or of maximal activation) within the structure containing the map. Other forms of representation are possible, however, and in Sects. 7.5 and 8.2.4 alternative models of spatial coding will be considered.

7.5 Auditory Response Characteristics and Spatial Representation in Deep Layers of Superior Colliculus

Anatomical evidence for projections to deep SC (SCD) from auditory brainstem structures is supported by electrophysiological evidence of acoustically responsive neurons in the deep layers (see Gordon 1975 for review). Such cells have been described in the cat (e.g., Gordon 1973; Harris et al. 1980; Hirsch et al. 1985; Rauschecker and Harris 1983; Stein and Arigbede 1972; Syka and Straschill 1970; Wise and Irvine 1983, 1984, 1985) and a variety of other mammals (e.g., rabbit: Aitkin et al. 1972a; Graham et al. 1981; Horn and Hill 1966; Perry 1974; mouse: Dräger and Hubel 1975a, b; chinchilla: Mast and Chung 1973; guinea pig: King and Palmer 1983; Palmer and King 1982, 1983; golden hamster: Chalupa and Rhoades 1977; Stein and Dixon 1979; Tiao and Blakemore 1976; echolocating bats: Jen et al. 1984; Poussin and Schlegel 1984; Shimozawa et al. 1984; Wong 1984; various monkeys: Allon and Wollberg 1978; Jay and Sparks 1984;

Updyke 1974). Acoustically responsive neurons are found throughout the rostrocaudal extent of SCD in the cat (e.g., Hirsch et al. 1985; Middlebrooks and Knudsen 1984; Wise and Irvine 1983, 1985), but appear to be restricted to the caudal region of SCD in rodents (mouse: Dräger and Hubel 1975b; guinea pig: King and Palmer 1983; golden hamster: Chalupa and Rhoades 1977). Although some acoustically responsive cells in SCD are polymodal, responding also to visual and/or somatosensory stimuli (e.g., Meredith and Stein 1983; see Gordon 1975 for review), a number of studies have reported a degree of segregation or clustering of neurons responsive to different modalities (e.g., Gordon 1973; Hirsch et al. 1985; mouse: Dräger and Hubel 1975b). This might reflect the "patchy" terminations characteristic of many projections to SCD (see Sect. 7.1.5). It seems probable that the function of the convergence of various forms of exteroceptive input into SCD might be to provide access to common premotor circuitry rather than to allow polysensory convergence per se (cf. Jay and Sparks 1984; Whittington et al. 1981).

Most of the earlier observations on acoustically responsive cells in SCD were made incidental to more detailed investigations of input in other modalities, with the consequence that data were generally obtained with free-field rather than dichotic stimulation. Despite a lack of precision in the control of acoustic stimuli in these early free-field studies, they established a number of basic features of the auditory responsiveness of SCD neurons. The most important of these were that most neurons responded best to spectrally complex acoustic stimuli rich in high frequencies and had large spatial RFs extending over much of the contralateral hemifield (Gordon 1973; Wickelgren 1971; mouse: Dräger and Hubel 1975a, b; golden hamster: Stein and Dixon 1979; Tiao and Blakemore 1976). Dräger and Hubel's (1975b) illustration of a large contralateral azimuthal RF of this sort in SCD of the mouse is reproduced in Fig. 51A. As noted previously, contralateral hemifield RFs of this sort would be expected to be generated by EI cells sensitive to IID; Dräger and Hubel (1975b) in fact inferred that the medial borders of such RFs were formed by ipsilateral inhibition from the fact that the RFs became omnidirectional when the ipsilateral ear was blocked (Fig. 51B). Gordon (1973; see also Wickelgren 1971) reported that many SCD neurons in the cat responded only (or best) to moving stimuli, but this aspect of their responses has not been examined in subsequent studies.

On the basis of these earlier studies, it is sometimes asserted that SCD contains RF-based maps of auditory, visual, and somatosensory space that are in register with one another (e.g., Harris et al. 1980). But in the case of auditory and visual input this assertion is not well founded in the early data. At best, it involves an *inference* from a *partial* correspondence in small numbers of cells (see King and Palmer 1983 for a detailed discussion of this point), at worst, a misinterpretation of the data. Many auditory RFs extend over the entire contralateral hemifield and are consequently much larger than visual RFs. This point is illustrated by Dräger and Hubel's (1975b) data in Fig. 51A, in which the larger and smaller arcs on the semicircle represent the horizontal extent of the visual RFs of an SCS neuron in the same penetration (small arc) and of the bimodal cell from which the auditory RF was obtained (large arc). The claim of "registration" on the basis of such data seems to be a loose generalization from the fact that RFs in both modalities are contralateral.

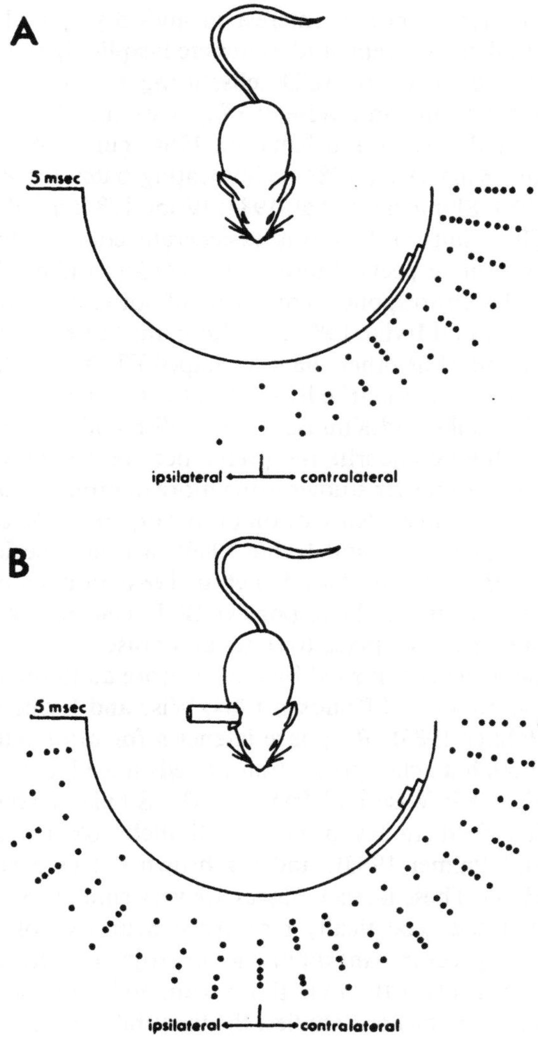

Fig. 51 A, B. Dot diagrams of responses of a trimodal cell in deep layers of right superior colliculus of mouse to click stimulus as a function of azimuth. *Line of dots* at each point on semicircle representing frontal half of space indicates temporal sequence of unit spikes elicited by acoustic stimulus at that azimuth (note time calibration at left). **A** With both ears open, cell responded over the entire contralateral hemifield, but not at all on the ipsilateral side (only anterior 180° are plotted), **B** when the ipsilateral ear was blocked mechanically, the cell responded regardless of azimuth (i.e., was omnidirectional). Large segment inside circle indicates azimuthal extent of visual receptive field (RF) of the cell whose auditory RF is plotted; small segment represents visual RF at surface of colliculus in same penetration. Cell was also excited by somatosensory stimulation of whiskers. (Dräger and Hubel 1975b)

One of the early studies did, however, describe an important correspondence between visual and auditory RFs. Gordon (1973; see also Wickelgren 1971) reported that the *medial edges* of auditory and visual RFs were in register and were topographically organized such that the position of the medial edge shifted from the median plane to larger contralateral azimuths as the recording site moved from rostral to caudal in SC. A similar topographic organization of medial edge position was described by Updyke (1974) in *Cebus* monkeys. This is an important observation and one that will be returned to in discussion of the form of the representation of auditory spatial location later in this section and in Sect. 8.2.4. For the moment, however, it is emphasized that auditory RFs in Gordon's study commonly extended over the contralateral hemifield, with indistinct boundaries other than at their medial edge, and that registration in medial edge position is not equivalent to registration of RFs.

The resurgence of interest in auditory spatial representation has prompted a number of recent studies directed explicitly to examination of auditory response characteristics of SCD cells, using precisely controlled dichotic (Hirsch et al. 1985; Irvine and Wise 1983a; Wise and Irvine 1983, 1984, 1985) or free-field (Middlebrooks and Knudsen 1984; guinea pig: King and Palmer 1983; Palmer and King 1982, 1983; echolocating bats: Jen et al. 1984; Poussin and Schlegel 1984; Shimozawa et al. 1984; Wong 1984; monkey: Jay and Sparks 1984) stimuli. The results of these studies generally confirm, but also extend in a number of important respects, those of the earlier studies. Thus, all acoustically responsive SCD cells respond to broad-band noise, but some are unresponsive to pure tones (Wise and Irvine 1983). Of those units responsive to pure tones, some have broad tuning while others have V-shaped FTCs, but the distribution of CFs for sharply tuned cells is markedly skewed toward high frequencies (Hirsch et al. 1985; Middlebrooks and Knudsen 1984; Wise and Irvine 1983). These characteristics undoubtedly underlie the preference for complex stimuli rich in high frequencies noted in earlier studies using more natural stimuli such as hisses, key jangles, etc. There is no evidence of tonotopic organization of SCD in either the cat (Hirsch et al. 1985; Wise and Irvine 1983) or other species (guinea pig: King and Palmer 1983; big brown bat: Jen et al. 1984; Poussin and Schlegel 1984).

Most acoustically responsive SCD neurons exhibit onset responses comprising one or a few spikes to tone- and noise-burst stimuli, the onset response in some cases being followed by one or more additional spikes (Hirsch et al. 1985; Middlebrooks and Knudsen 1984; Wise and Irvine 1983, 1985; guinea pig: King and Palmer 1983). Response latencies for most cells fall in the range 6–17 ms, although a small proportion of cells have longer latencies (up to 50 ms; Hirsch et al. 1985; Wise and Irvine 1983). Similar latency ranges have been reported in free-field studies in the cat (Middlebrooks and Knudsen 1984), guinea pig (King and Palmer 1983), and big brown bat (Jen et al. 1984; Poussin and Schlegel 1984). These latency ranges are very similar to those for IC neurons, and indicate that the acoustically responsive neurons isolated in these experiments receive their predominant input via the projections to SCD from IC and lower brainstem nuclei rather than via those from auditory cortex.

In recent dichotic studies, the binaural input characteristics and sensitivity to interaural disparities of SCD cells have been studied in detail. Data from three independent samples (Hirsch et al. 1985; Wise and Irvine 1983, 1985) are presented in Table 5. A striking feature of these data is the high proportion of EI cells and the low proportion of EE cells. Although this trend would be expected on the basis of the biased high-frequency representation in SCD, the proportion of monaural cells is much lower, and that of binaural cells showing facilitatory binaural interactions (viz., EO/F and OO/F) much higher, than the corresponding proportions for high-frequency cells in ICC (see Table 4B). As elaborated later in this section, EI cells and the two classes of facilitatory binaural cell show distinctive forms of IID sensitivity, with the consequence that SCD contains a much higher proportion of binaural cells with sensitivity to IIDs (viz., 70% – 80%) than any auditory nucleus other than LSO. The functional implications of this feature of auditory input to SCD will also be considered later in this section.

EI cells in SCD show the form of IID sensitivity characteristic of such cells in other nuclei, viz., maximal response over a range of positive IIDs and total sup-

Table 5. Percentages of neurons in deep superior colliculus of the cat showing various forms of binaural input

Study	Sample size	Binaural category					
		EO/mon	EE	EI	EO/F	OO/F	Other
Hirsch et al. (1985)	80	20	5	56	6	9	4
Wise and Irvine (1983)	165	1	8	58	20	7	5
Wise and Irvine (1985)	103	4	10	55	7	15	9
Combined samples	348	6	8	57	13	10	6

pression over a range of ipsilateral IIDs (e.g., Fig. 52). When IID functions are obtained for broad-band stimuli, the 80% dynamic ranges of most EI cells in SCD are relatively small, i.e., the functions are characterized by sharp cutoffs (Wise and Irvine 1985; Fig. 52). Although quantitative comparisons have not been made, these cutoffs appear to be generally sharper than those of the IID functions for pure tones of most neurons in ICC (D. R. F. Irvine and G. Gago 1984, unpublished work) or in SCD itself (Hirsch et al. 1985). IID functions of SCD neurons for broad-band noise stimuli also appear to be more invariant with changes in ABI (Wise and Irvine 1985) than do tone functions in either IC (Irvine and Gago 1985; gerbil: Semple and Kitzes 1984; kangaroo rat: Stillman 1972) or SC (Hirsch et al. 1985). If these impressions are confirmed by systematic quantitative comparisons, it will be important to determine whether sharpness and invariance of cutoff are characteristics of noise functions generally, and hence also exhibited by IC responses to broad-band stimuli, or whether they are specific to SCD responses. The former alternative finds some support in the invariance with intensity of the delay functions for broad-band stimuli of ICC neurons (Yin et al. 1986; Sect. 7.2.5).

As argued previously, the IID sensitivity of EI cells would be expected to generate large contralateral hemifield RFs, and differences between cells in the location of the IID cutoff would be expected to generate differences between cells in the position of RF medial borders. Gordon (1973) reported that medial border position was topographically organized in SCD, such that rostrally located cells had medial borders located close to the median plane while progressively more caudally located cells had medial borders at progressively greater contralateral azimuths. This topography, together with the hypothesis elaborated previously concerning the mechanisms by which the medial edges of hemifield RFs are generated, leads to the prediction of a rostrocaudal topography in the position of the cutoff in the IID-sensitivity functions of EI cells in SCD. Irvine and Wise (1983 a) and Wise and Irvine (1985) confirmed this prediction in experiments in which multiple dorsoventrally oriented penetrations into SCD were made along the rostrocaudal axis of the nucleus in individual animals. Representative data are presented in Fig. 52, illustrating both variation in cutoff position between cells and its topographic organization. The numbered IID-sensitivity functions in each panel of this figure were obtained from cells isolated in the correspondingly num-

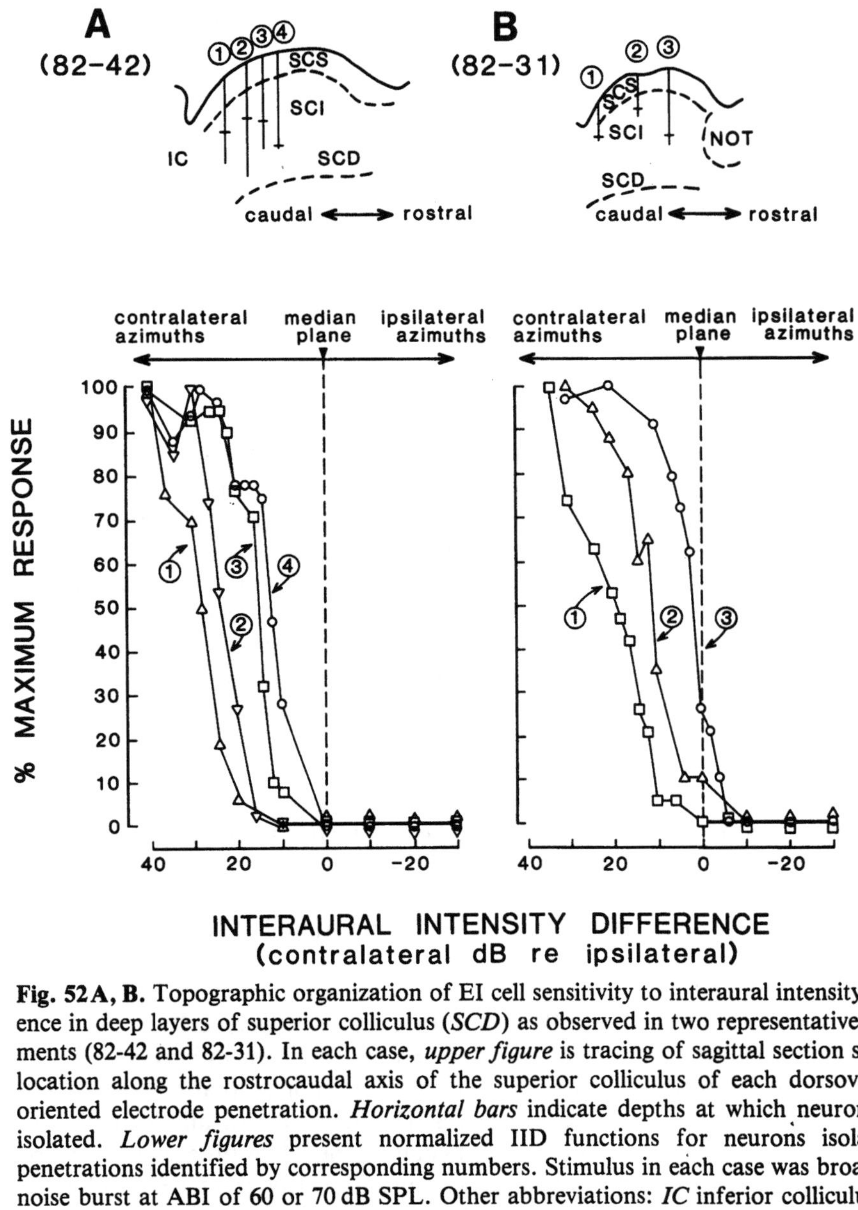

Fig. 52 A, B. Topographic organization of EI cell sensitivity to interaural intensity difference in deep layers of superior colliculus (*SCD*) as observed in two representative experiments (82-42 and 82-31). In each case, *upper figure* is tracing of sagittal section showing location along the rostrocaudal axis of the superior colliculus of each dorsoventrally oriented electrode penetration. *Horizontal bars* indicate depths at which neurons were isolated. *Lower figures* present normalized IID functions for neurons isolated in penetrations identified by corresponding numbers. Stimulus in each case was broad-band noise burst at ABI of 60 or 70 dB SPL. Other abbreviations: *IC* inferior colliculus; *SCS* superficial layers of superior colliculus; *SCI* intermediate layers of superior colliculus; *NOT* nucleus of optic tract. (Wise and Irvine 1985)

bered penetrations in the tracing of a sagittal section through SC shown in the upper panel. In each case, the cell isolated in the most rostral penetration shown has the cutoff closest to zero IID (the median plane), and the cutoff shifts progressively to larger positive IIDs for cells located progressively more caudal in the nucleus. Data in accordance with this topography were obtained in each of nine experiments in which functions were obtained for EI cells at more than one rostrocaudal position. Evidence suggestive of this topography was previously obtained from data pooled across animals by Wise and Irvine (1983) and by Hirsch et al. (1985). As noted in Sect. 7.2.4, a corresponding topography has been described in the large 60-kHz region of ICC in the mustache bat (Wenstrup et al. 1985), but topography of this sort has not yet been demonstrated within a frequency-band sheet in the cat's IC.

Neurons in SCD exhibiting facilitatory binaural interaction also show characteristic forms of sensitivity to IIDs (Hirsch et al. 1985; Wise and Irvine 1983, 1984, 1985). Representative functions for OO/F cells are presented in Fig. 53; the functions are typically sharply peaked and symmetric with maximum response at zero or small positive IIDs. The IID-sensitivity functions of EO/F cells are also peaked, but tend to be asymmetric, with a shallower slope declining to only about 50% of maximal response in the direction of larger positive IIDs (e.g., Fig. 54 full lines). For many EO/F cells, comparison of the IID-sensitivity function with the monaural contralateral intensity function (broken lines in Fig. 54) indicates that IID sensitivity in fact reflects mixed facilitatory and inhibitory interactions (Wise and Irvine 1984). Some EE cells characterized by facilitatory binaural interaction (EE/F cells) also have peaked IID-sensitivity functions (Wise and Irvine 1985), although the peaks are broader than those of OO/F and EO/F cells. Hirsch et al. (1985) have described similar forms of IID sensitivity based on facilitatory binaural interaction, but have argued that these should be regarded as variants of a single class. Although Wise and Irvine (1984, 1985) distinguished different classes of facilitatory binaural interaction, they have suggested that

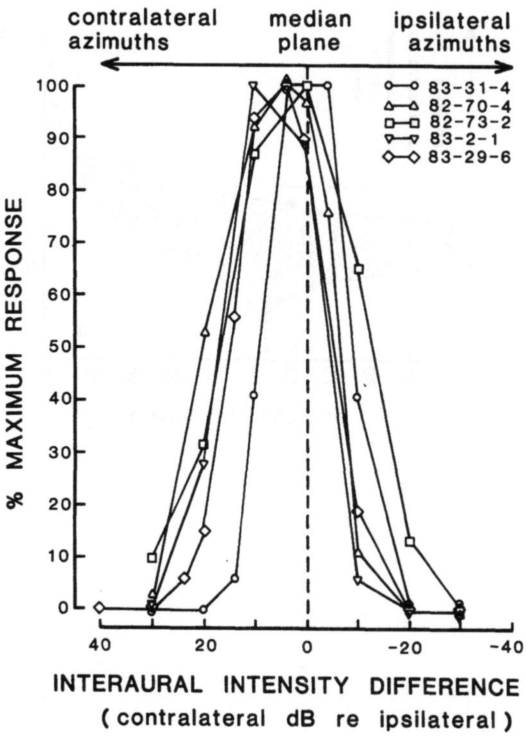

Fig. 53. Representative IID-sensitivity functions for five OO/F cells from deep superior colliculus of cat. Stimulus in each case was broad-band noise at ABI of 60 or 70 dB SPL. (Wise and Irvine 1985)

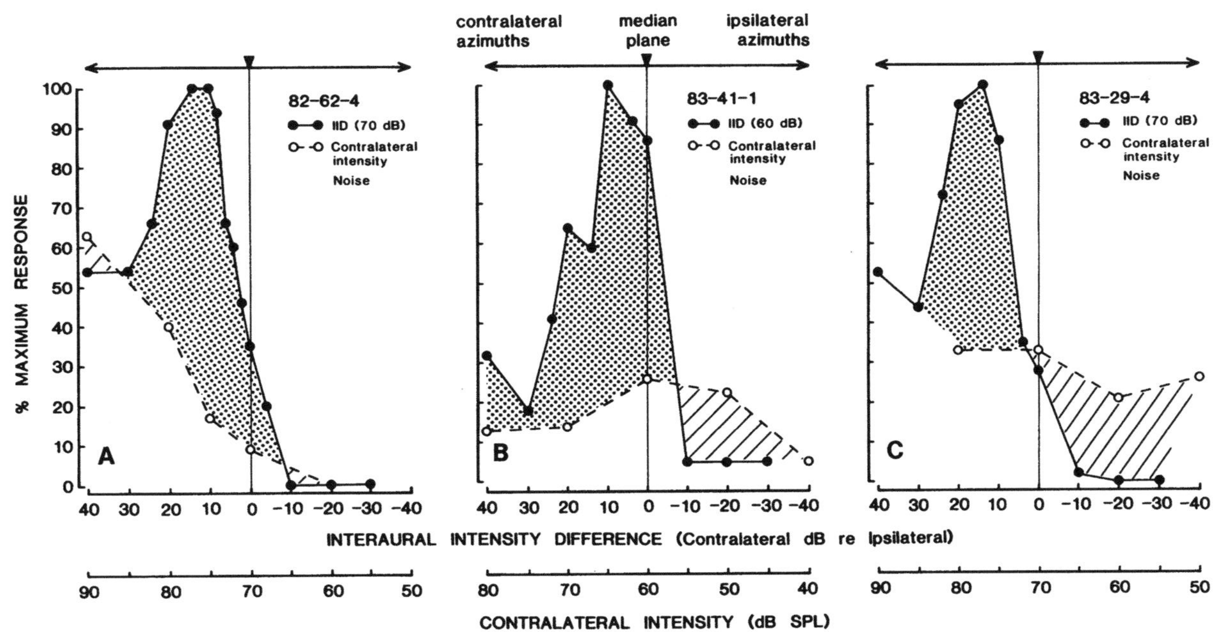

Fig. 54 A – C. Normalized interaural-intensity-difference (IID)-sensitivity functions and monaural contralateral intensity functions for three EO/F cells in deep superior colliculus of cat. In each case, *full circles* (*full line*) show binaural response as a function of IID (*upper abscissa*) for noise stimulus at average binaural intensity indicated, and *open circles* (*broken line*) show response to monaural contralateral stimulation over the range of intensities (*lower abscissa*) involved in generating the IIDs. Monaural response is expressed as a percentage of the maximum response to binaural stimulation. The region in which the binaural response exceeds the monaural response (i.e., region of facilitatory binaural interaction) is indicated by *stippling*; the region in which the binaural response is smaller than the monaural response (region of binaural inhibition) is indicated by *diagonal lines*. Cell in **A** appears to show purely facilitatory binaural interaction, while cell in **B** shows a wide area of facilitation, and inhibition only at large negative IIDs. Cell in *C* shows little or no evidence of binaural interaction at zero IID, but shows marked facilitation at positive IIDs and inhibition at negative IIDs. (**A** Redrawn from Wise and Irvine 1984; **B**, **C** L. Z. Wise and D. R. F. Irvine 1983, unpublished work)

there might in fact be a continuum between cells with sharply peaked IID-sensitivity functions based on binaural facilitation and cells with step-like functions based on binaural inhibition, with intermediate forms of sensitivity involving mixed binaural interactions. As noted previously, IID sensitivity based on facilitatory binaural interaction has also been described in primary auditory cortex of the cat (Kitzes et al. 1980; Phillips and Irvine 1981 b) and in IC of a number of species (Benevento et al. 1970; gerbil: Semple and Kitzes 1984; kangaroo rat: Stillman 1972).

In addition to the topographic organization of EI cell sensitivity to IIDs in SCD, Irvine and Wise (1983 a) and Wise and Irvine (1985) obtained evidence of topography *across* binaural classes. This topography is illustrated in Fig. 55; in all animals in which IID-sensitive cells of different types were isolated, OO/F cells were found in the most rostral penetrations (penetration 3 in Fig. 55 A and B) and EO/F cells were found in penetrations intermediate between those yielding OO/F and EI cells. There is thus a topography from sharply peaked IID-sensitivity functions to step-like functions with cutoffs at progressively larger positive IIDs along the rostrocaudal axis of SCD. Since OO/F cells show the strongest facilitation, and EI cells with lateral cutoffs the strongest inhibition, this topography in IID sensitivity reflects a topography in binaural interaction, from facilitation, through mixed facilitation/inhibition, to increasingly stronger inhibition. It has not yet been established whether this topography in SCD in the cat is unique to this structure or reflects a similar topography in one or all of the structures providing auditory input to SCD. The evidence cited in Sect. 7.1.5 suggesting a possible rostrocaudal topography with respect to the termination of afferents from different auditory brainstem structures provides some support for the former view. The implications of the topographic organization of IID sensitivity in SCD will be examined further in Sect. 8.2.4.

Hirsch et al. (1985) also studied the responses of SCD cells to ITDs. Only a small proportion of cells were sensitive to IPDs in low-frequency tones or in the envelopes of high-frequency AM tones. A larger proportion of cells were sensitive to onset ITDs (TTDs); these cells were generally also sensitive to IIDs and had similar (monotonic or nonmonotonic) functions for each form of disparity. Hirsch et al. (1985) proposed that these data support a "latency model" of IID sensitivity, and developed such a model in detail in a companion paper (Yin et al. 1985). This and other models of binaural interaction are considered in Sect. 8.2.2.

The biased high-frequency representation in SCD, and the fact that the vast majority of SCD units are sensitive to high-frequency cues to spatial location, have interesting functional implications. Given the role of SC in the control of orientation movements of the eyes and ears to auditory targets (see Sects. 7.1.5 and 7.6.2), the physiological data suggest that these components of the orienting response might only be elicited by stimuli with high-frequency components. Although informal observations provide some support for this suggestion, the effects of the spectral composition of the eliciting stimulus do not seem to have been examined in studies of eye and pinna movements to auditory targets. The question is worthy of examination.

Recent studies of SCD units with precisely controlled free-field stimuli have generally confirmed that most SC neurons have large contralateral RFs, but have also provided evidence for the existence of some form of space map in SCD, and

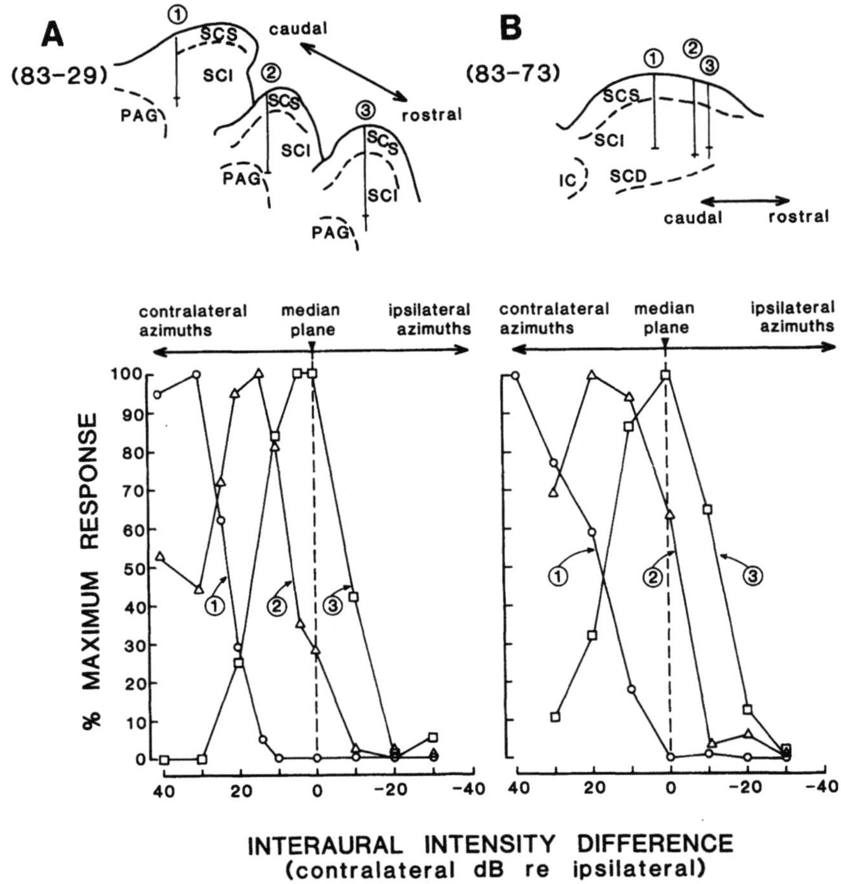

Fig. 55 A, B. Topographic organization of interaural-intensity-difference (IID) sensitivity across binaural cell classes in deep superior colliculus (*SCD*) of cat, as revealed in two representative experiments (83-29 and 83-73). Locations of electrode penetrations are shown in a set of tracings of frontal sections, arranged in caudorostral sequence, in upper part of **A**, and in tracing of sagittal section in upper part of **B**. *Horizontal bars* indicate depths at which neurons were isolated. *Lower figures* present IID functions for cells isolated in penetrations with corresponding numbers. See description of topography in text. Other abbreviations: *IC* inferior colliculus; *PAG* periaqueductal gray; *SCI, SCS* intermediate and superficial layers of superior colliculus. (Wise and Irvine 1985)

will therefore be examined in some detail. Large hemifield RFs have been described in SCD of the cat (Irvine and Wise 1983b; Middlebrooks and Knudsen 1984; Wise et al. 1982) and in echolocating bats (Jen et al. 1984; Poussin and Schlegel 1984; Shimozawa et al. 1984; Wong 1984). In the cat, Middlebrooks and Knudsen (1984) have presented evidence that although the RFs of SCD cells are very large, the centers of their best areas form a representation of auditory space analogous to that in the barn owl MLD (Sect. 7.3) and optic tectum (see below). A small proportion (8%) of their sample ($N = 90$) were omnidirectional, and the remainder exhibited one of two forms of spatial selectivity. Neurons located rostrally in SCD had "frontal" fields, which were located entirely within the frontal half of space, and had azimuthal extents ranging from 25° to 90°. More caudally located cells had even larger (azimuthal extent 100° – 200°) contralateral "hemifield" RFs (54% of sample) with geometric centers within 20° of the interaural axis. The "best areas" of these large RFs (defined by a response greater than 75% of maximum response) were themselves large, ranging from 10° to 50° in azimuthal extent, but the *geometric centers of the best areas* formed a topographic representation of auditory space. Best area centers shifted from frontal to contralateral azimuths along the rostrocaudal axis of SCD and from superior to inferior elevations along the mediolateral axis. The best area centers of bimodal neurons in the intermediate gray layer of SC were reported to correspond to the centers of their visual RFs, although the RFs themselves were not in register.

There is conflicting evidence on the question of topographic organization of spatial selectivity in bat SCD, and discussion is complicated by the variety of measures employed. In the big brown bat, there was no clear evidence of topographic organization in terms of either the azimuth at which unit discharge was maximal (Poussin and Schlegel 1984) or threshold was minimal (Jen et al. 1984; Shimozawa et al. 1984). The latter measure, however, would not be expected to reveal topography, since – as pointed out previously – the azimuth yielding lowest threshold for most units of a given CF is that of the pinna's acoustic axis at that frequency. In the little brown bat, Wong (1984) has presented preliminary evidence that the azimuth eliciting maximum discharge shifts from medial to lateral azimuths along the rostrocaudal axis of SCD, at least for intensities within 30 dB of threshold. Whether this topography would also be present in the geometric centers of best areas as defined by Middlebrooks and Knudsen (1984) cannot be determined from Wong's (1984) data.

The variation between units in the position of RF best areas described by Middlebrooks and Knudsen (1984) in SCD of the cat contrasts with the observation by Moore et al. (1984b) that for most high-frequency ICC units the best areas of RFs for tonal stimuli fell on the acoustic axis of the pinna (see Sect. 7.4). This contrast suggests that there must be significant transformations between responses in the primary pathway and those in SCD and/or between responses to tones and those to broad-band stimuli.

It seems probable that the rostrocaudal azimuthal topography described by Middlebrooks and Knudsen (1984) reflects the rostrocaudal topography in IID sensitivity described by Irvine and Wise (1983a) and Wise and Irvine (1985), although detailed comparisons are not possible because of the nonmonotonicity of the IID – azimuth relationship and the fact that broad-band stimuli provide different IIDs in different frequency bands. Nevertheless, it is clear that rostrally located OO/F and EO/F cells with sharply peaked IID-sensitivity functions would be expected to produce best areas of precisely the sort exhibited by rostrally located cells with "frontal" RFs, and that caudally located EI cells would generate "hemifield" RFs of the type exhibited by caudal cells in Middlebrooks and Knudsen's (1984) study. There is, furthermore, a close correspondence between the proportion of EI cells encountered in dichotic studies of SCD (57%; see Table 5) and the proportion of hemifield RFs reported by Middlebrooks and Knudsen (54%). Although the proportion of their cells with frontal fields (38%) exceeds the combined proportions of OO/F and EO/F cells (23%; Table 5), the correspondence is not unreasonable if the further 8% of EE cells, most of which have facilitatory binaural interactions and peaked IID functions (Hirsch et al. 1985; Wise and Irvine 1985; see earlier discussion) are included. The implications of these correlated free-field and dichotic observations for the manner in which auditory spatial location is represented by populations of SC neurons are examined in Sect. 8.2.4.

Harris et al. (1980) have also argued for a map of auditory space in cat SCD, but on the basis of data that differ significantly from those of Middlebrooks and Knudsen (1984) and of most other studies. They measured auditory RFs of SCD neurons in the awake cat to "quiet" (sic) 1.2-kHz tone bursts, and for a small sample of bimodal neurons reported highly restricted auditory RFs which coincided almost exactly with their visual RFs. These data are puzzling in a number of respects. In terms of all available data on auditory input to SCD, such a low-frequency stimulus would be expected to excite very few cells. Furthermore, the restricted auditory RFs they describe are at variance with the consistent reports in all other studies of extremely large RFs. The available evidence does not allow any resolution of these discrepancies. The paper by Harris et al. (1980) was addressed to an important issue concerning the relationship between visual and auditory spatial representations in SCD which is examined below in connection with the more recent results of Jay and Sparks (1984).

Yet another form of RF-based map of auditory space has been described by Palmer and King (1982) and King and Palmer (1983) in SCD of the guinea pig. They found that SCD neurons had restricted auditory azimuthal RFs for near-threshold noise stimuli, and that the azimuthal position of the RF moved from the median plane to larger contralateral azimuths as a function of rostrocaudal position in the nucleus. For most cells, however, restricted RFs of this sort were seen only within a few dB of threshold. They reported that for 63% of their cells the azimuthal response range expanded with intensity; in the examples they present, such cells have large hemifield RFs at intensities only 10 dB above threshold. Even in the 33% of cells they classified as not expanding with intensity, RFs could be quite large at suprathreshold levels (e.g., the cell for which data are presented in Fig. 5B of King and Palmer 1983 has a RF 45° wide approximately 15 dB above threshold). Auditory RFs at suprathreshold levels in the guinea pig SCD thus resemble those described in most studies of SCD in the cat. In a subsequent series of experiments, Palmer and King (1983) established that the restricted RFs at near-threshold levels reflected monaural, pinna-based effects. Occlusion of the ear ipsilateral to the recording site, or destruction of the ipsilateral cochlea, had no effect on azimuthal RFs at near-threshold levels, but resulted in further enlargement of suprathreshold RFs. Under monaural conditions, removal of the contralateral pinna and concha eliminated the map at near-threshold levels, all units having RFs opposite the contralateral meatus. Although these data are of interest in demonstrating the existence of monaural pinna cues for the azimuthal location of low-intensity broad-band signals in the guinea pig, their implications for more general spatial processing are unclear. Accurate localization of, and orientation to, auditory targets is exhibited by most mammals over a wide range of suprathreshold intensities. Although the effects of SPL on azimuthal acuity appear not to have been determined, it seems unlikely that performance at near-threshold levels is superior to that at suprathreshold levels, and not improbable that it is inferior. Auditory space must therefore be represented in the nervous system in a manner that allows good localization performance at levels well above threshold.

The optic tectum of the barn owl (which is homologous to the mammalian SC) contains a best area map of auditory space (Knudsen 1982, 1984a, b) derived via point-to-point projections from the space-mapped lateral region of MLD (Knudsen and Knudsen 1983). In contrast to the situation in mammalian SCD, acoustically responsive neurons were found in both the superficial and deep layers of the owl's optic tectum, and most neurons were responsive to both auditory and visual stimuli. Although the auditory RFs (and best areas) of bimodal units were much larger than their visual RFs, there was generally a close alignment between the *centers* of auditory best areas and visual RFs, so that the visual and auditory maps were in register except at the nuclear margins where the most peripheral regions of space were represented. Knudsen (1983b) subsequently reported that auditory and visual maps were also in register in birds reared with an attenuating plug in one ear. When the plug was removed, the auditory RFs and the auditory map were shifted. This remarkable plasticity indicates that the spatial location of a tectal neuron's auditory RF must be determined by that of its visual RF, and that the neural connections necessary to produce an auditory RF corresponding to its visual RF are modifiable.

Correspondence between visual and auditory spatial representations, of whatever form, creates a paradox which was pointed out by Pöppel (1973) and has hence been termed "Pöppel's paradox". The paradox arises from the fact that visual RFs are defined within a retinotopic coordinate system, whereas auditory RFs are defined within a head-centered coordinate system. In the barn owl, the eyes are immobile, and these two coordinate systems are equivalent. In animals capable of eye movements, however, the correspondence between auditory and visual RFs and of neural maps based on these different coordinate systems holds only when the eye is centered in the orbit, and is disrupted by eye movements. One solution to this problem would be to convert different sensory inputs into a common coordinate system, thus allowing them to share a common premotor circuit. Sparks and his colleagues (Mays and Sparks 1980; Sparks and Mays 1983; Sparks and Porter 1983) have presented evidence that primate SCD is organized in motor, not sensory, coordinates, and Jay and Sparks (1984) have recently presented evidence for transformation of the coordinate systems of auditory RFs in SCD. They measured the medial edges of auditory RFs in SCD of awake monkeys trained to maintain fixation on visual targets at different azimuthal positions. For all units analyzed ($N = 116$), the position of the auditory RF medial edge was found to shift as a function of eye position. In some cases, the RF edges for a given unit at different eye positions became almost coincident when they were replotted in terms of motor error (i.e., in terms of the horizontal component of the eye movement necessary to shift gaze to the auditory stimulus), indicating a transformation into motor coordinates. An interesting feature of Jay and Sparks's (1984) data is that the mean latency of the responses of these cells to intense (90 dB) acoustic stimulation was about 50 ms. These latencies are much longer than those of most acoustically responsive cells in other studies (viz., 6–20 ms), suggesting that the integration of auditory spatial and eye position information on which the transformation into motor coordinates must depend involves processing at levels well beyond the initial auditory input to SCD. Whether this processing takes place in SCD itself, or in some other structure, is unknown.

The major aim of the study by Harris et al. (1980) discussed earlier was to investigate Pöppel's paradox in the cat, but in contrast to Jay and Sparks (1984) they found no evidence for changes in auditory RFs as a function of eye position. They also recorded eye movements and observed that the cat does not maintain eccentric eye positions, each saccade being followed by a head movement that recenters the eye in the orbit. They therefore concluded that the paradox does not arise for the cat and that transformation of auditory space maps is consequently not required. As noted previously, however, the IID–azimuth relationship and hence the location of those auditory RFs formed by IID sensitivity depend on pinna position, and correspondence between auditory and visual representations would also be disrupted by pinna movements. It seems likely that purely auditory information must be combined with information about pinna position (see discussion in Sect. 8.2.4), but it is also possible – as suggested by some informal observations (e.g., Middlebrooks and Knudsen 1984; W. M. Jenkins, personal communication) – that the alert cat tends to maintain a fixed pinna position.

7.6 Behavioral/Lesion and Electrical Stimulation Evidence on the Role of Auditory Midbrain Structures in Spectral and Spatial Processing

7.6.1 Lesions of Inferior Colliculus or Its Brachium

Evidence on the effects of lesions of IC or BIC on frequency discrimination was reviewed in detail by Neff et al. (1975), and there appears to have been no research on this topic subsequent to their review. Goldberg and Neff (1961 a, b) studied the effects of auditory system lesions on frequency discrimination in cats, using an avoidance conditioning task in which the warning signal was a change in the frequency of an ongoing train of tone pulses from 800 to 1000 Hz. Complete bilateral transection of BIC completely abolished this discrimination; in animals with incomplete sections, the time taken to relearn the discrimination was related to the extent of the section (Goldberg and Neff 1961 b). Since this discrimination can be relearned after total bilateral ablation of all auditory cortex (Butler et al. 1957; Goldberg and Neff 1961 a; Thompson 1960), which results in retrograde degeneration of all or most of auditory thalamus, it appears that transmission of information via BIC is necessary for such discrimination, but that the discrimination can be mediated by structures other than primary auditory thalamus and cortex to which fibers interrupted by a BIC lesion convey auditory input (Goldberg and Neff 1961 b; Neff et al. 1975). Abolition of this discrimination by BIC lesions does not, however, indicate whether IC plays a role in *processing* spectral information, or simply conveys the information to these other structures.

Frequency discrimination under conditions where *identification* of the frequency of a stimulus presented against a silent background is required (i.e., *absolute* discrimination), rather than simply detection of a change in frequency, is abolished in cats with bilateral destruction of IC (Goreva and Kalinina 1967, cited by Neff et al. 1975), and in cats with lesions of all auditory cortex (Thompson 1960; see Aitkin et al. 1984b; Neff et al. 1975 for reviews). Although these data indicate that absolute frequency discrimination depends on the integrity of higher levels of the primary auditory pathway, they once again throw no light on the question of whether significant processing of frequency information takes place in IC. Evidence of *differential* effects of subtotal lesions of IC's afferent and efferent pathways – of the type provided by Masterton et al. (1968) on SOC's role in ITD processing (see Sect. 5.3) – would be necessary to establish that frequency information undergoes some form of transformation in IC.

Although the more extensive data on the effects of IC lesions on auditory localization also fail to establish whether significant processing of spatial information takes place at this level, recent data have provided important information on the nature of auditory spatial representation in the CNS. In earlier studies using the Neff apparatus with two speakers (Sect. 5.3; see Fig. 33), it was found that bilateral section of BIC completely abolished localization performance (Casseday and Neff 1975; Neff 1968; Strominger and Oesterreich 1970). Comparable results were obtained by Masterton et al. (1968) in the lateralization task described in Sect. 5.3: large bilateral lesions of IC abolished discrimination based on ITD (task 3) and severely impaired the simple left – right discrimination (task 2). In the Neff situation, unilateral section of BIC resulted in an initial impairment of performance, but the task was subsequently relearned with elevated angular discrimination thresholds (Casseday and Neff 1975; Strominger and Oesterreich 1970). Strominger and Oesterreich (1970) noted, however, that unilateral section resulted in an increase in errors *contralateral* to the lesion. They suggested that spatial location in one half of space might be represented in the contralateral side of the brain, and that the post-lesion improvement in animals with unilateral lesions reflected the animals' learning of new strategies.

The hypothesis of contralateral spatial representation has been strikingly confirmed in recent experiments by Jenkins and his associates (Jenkins and Masterton 1982; Jenkins and Merzenich 1984) using the seven-speaker situation described previously (Sect. 5.3; see Fig. 33C) and either single click or short tone-burst stimuli (to prevent scanning head movements). Complete unilateral lesions of LL, IC, BIC, MGB, or auditory cortex resulted in total or near-total loss of localization ability for sounds in the contralateral hemifield and little or no impairment for sounds in the ipsilateral hemifield. Representative data for animals with LL and IC lesions are presented in Fig. 56A and B. In each of these cases, as in all other cases with unilateral lesions, no impairment was apparent when tests were made using only two speakers at a fixed 60° separation (data points indicated by stars). In one interesting case in which a large IC lesion spared the ventromedial portion of ICC, no localization deficit was observed. This result indicates that a portion of high-frequency ICC is sufficient to support performance of a task requiring localization of clicks with 30° acuity. This result is not particularly surprising, since the evidence on binaural properties in IC indicates that EI cells sensitive to IID and TTD are concentrated in ventral ICC. The contralateral representation of sound source location is also established by Thompson and Masterton's (1978) evidence that large unilateral lesions at the level of LL or BIC abolish the accuracy of reflexive head orientation to broad-band stimuli presented contralaterally. This reflex response differs from perceptual localization, and

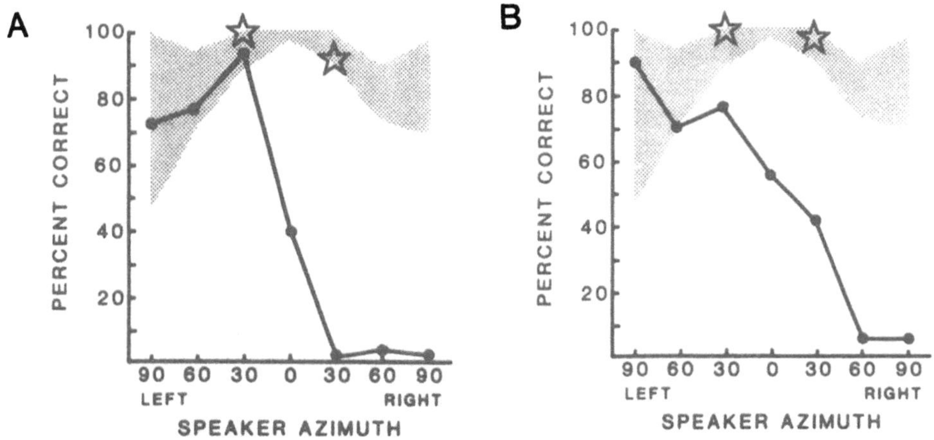

Fig. 56 A, B. Sound localization performance in seven-speaker localization test of two cats with unilateral lesions. **A** Cat with complete lesion of left lateral lemniscus, **B** cat with large lesion of left inferior colliculus and dorsal nucleus of lateral lemniscus. In each panel, *stippled area* indicates range of performance for five normal cats, and *stars* indicate performance levels achieved by the lesioned animals on the two-speaker test. In each case, performance on the seven-speaker task indicates a profound localization deficit in the hemifield contralateral to the lesion, but normal or near-normal localization in the ipsilateral field. Performance on the two-speaker task gives no evidence of any deficit. (Jenkins and Masterton 1982)

Thompson and Masterton's (1978) results indicate that it is mediated in part by nonlemniscal auditory pathways (see further discussion in Sect. 7.6.3).

A detailed account of the effects of auditory cortical lesions on localization is beyond the scope of this review (see Aitkin et al. 1984b; Jenkins and Merzenich 1984; Neff et al. 1975 for reviews). The recent studies by Jenkins and Merzenich (1984) require consideration, however, because they have important implications for auditory spatial representation at lower levels. They examined the effects of cortical lesions in cats on the localization of short tone bursts in the seven-speaker situation. Lesions of a restricted region of the frequency representation in primary auditory cortex (area AI) produced profound localization deficits in the contralateral hemifield for frequencies corresponding to the area of the frequency representation destroyed by the lesion, while localization of other frequencies was unaffected. Conversely, in one animal in which a large lesion was made of all auditory cortex except for a restricted frequency region of AI, contralateral localization deficits were seen at all frequencies except for those represented in the spared region. These results indicate that the region of AI in which a particular frequency is represented is both *necessary* and *sufficient* for the localization of sounds of that frequency in the contralateral hemifield. Given the strict topography that characterizes the tonotopic core of the auditory pathway (as described in earlier sections), the strong implication of these data is that spatial representation occurs *within* frequency channels (cf. Merzenich et al. 1984). If this is the case, it would indicate a fundamental difference between the cat and the barn owl, in which there is a segregation of tonotopic and space-mapped regions in MLD

(Knudsen and Konishi 1978b; Sect. 7.3). The available data do not allow any judgment as to whether the representation in AI of the cat is derived from specifically cortical processing or reflects processing that has already occurred in either ICC or MGB. The latter alternative derives some support from the fact that, at least for current levels of analysis, there do not appear to be any major differences in the sensitivity to interaural disparities or in spatial RF properties of ICC and AI neurons (Sects. 7.2.4, 7.2.5, and 7.4; Aitkin et al. 1984b; Middlebrooks and Pettigrew 1981). Whichever of these alternatives turns out to be the case, it appears that an individual frequency-band sheet in ICC must contain either a representation of contralateral spatial location for that frequency band or the precursors of such a representation. The form of these representations, wherever they might be located, is discussed in Sect. 8.2.4.

It remains to consider a final study of IC lesions which, although is does not relate to either spectral or spatial processing, bears on the functions of the different divisions of IC. Jane et al. (1965) trained cats in a shock-avoidance task, using a combined auditory (300 Hz tone approximately 25 dB above behavioral threshold) and visual (flashing light) stimulus as warning signal, and subsequently observed performance on test trials on which either the visual or auditory stimulus alone was presented. Three of four normal animals responded consistently to the auditory stimulus alone, but almost never to the visual stimulus alone, while the fourth responded consistently to the compound stimulus, but almost never to either stimulus alone. The former pattern of response defines what the authors termed "prepotency" of the auditory stimulus under these test conditions. In the experimental animals, training and testing were carried out *after* bilateral lesions at various levels of the auditory pathway. The absence of pre-lesion training is a critical feature of the experimental design, since the aim is to establish the effects of lesions on determination of which element of the compound stimulus acquires prepotency during training. The performance of animals with large bilateral ablations of auditory cortex or with transections of TB was characterized by prepotency of the auditory stimulus indistinguishable from that seen in the four normal animals. In cats with bilateral lesions of BIC or of the dorsal part of IC, however, the *visual* stimulus was prepotent, almost no responses being made to the auditory stimulus alone. These animals could subsequently be trained to normal performance on the task using an auditory stimulus alone, and had normal or near-normal thresholds, indicating that the loss of prepotency did not reflect a change in auditory sensitivity. It thus appears that prepotency of one or the other modality reflects some form of differential *attention* to that modality. The contrasting effects of midbrain and cortical lesions suggests that this form of attention to auditory stimulation depends on dorsal IC and its projections via BIC (or adjacent pathways affected by the BIC lesion) to areas other than primary auditory cortex or thalamus. The similarity to the structures sufficient for *relative* frequency discrimination, itself perhaps a basically attentional task, is striking. Although the extent of the IC lesions with respect to cytoarchitectonic subdivisions was not described, the reconstructions indicate that they included most or all of ICP and some of dorsal ICC of the traditional parcellation. Jane et al. (1965) reported that the IC-lesioned animals in which auditory prepotency was abolished had normal performance on the ITD-based lateralization task of Masterton et al. (1967, 1968), and Jenkins and Masterton (1982) reported no localization deficit in an animal with a similar dorsal lesion. Taken together, these data suggest a segregation between regions of IC involved in processing and/or transmission of information about spatial properties of the acoustic stimulus and regions involved in attention to the auditory component of a bimodal stimulus, and imply involvement of the areas comprising ICP of the standard parcellation in the latter function. These issues will be considered further in Sect. 7.7.

7.6.2 Electrical Stimulation of Superior Colliculus

The SC was one of the first brain structures to which the technique of electrical stimulation was applied (Adamük 1870), and there is now a vast literature on the effects of such stimulation (see Sparks and Pollack 1977 for review). Most of the literature concerns SC's role in the control of saccadic eye movements and the manner in which saccades are coded (e.g., Robinson 1972; Schiller and Stryker 1972; Sparks and Mays 1983), and would thus appear to be outside the scope of this review. However, as noted previously, accurate saccades are made to auditory targets (human: Zahn et al. 1978; Zambarbieri et al. 1981, 1982; monkey: Whittington et al. 1981). Moreover, electrical stimulation of SC elicits movements of the head and pinnae in addition to eye movements in the cat (Harris 1980; Schaefer 1970; Stein and Clamann 1981; Syka and Radil-Weiss 1971) and in other mammals (rat and hamster: McHaffie and Stein 1982; rabbit: Schaefer 1970; Schaefer and Schneider 1968), and the various movements appear as components of a coordinated orienting response of the type elicited by auditory and other stimuli. Although these observations establish the relevance of the SC electrical stimulation evidence to the brainstem mechanisms of auditory spatial processing, it will not be possible to give more than a cursory review of the relevant literature here.

In the monkey, discrete electrical stimulation in SCD produces contralaterally directed saccades whose amplitude and direction depend on the site of stimulation and are largely independent of initial eye position (e.g., Robinson 1972; Schiller and Stryker 1972). The saccade elicited by stimulation at a given site brings the fovea to the point in the visual field occupied by the visual RFs of units at that site prior to stimulation ("foveation"; Schiller and Stryker 1972). Thus, the horizontal extent of the elicited saccade increases as the stimulation site is shifted from rostral to caudal in SC, and stimulation at medial and lateral sites elicits saccades with up and down components, respectively. This correspondence between the visual and visuomotor maps suggests that saccades are coded solely in terms of retinal error. However, although the vector of an *electrically elicited* saccade in monkeys is independent of initial eye position, recent evidence has demonstrated that saccades to *visual targets* are based on both retinal error and eye position information (Mays and Sparks 1980; Schiller and Sandell 1983; Sparks and Mays 1983; Sparks and Porter 1983).

There is conflicting evidence concerning the nature of saccades elicited by SC stimulation in cats. Stein et al. (1976) reported that the saccades elicited from a given locus in SC were, like those in the monkey, independent of eye position. In contrast, Straschill and Rieger (1973) reported that the direction and amplitude of the saccades elicited from a given stimulation site changed with changes in initial eye position so as to bring the eyes to a constant position (i.e., the eye movements were "goal-directed"). Yet a third alternative was suggested by Roucoux and Crommelinck (1976) and Crommelinck et al. (1977), who reported that saccades elicited by stimulation in the rostral half of SC were independent of eye position, but that those elicited from the caudal half were goal-directed. In all of these experiments, the cat's head was fixed so that only eye movements were possible. As Stryker and Blakemore (1972) have shown, however, cats do not make saccades greater than 20°, and changes of gaze are usually associated with head movements. Harris (1980) examined the effects of SC stimulation in cats whose heads were free to move. Stimulation in the rostral part of the colliculus (representing the central 25° of the visual field) elicited saccades whose amplitude and direction depended on stimulation site and were independent of eye position. These results are thus in accord both with those of Stein et al. (1976), and with those of Roucoux and Crommelinck (1976) and Crommelinck et al. (1977). Stimulation at more caudal sites elicited coordinated eye *and* head movements, but goal-directed eye movements were never seen. Harris's (1980) data accord well with the characteristics of cat oculomotor behavior described by Stryker and Blakemore (1972), but the factors responsible for the conflicting results of earlier studies remain unclear.

In studies in which orientation movements of the pinnae have been elicited by electrical stimulation of SCD, it has generally been reported that the pinnae were directed to the same region as the eyes, i.e., that the pinna movement and eye movement maps are in register (Schaefer 1970; Stein and Clamann 1981; Syka and Radil-Weiss 1971; rat and hamster: McHaffie and Stein 1982). Similar observations were made by Syka and Straschill (1970) with IC stimulation that was shown to produce activation of neurons in SCD. The claim of correspondence between the direction of eye and pinna movements is difficult to interpret given the conflict concerning the nature of the eye movements elicited by SC stimulation. Furthermore, it is not clear that visual or photographic evidence provides an adequate measure of the direction of pinna orientation. It is probable that the function of orienting movements of the ear is to bring the target onto the acoustic axis of the pinna. For targets with high-frequency components, the amplification provided by the pinna for on- or near-axis targets would provide enhanced resolution, so that this process can reasonably be considered analogous to foveation. If this is the case, measurement of naturally occurring or electrically elicited pinna movements requires that the change in the direction of the pinna *axis* be established. Current methods of determining the pinna axis (measurement of tympanic membrane SPL by means of either the cochlear microphonic response or probe microphones) are not suited to dynamic situations, and it would therefore be useful to determine whether any physical feature of the pinna indicates its axial direction.

7.6.3 Lesions of Superior Colliculus

The literature on the effects of lesions of SC is of comparable magnitude to that on SC stimulation (see Goldberg and Robinson 1978; Sparks and Pollack 1977 for reviews) and can also be only summarily treated here. Studies in primates have been concerned almost exclusively with the effects of lesions on saccades to visual targets. Lesions of SC in monkeys affect only the latency of visually elicited saccades (e.g., Mohler and Wurtz 1977), a result apparently at variance with the ascription to SC of a major role in oculomotor control. However, when lesions of SC are combined with lesions of striate cortex (Mohler and Wurtz 1977) or the frontal eye fields (Schiller et al. 1980), which in isolation also have only minor and transient effects, there is a permanent loss of visually guided saccades. These observations suggest that there are two parallel pathways for controlling saccades, and that in normal animals the faster SC pathway is employed (Schiller et al. 1980; Sparks and Mays 1981). There has been no investigation of the effects of lesions on saccades to auditory targets, but it would be of great interest to determine whether there is a parallel pathway for acoustically elicited saccades and, if there is, the cortical or other areas involved in this pathway.

Studies of the effects of unilateral SC lesions in a variety of nonprimate mammals have described a visual neglect of stimuli presented in the contralateral field and a failure to orient to contralateral stimuli of any modality. Most of these studies have focused on visual and visuomotor deficits, and have made only incidental observations on responses to auditory stimuli. There is general agreement, however, that following unilateral SC lesions auditory stimuli on either side of the body produce an arousal reaction, but that there is a failure to orient toward contralateral sounds and, in some cases, a loss of mobility of the contralateral pinna to sound (Sprague et al. 1963; Sprague and Meikle 1965; tree shrew: Casagrande and Diamond 1974; hamster: Schneider 1970). Sprague and Meikle (1965) reported that some time after the lesion there was considerable recovery of orientation to contralateral sounds, but no recovery of pinna mobility.

Although these results accord well with the other data indicating SC's involvement in orientation of receptors to auditory targets, two recent studies concerned explicitly with auditory orientation have given apparently conflicting results. Czihak et al. (1983) reported that unilateral lesions of SC resulted in a massive reduction in what they termed "targeting reactions" to contralateral tone bursts. Although they defined the targeting reaction as a directed movement of ears, eyes, and head toward the stimulus, they did not specify the response measure used. Their results are, however, clearly in accord with the earlier observations. In contrast, Thompson and Masterton (1978) reported that unilateral lesions of SCD had no effect on the accuracy of reflex head orientation to contralateral sound (white-noise burst). Thompson and Masterton's (1978) experiments involved far more precise response measurement than any previous study, and careful delimitation of the extent of the lesions, so this negative result – although isolated – cannot be discounted. It is possible that the orienting response is not in fact a unitary response and that different mechanisms subserve orienting movements of the head and those of the eyes and pinnae. It is also possible that differences in the extent of the lesions or in the lesion-test recovery interval (cf. Sprague and Meikle 1965) might have contributed to the conflict between Thompson and Masterton's data and the results of other studies. Until this conflict is resolved, the nature of SCD's involvement in orientation to auditory stimuli remains unclear.

Tunkl (1980) examined the effects of SC lesions on auditory localization in cats, using the Jenkins and Masterton (1982) seven-speaker version of the Neff apparatus. His lesions were restricted to the superficial layers and had little or no effect on performance, an unsurprising result in view of the exclusively visual connections of these layers.

7.7 Hypotheses Concerning Parallel Processing in Inferior Colliculus

The question of parallel processing at the level of the IC involves two separate issues. The first concerns the segregation within the laminated central nucleus of the projections of the putative parallel processing channels in the lower brainstem that were discussed in previous sections (Sects. 4.3 and 5.4). The second concerns the possibility that the three major divisions of the standard parcellation of IC might be the origins of separate, and to some extent parallel, forebrain auditory systems. These two issues will be considered in turn.

As discussed previously, the properties and connections of neurons in some of the different subdivisions of CN and SOC support a broad distinction between a system comprising rostral AVCN and MSO that is specialized for IPD processing and a system comprising caudal AVCN and MNTB/LSO that is specialized for IID and TTD processing. The projections of these systems to ICC are largely segregated. This segregation is in part a direct consequence of tonotopic organization and the different frequency biases of the two systems. There is also evidence, however, of a degree of segregation of IID- and IPD-sensitive neurons in ICC beyond that consequent on frequency organization (e.g., Semple and Aitkin 1979; Sect. 7.2.3).

ICC also contains a high proportion of monaural cells (Table 4), reflecting the fact that many CN neurons project to IC directly or via VNLL. The function of this large number of monaural cells is not known. As noted in Sect. 4.3, Evans and his colleagues suggested that the projection from DCN to IC was the basis of a pattern recognition pathway; this suggestion was based partly on the complex response properties of DCN neurons and partly on the fact that their projections bypass the binaural integration centers in SOC. However, neurons in many other divisions of CN (almost all except the area of large spherical/bushy cells) also project directly to ICC, and these neurons generally do not have complex response properties. Furthermore, it is by no means certain that significant spectral information is extracted in CN; a number of models that are considered in more detail in Sect. 8.1 suggest that spectral discrimination involves processing of the pattern of synchronized rate across the tonotopic array, and that this processing takes place central to CN, perhaps in SOC.

It is, of course, entirely possible that the large group of cells characterized as monaural is functionally heterogeneous, in ways that relate to their diverse sources. To state this is merely to acknowledge that we have no real evidence on the functional significance of monaural cells or of those EE cells that are not delay sensitive. Although it is tempting to assume that these cells have a specialized role in spectral processing, it is not clear that IID- and delay-sensitive cells could not also contribute to the spectral representation of the contralateral sounds by which they are most strongly driven. Information on the degree of independence of spectral and spatial processing in the auditory system – perhaps provided by some form of selective adaptation procedure (e.g., Regan and Tansley 1979) – would be extremely valuable as a basis for consideration of the functions of those cells that do not appear to contribute to spatial coding. It is not clear how this question could be readily resolved by lesion studies. For example, even if spectral processing were performed *only* by monaural and delay-insensitive EE cells, unilateral lesions would be expected to produce a greater loss of spectral discrimination of sounds in the contralateral than the ipsilateral hemifield, since most monaural and EE cells receive their sole or stronger input, respectively, from the contralateral ear. It must therefore be concluded that evidence from ICC provides no information additional to that from lower levels concerning the existence of parallel processing channels in the brainstem. These issues are considered more generally in Sect. 8.

The second aspect of the question of parallel processing in IC derives from the marked differences in neuronal architecture, afferent and efferent connections, and physiological response properties between the three divisions of the standard parcellation (cf. Aitkin 1979). The laminated central nucleus is the major recipient of ascending lemniscal axons, exhibits precise tonotopic organization, and projects primarily to the laminated ventral division of MGB, which in turn is the major source of input to the tonotopic auditory cortical fields. ICC is also the major midbrain source of projections to nuclei in the lower brainstem auditory pathway. Conversely, ICP apparently receives no ascending lemniscal auditory projections, but derives its input predominantly via commissural projections from contralateral IC and descending projections from auditory cortex. It projects most strongly to the dorsomedial division of MGB, which in turn projects predominantly to the nontonotopically organized cortical area AII. Although ICX receives sparse lemniscal afferents, its major auditory input is derived from ICC and it also receives somatosensory afferents. Its major ascending projections are to the medial and deep dorsal divisions of MGB. The descending projections of ICP and ICX are to structures outside the auditory brainstem pathway, particularly to the pontine nuclei and to SCD.

Although the different divisions of IC cannot strictly be regarded as acting in parallel, since ICX derives its major input from ICC, and ICP its major input from auditory cortex, the differences summarized here strongly suggest that they are the sources of largely distinct ascending and descending systems. Merzenich and his colleagues have emphasized this in proposing a distinction between "cochleotopic" and "diffuse" systems in the auditory forebrain (Andersen et al. 1980a, b; Merzenich et al. 1979), a distinction that is in many ways a refinement of Graybiel's (1972, 1973) earlier "lemniscal line/lemniscal adjunct" distinction. In simplified form, the cochleotopic system comprises ICC, ventral MGB, and the tonotopic cortical fields, and is characterized by sharp frequency tuning and strict cochleotopic (tonotopic) organization. The diffuse system comprises ICP, dorsocaudal MGB, and the nontonotopic cortical fields, and is characterized by broad frequency tuning and the absence of cochleotopic organization. The functional significance of this distinction remains unclear, however, primarily because of the lack of knowledge of the functions of the nontonotopic cortical fields and of the "diffuse" system in general. Jenkins and Merzenich's (1984) lesion evidence indicates that AI, and thus the cochleotopic system in general, is critical for the localization of pure-tone stimuli. It also seems likely that the cochleotopic system is necessary for *absolute* spectral discrimination, although the effects of lesions restricted to physiologically defined AI and to frequency regions within it have not been determined (see Aitkin et al. 1984b for discussion of the need for physiological definition of auditory cortical fields and of the paucity of data on the effects of lesions of individual fields so defined). It is therefore generally accepted that the tonotopically organized core of the forebrain auditory system which has its origin in ICC is *necessary* for (at least) basic spectral and spatial processing.

The only clue to the function of the diffuse system is the evidence provided by Jane et al. (1965) on the effects of dorsal IC lesions on the "prepotency" of auditory stimuli in the cat (see Sect. 7.6.1). These data suggest that ICP (and perhaps the diffuse system in general) might be involved in some aspect of attention to auditory stimuli (cf. Aitkin 1979), but this suggestion carries with it all the difficulties of more precise formulation that attach to the concept of attention itself.

The various regions comprising ICX of the standard parcellation are not considered within the cochleotopic–diffuse classification. The ascending projections of ICX are primarily to the medial and deep dorsal divisions of MGB. In the cat and anthropoid primates, the medial division projects mainly to layer I of all auditory cortical fields (Jones and Burton 1976; Niimi and Naito 1974; Wilson and Cragg 1969), in a way that resembles the nonspecific projections of the intralaminar nuclei (e.g., Jones 1981). Although these observations serve to distinguish the system arising from ICX from those with their origins in the other major divisions, they throw little light on the functions of the system.

The fact that ICX is the major source of auditory projections to the pontine nuclei and also projects to SCD suggests a role in the mediation of auditory reflexes and of orienting responses to auditory targets (cf. Aitkin 1979). In the barn owl, in which such a function is particularly clear, spatial selectivity and the space map appear first in the midbrain division that Knudsen (1983a) has suggested might be homologous to ICX, and the spatial representation there is the basis of that in the optic tectum (see Sect. 7.3). The evidence so far available for mammals does not indicate a higher degree of spatial selectivity in ICX than in ICC, but this issue has not yet been thoroughly explored (see Sect. 7.4). It was suggested in Sect. 7.5 that the IID functions of SCD neurons appear to have sharper cutoffs and to be more stable with variations in other stimulus parameters than those of ICC neurons, and it is possible that such changes (if confirmed) might be mediated in part by changes in ICX. It therefore remains a possibility that ICX plays a role in the cat in the derivation of the representation of auditory spatial location on which orientation to acoustic targets is based. The function of somatosensory input to ICX is not encompassed by this potential function (since SCD itself receives direct somatosensory projections) and remains to be established.

It is clear from this account that although the differences in the neuronal architecture, connections, and physiological response properties of the major divisions defined by the traditional parcellation of IC undoubtedly reflect differences in function, the available evidence does not allow any but the sketchiest account of what these functions might be. Furthermore, each of these major divisions contains multiple subdivisions, defined either morphologically (Morest and Oliver 1984) or, in the laminated central nucleus at least, by the partial segregation of neurons with different patterns of binaural input and different response properties. Clarification of the question of parallel processing in IC will depend, in part, on evidence establishing correlations between morphologically and physiologically defined regions, and on elucidation of the functional significance of different physiological response characteristics.

8 Auditory Brainstem Processing: Integration and Conclusions

8.1 Spectral Processing

Evidence reviewed in Sect. 2 indicates that information about frequency spectrum is conveyed to the auditory central processor in both the spatial and temporal pattern of AN fiber activity. For high-frequency stimuli, it would appear that the only information provided to the CNS is in terms of the distribution of *discharge rate* across the fiber array. At low frequencies, the phase-locking of AN fibers means that temporal information is also available to the CNS, and it seems likely that critical information on the spectral composition of low-frequency sounds with steady state spectra is provided by the distribution of *synchronized activity*. However, despite the extensive anatomical and physiological evidence on auditory brainstem mechanisms examined in the body of this review, there is little that can be said with certainty about the manner in which the information about frequency spectrum provided by the auditory nerve is processed in the brainstem.

One of the basic features of the brainstem pathways examined here, and of higher levels of the auditory system (e.g., Aitkin et al. 1984b), is the preservation of sharp frequency tuning and tonotopic organization throughout the "core" nuclei of the pathway. Although it was once considered that frequency tuning became sharper at higher levels (Katsuki 1961), quantitative evidence indicates no significant difference in sharpness of excitatory tuning curves between VCN, ICC, ventral MGB, and cortical area AI (Calford et al. 1983). Precise tonotopic organization in the core nuclei of the subcortical pathways is generally associated with fibrodendritic lamination (see particularly Sects. 5.2.2 and 7.2.1), as originally proposed by Morest (1964c). Although tonotopy and sharp frequency tuning are sometimes discussed as if their mere existence resolved the problem of spectral processing, it is clear that they do no more than provide the framework within which spectral representation takes place. This point is established unequivocally simply by considering how little the fact of tonotopy in AN tells us about the manner in which frequency is represented in the tonotopic fiber array.

Although the available data do little to elucidate the *mechanisms* by which AN frequency spectrum information is processed in the auditory brainstem, they do establish significant constraints on the *sites* at which such processing must occur. These constraints are most clear in the case of place–periodicity information. The phase-locking of AN fibers is preserved in the discharge of AVCN primary-like units (Sect. 4.2.4) and apparently, though the data base is small, in MSO (see Sect. 5.2.2). In ICC, however, there is a dramatic decrease both in the proportion of neurons exhibiting phase-locking and in the upper frequency limit of its occurrence (Sect. 7.2.2). There would seem to be little doubt that information conveyed in the pattern of synchronized discharges across the tonotopic array must be converted into some other code at levels below IC. The position with respect to place–rate information is less clear. The presence in ICC of a large proportion of neurons with sharply nonmonotonic rate–intensity functions at CF, and of cells sensitive to interaural disparities, would have the consequence that the distribution of activity elicited by a given frequency would vary dramatically with changes in intensity and in the azimuthal location of the source. It is conceivable that the central auditory processor might contain mechanisms capable of retrieving frequency information despite lack of invariance in the pattern of activity associated with a particular frequency. Alternatively, frequency spectrum information might be conveyed only in the pattern of discharge rate across those neurons that are insensitive to interaural disparities and have monotonic rate–intensity functions. Only in the case of information coded in terms of synchronized activity, therefore, do the data unequivocally indicate that a major transformation must take place within the brainstem caudal to IC.

As noted in Sect. 3, a major problem for periodicity theories of frequency coding has been to understand how the periodicity could be detected and processed in the CNS. Most models have postulated some form of temporal cross-correlation process, in which the output of a coincidence detector is determined by the temporal correlation between the phase-locked input and a delayed replica of it. The physiological plausibility of such models is established by the evidence that cells in MSO and ICC that are sensitive to IPDs function as coincidence detectors and perform what is essentially a cross-correlation process (Chan et al. 1985; Goldberg and Brown 1969; Yin and Kuwada 1983b; Sects. 5.2.4 and 7.2.5). In the binaural case, the delay between the phase-locked ipsilateral and contralateral inputs to the coincidence detector is determined jointly by the physical ITD and by different delays in the pathways from the two ears to the MSO neuron. Temporal cross-correlation models of frequency coding, however, must postulate some form of parallel delay line to provide the delayed version of a monaural signal. As pointed out by Whitfield (1970), the alternative possibility of cross-correlation with input derived from an internal clock is implausible given that the precision and variety of clock rates required greatly exceed those observed in regularly firing spontaneous units. Some neural cross-correlation models (e.g., Licklider 1951) have postulated that the necessary delays arise from synaptic or axonal delays within chains of neurons, without speculating on their specific locus. Two recent models, however, have suggested particular brainstem mechanisms which might give rise to the required delays. Langner (1981, 1983) has described neurons in MLD (the avian homolog of IC; see Sect. 7.3) of the guinea fowl that exhibit intrinsic oscillations in their responses to constant-amplitude and AM tone signals, and has argued that these oscillations are the source of delayed versions of the response pattern. He has suggested that chopper units in mammalian PVCN might serve a similar function. Loeb et al. (1983) have proposed that delayed versions of a phase-locked signal could be derived via projections to individual coincidence detector neurons from different points on the basilar membrane. In this model, therefore, the basilar membrane acts as a tapped delay line. One version of their model proposes that the MSO neurons known to be sensitive to IPD might also function as coincidence detectors for monaural delayed signals generated in this fashion. An interesting feature of this proposal is therefore a lack of separation of spectral and spatial processing in the lower brainstem pathways (cf. earlier discussion in Sect. 5.4).

Although both of these models invoke known characteristics of auditory brainstem neurons, they remain highly speculative and their physiological implications have yet to be examined in detail. Furthermore, these models are directed only to the mechanisms by which periodicities in the input to an array of coincidence detectors could be converted to a discharge rate distribution across the array. As Loeb et al. (1983) point out, the output of the coincidence detector array would itself have to be analyzed by a pattern recognition network of some kind, i.e., a network of the type required to analyze the discharge rate pattern produced by high-frequency sounds. In terms of the "central spectrum" model of Goldstein and his colleagues (e.g., Goldstein and Srulovicz 1977; Srulovicz and Goldstein 1983) that was discussed in Sect. 3.5, the cross-correlation models deal only with the manner in which the central spectrum is generated. In their model, the pattern recognition process required to identify the frequency spectrum of the sound involves a template matching procedure. We do not, as yet, have even speculative physiological models of the template matching process. However, the psychophysical evidence that the pitch sensation from a two-tone complex can be elicited when the tones are presented separately to the two ears (Houtsma and Goldstein 1972) indicates that the template matching (or other form of pattern recognition) must take place at a level beyond that at which spectral information from the two ears is combined. A further and more general point is that pattern recognition processes do not necessarily involve the postulation of higher-order feature detector neurons of great selectivity. The "recognition" process might simply involve the way in which differences in the distributed pattern of activity are transformed into different patterns of activity in premotor systems or in the way in which they map onto different perceptual experiences (cf. McIlwain 1976). One example of the way in which the former transformation might occur is considered in Sect. 8.2.4.

8.2 Spatial Processing

8.2.1 Are There Parallel Pathways for Spatial Processing in the Brainstem?

At various points in this review, evidence has been examined that indicates a considerable degree of segregation of the brainstem pathways within which interaural phase and intensity differences are processed in the cat (see particularly Sects. 5.4 and 7.7). The initial processing of IPDs is carried out by neurons in MSO which function as coincidence detectors (Sect. 5.2.4) and derive their major input from the large bushy/spherical cells in rostral AVCN which preserve with high fidelity the phase-locked information provided by AN fibers (Sects. 4.2.4 and 5.1.3). Conversely, IE neurons, which are the major processors of IIDs, are concentrated in LSO (Sect. 5.2.3), and LSO and MNTB derive their major input from bushy/spherical and modified bushy/globular cells in AVCN (Sects. 5.1.2 and 5.1.4). In ICC, the ascending projections of MSO and LSO, and consequently the neurons sensitive to IPD and IID, are largely segregated (Sects. 7.1.3 and 7.2.3). Although these data provide strong support for the processing of IPDs and IIDs in largely separate and parallel brainstem pathways, a number of interrelated qualifications must be considered.

The first is that it does not follow from the fact that these disparities are processed in separate pathways that the pathways themselves can be regarded as dedicated IPD and IID processing channels. The evidence examined in the previous section indicates strongly that the preservation of temporal information in the projections from rostral AVCN (and perhaps even the coincidence detector mechanisms of MSO) are involved in spectral processing of low-frequency sounds with steady state spectra. Thus, it would seem possible that the pathway passing from rostral AVCN through MSO to the midbrain should be viewed as a pathway specialized for the processing of both spatial and spectral features of low-frequency sounds. (A similar conclusion was reached in Sect. 5.3 in considering the interpretation of the hedgehog's inability to localize low-frequency sounds.) Similarly, transient ITDs are processed in SOC and IC by the same classes of neuron, and in some cases by the very same neurons, by which IIDs are processed. Indeed, some binaural interaction models (see following section) propose that sensitivity to time differences is the basic mechanism underlying IID sensitivity. Thus, the pathway within which IIDs are processed is not exclusively devoted to this function. As noted earlier, it is not known whether EI cells play a significant role in the spectral processing of high-frequency sounds, but it is at least possible that the pathways passing from AVCN through MNTB and LSO to the midbrain have a role in high-frequency processing that is more general than the coding of IIDs and transient ITDs.

The arguments developed here in connection with the characterization of these pathways have also brought out a related point. Although the binaural levels of these two pathways are characterized by different binaural processing mechanisms, the high degree of segregation between them in large part reflects the frequency organization of the brainstem auditory system and the fact that IID and IPD function as cues for localization in different, and largely nonoverlapping, frequency ranges. In the cat (and other mammals in which both MSO and LSO are well developed – see Sect. 5.1.1), the segregation is sharpened by the partial segregation of high and low frequencies to different SOC subdivisions. It is unfortunate that there appear to be no physiological data on SOC in those primates in which LSO and MNTB are small or absent. It seems likely that in such cases IID and IPD are processed by different classes of cell in a single nucleus (MSO) and are segregated primarily by virtue of their different frequency selectivity.

The only clear evidence for segregation of IPD and IID processing independent of frequency segregation is provided by the barn owl (see Sect. 7.3). The owl uses IPD and IID in the *same* frequency range (5 – 9 kHz) to determine accurately the azimuth and elevation, respectively, of a sound source. Recent evidence (Moiseff and Konishi 1983 a, b; Sullivan and Konishi 1984; Takahashi et al. 1984; Sect. 7.3) indicates that these cues are processed in separate pathways in the brainstem and that these pathways converge at the MLD to produce neurons selective for *both* IPD and IID. As noted in Sect. 7.3, the neurons in the owl's magnocellular nucleus that have been identified as the origin of the IPD-processing pathway have many physiological properties in common with the bushy/large spherical cells in rostral AVCN of the cat that project to MSO and are the origin of the mammalian IPD-processing pathway. These similarities support Sullivan's (1985) proposal that the magnocellular nucleus in the barn owl is analogous to the bushy/large spherical cell area of mammalian VCN. Analogies between the neurons in the angular nucleus that have been identified as the origin of the IID-processing pathway in the owl and cell types in mammalian VCN are less clear.

On the grounds that they exhibited chopper PSTHs and little or no phase-locking, Sullivan suggested that they might be analogous to "chopper" units in VCN, which have been identified with stellate cells. But "choppers" in PVCN do not have the extended dynamic range characteristic of angular neurons (see Sect. 4.2.4), and stellate cells appear to make only a minor contribution to the projections from CN to the SOC nuclei primarily responsible for IID processing in the cat (viz., LSO and MNTB) (see Sects. 5.1.4 and 5.2.2). The cat CN neurons whose properties most resemble angular nucleus neurons would appear to be the "on-type C" units with extended dynamic range that were discussed in Sects. 4.2.4 and 4.2.5, but there is no evidence that these cells project to LSO in the cat.

Reference to the neural coding of elevation in the barn owl emphasizes the lack of evidence on neural mechanisms for elevation coding in mammals. In humans, the major cue for elevation is provided by spectral transformations of high-frequency complex sounds by the pinna. The fact that the amplification produced by the cat's pinna is characterized by frequency-dependent vertical directionality indicates that spectral transformations in complex stimuli would almost certainly provide a cue for elevation. Grinnell and Grinnell (1965) and Fuzessery and Pollack (1984) have pointed out in reference to echolocating bats that such frequency-dependent directional amplification would produce not only monaural spectral transformations, but also interaural spectral differences as cues to elevation. It seems reasonable to assume that information about spectral transformations of this sort would be processed within the brainstem pathway (or pathways) for the spectral processing of high-frequency sounds and would involve the same pattern recognition or template matching processes. Determination of the acuity of elevational localization in the cat and other nonprimates, and of the cues on which it is based, would provide a foundation for investigation of the brainstem pathways involved and their relationship to other pathways.

8.2.2 Models of Binaural Interaction

Models of binaural interaction, and their ability to describe the *psychoacoustic* data, have been examined in detail in the excellent review by Colburn and Durlach (1978). Only a brief account of some general issues relating to models of lateralization (specifically of processing of interaural time and intensity disparities), and of their relationship to *physiological* data, will be provided here. Although most recent models have been developed at the psychoacoustic level, many postulate mechanisms that have relatively clear implications at the single-neuron level.

As Colburn and Durlach (1978) point out, almost all models postulate some form of cross-correlation mechanism, most often realized as a coincidence detector, for the processing of ITDs. This postulate is well supported by physiological data indicating that IPD-sensitive neurons in MSO (Goldberg and Brown 1969; Sect. 5.2.4), DNLL (Brugge et al. 1970; Sect. 6.2), and ICC (Kuwada et al. 1984; Sect. 7.2.5) function as coincidence detectors. The recent evidence of Chan et al. (1985; Sect. 7.2.5) on the sensitivity of low-frequency ICC neurons to the degree of correlation of interaurally delayed noise stimuli supports the view that these neurons, or the neurons from which their input is derived, carry out a cross-correlation process.

Cross-correlation models require an additional mechanism, however, to deal with the effects of IIDs, because the cross-correlation operation is symmetric with respect to input amplitude imbalances (see Colburn and Durlach 1978 for detailed discussion). The major differences between models arise with respect to the nature of the additional mechanisms they postulate. Two general approaches can be distinguished. In his original formulation of the coincidence detection model, Jeffress (1948) postulated that IIDs were converted into time differences in the signals from the two ears as a consequence of the effects of intensity on the latency of AN responses. According to what has subsequently been termed the "latency" hypothesis, time–intensity trading in auditory lateralization therefore reflects an intensity-to-time conversion (or "trade") in the peripheral auditory system (e.g., Deatherage and Hirsh 1959; Hafter and Carrier 1972). Deatherage and Hirsh (1959) offered indirect evidence for this model by demonstrating that psychophysical time–intensity trading relationships for transient stimuli exhibited nonlinearity and intensity dependence that corresponded to those of the latency–intensity function of the AN gross action potential. More direct physiological evidence is provided by the fact that neurons at a number of levels of the auditory pathway have been shown to exhibit similar forms of sensitivity to IID and TTD. Such neurons have been described in SOC (Caird and Klinke 1983; Hall 1965; Sect. 5.2.4), IC (Benevento et al. 1972; Sect. 7.2.5), SCD (Yin et al. 1985; Sect. 7.5), and primary auditory cortex (Kitzes et al. 1980). As noted in discussion of Caird and Klinke's (1983) data in Sect. 5.2.4 (see Fig. 31), and as emphasized by Yin et al. (1985), the ITD range over which some of these cells exhibit sensitivity greatly exceeds the physiological range, but is comparable to the range over which latency varies with intensity change. Such cells would presumably not contribute to the processing of naturally occurring TTDs, and their sensitivity to large TTDs in dichotically presented stimuli could simply reflect the mechanisms underlying IID sensitivity. Yin et al. (1985) have recently proposed a qualitative model of single-neuron sensitivity to IIDs based on data obtained from cells of this type in SCD. According to this model, IID sensitivity depends on temporal coincidence of the inputs from the two ears, in accordance with the latency hypothesis, and not on the amplitude (or rate) of the inputs. If this were the *sole* mechanism of IID sensitivity, *all* neurons sensitive to IIDs would also be sensitive to ITDs over delay ranges reflecting neural latency–intensity relationships. That this is not the case is indicated by the data of Benevento et al. (1970; Sect. 7.2.5) on IC unit responses to interaural disparities in dichotically presented clicks. Of 70 units tested for sensitivity to both IIDs and ITDs (up to values of 3000 μs), 40% were sensitive to both, while 60% were sensitive to one, but not the other (40% ITD only; 20% IID only). Similar data on dissociation of IID and ITD sensitivity have been presented by Altman (1978). These data suggest that while temporal coincidence and summation may underlie IID sensitivity in some neurons, they cannot be the sole mechanisms of such sensitivity.

The alternative to the latency hypothesis is to postulate that IIDs are processed *independently* of ITDs, and that information about IIDs is used to weight the output of the cross-correlation ITD processor. Weighting functions of this sort were originally proposed by Sayers and Cherry (1957) in the first quantitative cross-correlation model of binaural interaction. The apparent inability of the latency hypothesis to deal with all available lateralization data (e.g., Blauert 1983;

Stern and Colburn 1978) has prompted the postulation of independent IID processing and weighting functions in the more recent cross-correlation models of Colburn and his colleagues (Colburn and Latimer 1978; Stern and Colburn 1978) and of Blauert (1980, 1983). The general structure of such models is illustrated by Blauert's (1983) functional diagram in Fig. 57. Note that in this generalized model the output of the binaural coincidence processor in each frequency channel is passed to a "binaural level difference weighting" system. The pattern of outputs across an array of frequency-specific weighting devices is in turn analyzed by some form of pattern recognition process.

Although physiological mechanisms or sites for an independent IID processor have not been proposed in these models, Colburn and Moss (1981) have proposed a neural model of IID processing by IE (or EI) cells. Central to their model is the assumption that the sensitivity of the model neuron to IIDs reflects the relationships between the (intensity-dependent) *rates* of inhibitory and excitatory inputs to the cell from the two ears. This model successfully predicts some of the features of the IID-sensitivity functions of IE neurons in LSO. In contrast to the Yin et al. (1985) model, which ignores the effects of intensity on the rate or amplitude of the inputs from the two ears, the Colburn and Moss (1981) model ignores the effects of intensity on the timing of the inputs. Further evidence is required to determine whether development of one or the other of these models will be able to account for all the data, or whether a model incorporating the effects of IIDs on *both* the rate and timing of inputs from the two ears will be necessary.

The discussion of the separation of IID and ITD processing in models of binaural interaction obviously has many parallels to the discussion in the preceding section of the anatomical and physiological evidence for parallel processing channels in the brainstem. The fact that the most recent and sophisticated psychoacoustically based models (those of Blauert and of Colburn and his colleagues) postulate separate processors for IIDs and *ongoing* ITDs is in accord with the

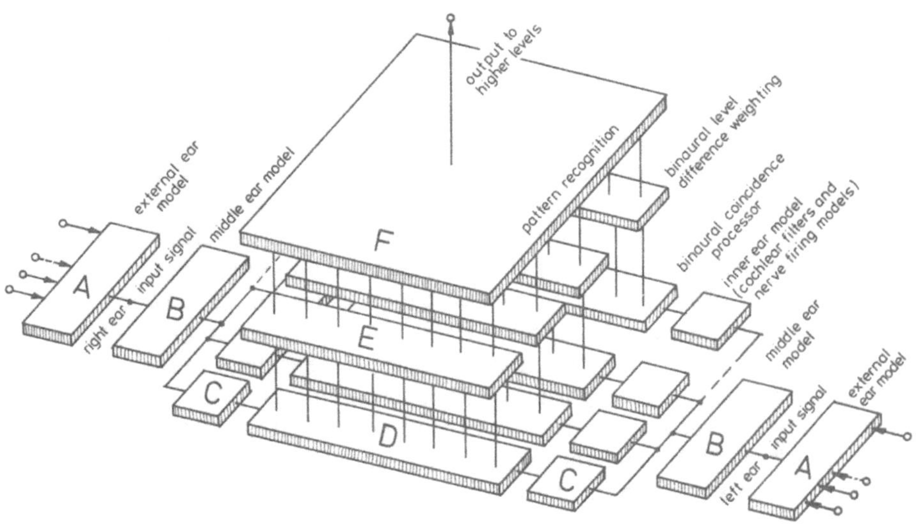

Fig. 57. Functional diagram of a model of signal processing in binaural localization and signal detection. See discussion in text. (Blauert 1983)

evidence for processing of these disparities in separate brainstem pathways. Conversely, and as already noted, the latency hypothesis is in accord with the physiological evidence that IIDs and *transient* ITDs might be processed within a single pathway.

One class of model that has not been considered in the preceding discussion is that described by Colburn and Durlach (1978) as "count-comparison" models, the central premise of which is that the lateral position of a sound source (or of the lateralized image generated by dichotic stimulation) is determined by comparison of the activity levels in two neural populations. The best known of these models is van Bergeijk's (1962) refinement of the earlier model of von Békésy (1930). In terms of the alternative approaches to the processing of IIDs already discussed, both of these models postulate *independent* effects of IID and ITD. The effect of ITD on the amount of activation is mediated by the location within the population of neurons at which inputs from the two ears arrive simultaneously (i.e., by what is essentially a coincidence detection mechanism), while the effect of IID is mediated by the magnitude of the input from the two ears (i.e., by what is essentially a weighting function). In van Bergeijk's version of the model, the two populations were identified with EI cells in MSO on the two sides, and the amount of activation in the two nuclei was said to be compared at some higher level. These aspects of the model are clearly at variance with more recent physiological and behavioral/lesion evidence, most notably with the facts that unilateral lesions at levels above SOC result in localization deficits only in the contralateral half of space, and that each side of the brain at these levels contains a representation of contralateral spatial locations (see Sects. 7.6.1 and 8.2.5). These features are specific to van Bergeijk's model, however, and are not generic characteristics of count-comparison models. Evidence on spatial representation in the auditory brainstem considered in Sect. 8.2.4 suggests that IID and auditory azimuth may be coded in terms of the *relative activation* of different subpopulations of neurons (i.e., by what is essentially a count-comparison process). More generally, as Blauert's functional diagram (Fig. 57) indicates, the output of the cross-correlation processor and IID weighting system in a generalized binaural processing model takes the form of a distribution of discharge rates across an array of processing channels and must also finally be subjected to some form of pattern recognition (count-comparison) process.

8.2.3 The Coding of Interreceptor Disparities in the Visual and Auditory Systems

Fundamental differences in the ways in which the auditory and visual systems derive representations of sensory space are emphasized at a number of points in this review (see Sect. 2.3 and later discussion in Sects. 8.2.4 and 8.2.5). Whereas the azimuth and elevation of a visual stimulus are given by the retinal locus of activation (together with information about eye and head position), those of an auditory stimulus must be derived computationally from interaural disparities and other cues. Information about the *third* dimension of the visual world (viz., depth or distance) is not available directly, however, and must also be derived computationally. The major cue for depth, and the basis for the unique experience of stereoscopic depth perception, is horizontal disparity in the positions of the retinal images in the two eyes, an interreceptor disparity analogous to those on which auditory azimuthal localization is based. Recent evidence on the sensitivity to binocular disparity of neurons in striate cortex of the monkey (Poggio 1979; Poggio and Fisher 1977; Poggio and Poggio 1984; Poggio and Talbot 1981) and cat (Ferster 1981) indicates the existence of mechanisms that bear a striking resemblance to those by which IIDs are processed, as noted previously by Irvine and Wise (1983b).

The two main types of binocular-disparity-sensitive cells described in these studies are illustrated in Fig. 58 by data reported by Poggio (1979) from area 17 in awake rhesus monkeys. The class of cells represented by the unit in Fig. 58A were designated "tuned excitatory", and had sharply peaked disparity sensitivity functions with maximum response at or near zero disparity. As indicated by the level of the monocular responses (circled *L* and *R*), tuned excitatory cells were characterized by marked binocular facilitation. The second major class of cells (designated "near" and "far" cells) is illustrated in Fig. 58B by a "far" cell, which responds maximally over a range of disparities corresponding to stimulus distances farther away than the fixation distance, and minimally over a range of disparities corresponding to nearer distances. "Near" cells have reciprocal sensitivity, and the sensitivity of both variants of this class reflects inhibitory, or mixed inhibitory and facilitatory, binocular interactions.

The similarity of these cell classes to those exhibiting the major forms of sensitivity to IIDs described in this review (Sects. 5.2.3, 7.2.4, and 7.5) are obvious and are illustrated in Fig. 58C and D by data from SCD. Neurons classified as OO/F are characterized by sharply peaked functions with maximum response at or near zero IID that are produced by binaural facilitation (Fig. 58C); in both respects they are remarkably similar to tuned excitatory cells. Similarly, neurons classified as EI have monotonic functions with maximal response over a range of IIDs corresponding to contralateral azimuths and marked suppression at IIDs corresponding to ipsilateral azimuths (Fig. 58D). The form of these functions is the product of inhibitory, or in some cases mixed facilitatory and inhibitory, interactions. Neurons responding maximally to IIDs corresponding to the left and right halves of auditory space are located on separate sides of the brain, but the similarity of the functions and the mechanisms by which they are produced to those of near/far neurons is striking.

Undoubtedly there will prove to be differences between the binocular disparity and IID coding systems. For example, the suggestion that there is in fact a continuum from sharply peaked (purely facilitatory) to step-function (purely inhibitory) IID-sensitivity functions, with intermediate forms produced by various mixtures of facilitation and inhibition (Wise and Irvine 1984; see Sect. 7.5) does not appear to have arisen in the visual case. Whatever differences may emerge, the general similarities described here strongly suggest that the visual and auditory systems have evolved very similar mechanisms for the processing of interreceptor disparities. In terms of the models of IID sensitivity considered in the preceding section, it would be interesting to examine the possibility that sensitivity in both systems involves some form of coincidence detection process.

A final point arises from the fact that the classes of binocular disparity detector described in the studies of cat and monkey striate cortex were originally predicted

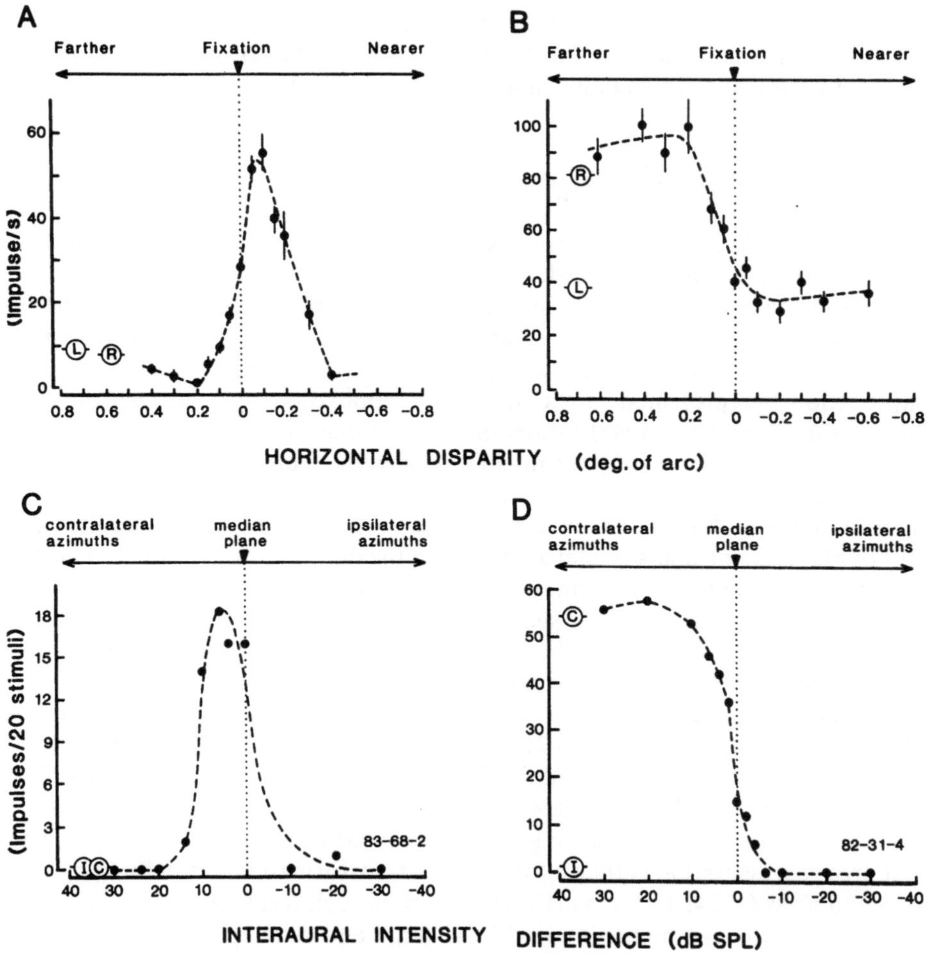

Fig. 58. A, B Depth response profiles of two neurons in primary visual cortex (area 17) of rhesus monkey, illustrating two major classes of binocular-disparity-sensitive neuron. In each case, binocular neural response (mean ± standard error of mean) is plotted against horizontal disparity of the images in the two eyes. Negative values indicate crossed (or convergent) disparities, associated with objects nearer than fixation; positive values indicate uncrossed or divergent disparities. Magnitude of monocular responses is indicated by *encircled L* (left eye) and *R* (right eye). **A** shows the profile characteristic of a "tuned excitatory" neuron; the monocular responses are weak, but there is strong binocular facilitation that is restricted to a narrow region about zero disparity. **B** shows the response profile of a "far" neuron, which is activated by stimuli farther than the plane of fixation (positive disparities) and inhibited by nearer stimuli (negative disparities). (Redrawn, in mirror reversed form to allow comparison with **C** and **D**, from Poggio 1979). **C, D** Representative interaural-intensity-difference (IID)-sensitivity functions for **C** an OO/F and **D** an EI neuron in deep superior colliculus of cat. In each case, magnitude of monaural responses is indicated by *encircled C* (contralateral ear) and *I* (ipsilateral ear). (L. Z. Wise and D. R. F. Irvine 1983, unpublished work)

by Richards (1970, 1971) on the basis of psychophysical studies of anomalies of stereopsis in human subjects. The possibility therefore arises that there might be analogous anomalies in IID discrimination.

8.2.4 The Nature of Spatial Representation in the Auditory Brainstem

The evidence examined in the main body of this review has provided a great deal of information on the way in which binaural cues are processed by individual neurons in the mammalian auditory brainstem, and more limited, but generally concordant, information on the spatial receptive fields of individual neurons. There is very little information, however, on the manner in which auditory spatial location is represented in the activity of populations of such neurons, although since Jeffress's (1948) influential paper it has been widely assumed that the spatial location of a stimulus might be coded in terms of the neural locus of activity according to some form of "place" code. Evidence on the spatial representation in the midbrain of the barn owl (Sect. 7.3) has both supported the concept of place coding and stimulated interest in the question of the form of auditory spatial representation in mammals. Inextricably connected with this issue is the question of the coordinate system within which auditory spatial location is represented. These two issues will be briefly examined in this section, in the context of the evidence presented in the body of the review.

In the space-mapped region of MLD, the receptive fields of individual neurons are relatively large (mean: 70° ×25°), but contain more restricted "best areas" defined by maximal discharge rate within the RF (Sect. 7.3; see Fig. 48B). The best areas represent the contralateral field in a topological manner, neurons at adjacent locations in MLD having best areas in adjacent spatial locations. The best area representation is thus a "first-order transformation" as defined by Allman and Kaas (1974) and Kaas (1977). Although such transformations are commonly described as "point-to-point", the large and overlapping RFs of MLD neurons have the consequence that a sound from a point source in space would activate neurons over a large area of the nucleus. The location of the sound is represented not by activation at a point in the nucleus, but by the point of *maximum* discharge within a large region of activation. Detection of the point of maximum discharge requires some form of comparison of discharge rates across the activated population, that is, it requires a form of pattern recognition analogous to those required to analyze the place–rate pattern for spectral coding[24].

Within the mammalian auditory system, neurons sensitive to interaural disparities generally respond over wide IID and ITD ranges (i.e., are very broadly tuned for IID and ITD), and correspondingly have large azimuthal RFs. Within the primary pathway, best areas have been examined only for neurons in ICC; for many high-frequency neurons, the best area coincides with the acoustic axis of the pin-

[24] Although this argument has been couched in terms of MLD, it applies to all first-order transformations (see McIlwain 1976). Even in those representations based on very restricted RFs (e.g., the striate cortex representations of the fovea or area centralis) there is considerable overlap between the RFs of neurons at adjacent recording sites (cf. Mountcastle's 1978 discussion of partially shifted overlap).

na, and there is little evidence to indicate any topographic organization of best areas (Sect. 7.4). In SCD of the cat, however, Middlebrooks and Knudsen (1984) have reported that although RFs and even best areas are large, the *centers* of best areas provide a first-order representation of contralateral auditory space (Sect. 7.5). As Middlebrooks and Knudsen (1984) have argued, their data suggest that the location of a sound source is represented in SCD by the location of a relatively large region of maximum activity within a broad area of activation.

An alternative "place" model of IID coding in SCD, which can be generalized (with some qualifications) to azimuthal coding, has recently been proposed (Irvine and Wise 1983a; Wise and Irvine 1985). The first stage of this model is based on the variation between EI cells in the position of the cutoff in IID-sensitivity functions, and can be illustrated by reference to Fig. 59, in which normalized functions for a sample of cells are presented. Consider the effects of variation in IID on the response of a population of cells of which this was a representative sample. At zero IID, only the small proportion of cells whose responses are not totally suppressed at all negative IIDs would respond. At progressively larger

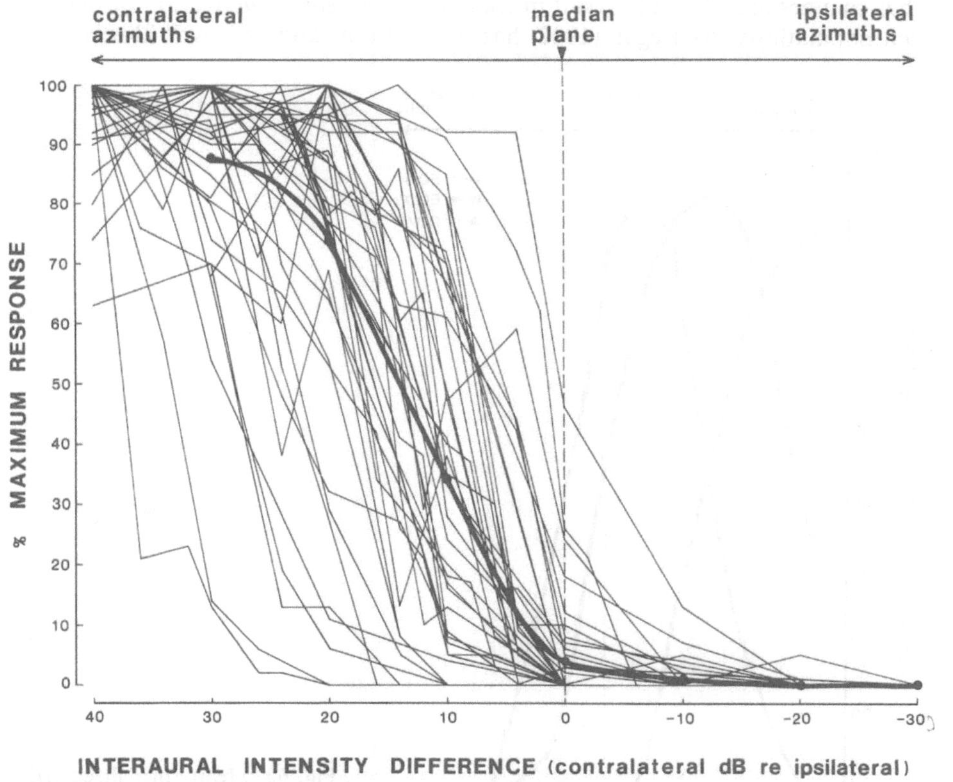

Fig. 59. Superimposed normalized interaural-intensity-difference (IID)-sensitivity functions (*light lines*) for a sample of 43 EI cells in deep superior colliculus of cat. *Full circles* indicate mean normalized responses, and *bold line* is fitted to mean values by eye. For an EI cell population of which the sample was representative, the mean curve would indicate the variation in level of activation across the population as a function of IID. (Wise and Irvine 1985)

positive (contralateral) IIDs, cells already responding would be driven to saturation and there would be a progressive recruitment of cells with more lateral cut-offs, with the result that there would be a progressive increase in the *extent of activation* of the EI cell population. This progressive increase is shown by the mean curve, which would represent the change in response of a population of which this was a representative sample. The topographic organization of EI cell sensitivity in SCD (Sect. 7.5) means that this increase in the extent of activation would involve a caudal extension of the EI cell region activated.

The output of the EI cell population changes very little with IID changes in the region of zero (as reflected in the shallow slope of the mean curve near zero in Fig. 59), so that discrimination based solely on activation of these cells would be poor in this region. As described in Sect. 7.5, however, other classes of IID-sensitive cells (viz., OO/F and EO/F cells) have different forms of sensitivity involving marked changes in response in the region of zero IID (see Figs. 53 and 54). The mean functions for all three cell types in Wise and Irvine's (1985) SCD sample are presented in Fig. 60. It is apparent that in those regions in which the slope of the EI curve is shallow, the slopes of the OO/F and EO/F curves are steep, so that changes in IID would result in changes in the *relative activation* of the three cell populations. As Regan (1982) has pointed out, and as is well illustrated by

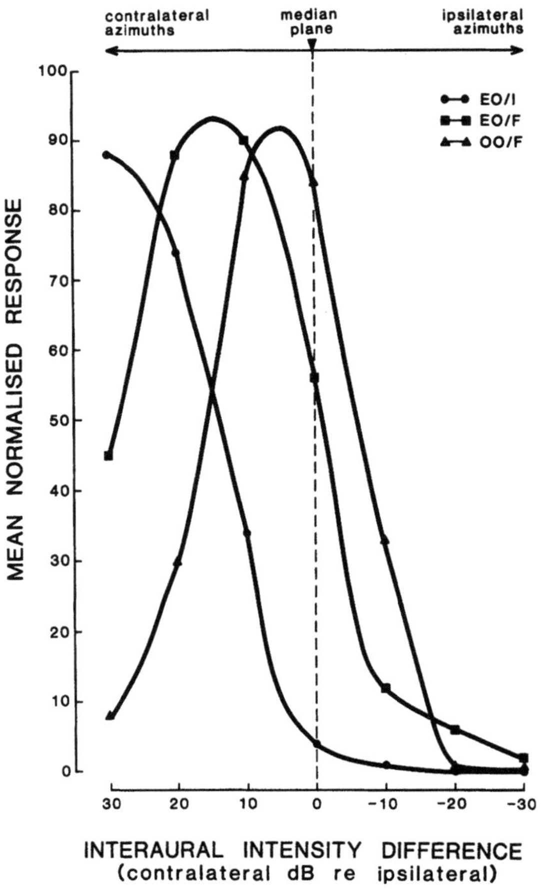

Fig. 60. Mean normalized IID-sensitivity functions (fitted by eye) for samples of EI (here designated *EO/I*), *EO/F*, and *OO/F* cells in deep superior colliculus of cat. (Wise and Irvine 1985)

the familiar example of the retinal coding of wavelength, precise coding and fine discrimination along a sensory dimension can be achieved in terms of the relative activation of subpopulations of neural elements (or, more generally, of filters or channels) with broad and overlapping sensitivity. Richards (1971) has made the same point in arguing that retinal disparity is coded in terms of the relative activation of populations of those types of disparity-sensitive neurons whose similarity to the classes of IID-sensitive cells was discussed in Sect. 8.2.3.

Although the azimuthal RFs of many SCD neurons and high-frequency ICC neurons resemble the IID functions of EI cells (Sects. 7.4 and 7.5), a number of factors preclude the direct generalization of this model of IID coding in SCD to the coding of azimuthal location. The major factors are the complex, frequency-dependent nonmonotonic relationship between IID and azimuth in the cat that was discussed in Sect. 2.3, and the fact that azimuthal receptive fields under free-field conditions can be determined by other factors in addition to IIDs. Nevertheless, the model provides an illustration of the way in which azimuthal location might be represented very precisely in terms of the *relative activation* of subpopulations of neurons with large and overlapping RFs of the types that are observed in the mammalian auditory system.

In this context, it is of interest to note the fact that the peaks of delay sensitivity functions of low-frequency neurons are not uncommonly at delays outside the physiological range (e.g., Kuwada and Yin 1983; Stillman 1971a). Yin and Kuwada (1983b) suggested that the most "characteristic" feature of the delay sensitivity functions of ICC neurons might be the location of the sloping portion of the function. These observations suggest the possibility that IPD might be coded in terms of the extent of activation of the population of delay-sensitive cells.

Both the best area model and the relative activation model require that the distribution of activity over an ensemble of neurons be analzyed by some form of pattern recognition process. The difference between the models is in the feature of the pattern to be recognized. For a simple best area model, it is the region of maximum discharge; since this can only be identified relative to the discharge level elsewhere in the ensemble, the pattern recognition task is essentially one of detecting relative activation *within* a single population. For the alternative model proposed by Irvine and Wise, what has to be detected is the relative activation *across* different subpopulations. Both are therefore "relative activation" models. An advantage of multi-population models is the precision of coding they make possible in regions in which the slopes of the sensitivity functions of the different subpopulations are different (see Regan 1982 for elaboration). But the existence of an advantage of this kind has no implications for the reality of the model.

SCD is one of the few structures in which some hint as to the nature of the pattern recognition process is available. The problem of precise coding despite large regions of activation is also posed by the contrast between the precision of visually and electrically elicited saccades in primates and the large regions of SCD activated by a visual target (because of large RFs) or by microstimulation. This problem has been discussed by McIlwain (1975, 1976, 1982) and by Sparks and his colleagues (e.g., Sparks et al. 1976; Sparks and Pollack 1977), who have argued that saccades must be coded in terms of the distribution of activity across a large number of neurons and that their precision results from the "summation of the movement tendencies produced by the population of neurons" (Sparks and Pollack 1977, p. 209). Thus, according to these proposals, the pattern recognition process is achieved by the superimposition of the spatial pattern of activity onto a distributed system of pulse generators for the oculomotor system (McIlwain 1975, 1976, 1982). The nature of the pattern recognition process in the core auditory system is unknown. As noted in discussion of this issue with respect to spectral processing (Sect. 8.1), it is not necessary to suppose that the output of the pattern recognition process involves single neurons with higher-order feature detector properties.

A final issue, the significance of which is independent of the *form* of auditory spatial representation, concerns the coordinate system within which auditory spatial location is represented. The location of an auditory stimulus is first coded in terms of the properties of the acoustic signals at the eardrums (i.e., in what might perhaps, somewhat fancifully, be termed a tympanocentric coordinate system). The characteristics of the signals reaching the two eardrums from a given sound source vary, however, with the position of the head and body, and – in animals with mobile pinnae – with the positions of the ears. Information about ear, head, and body position must therefore be combined with the original representation to produce a representation of sound source position in spatial coordinates. These processes are analogous to those involved in the transformation of the representation of the location of a visual stimulus in retinotopic coordinates to a representation in spatial coordinates by the addition of information about eye and head position. Nothing is known about where in the auditory system the integrations necessary for the transformation to spatial coordinates take place. The only evidence we have concerning coordinate transformations is the report by Jay and Sparks (1984) of changes in the position of the borders of auditory RFs in SCD of the monkey as a function of changes in eye position. These changes presumably reflect a further transformation specific to SCD and required by the fact that saccades in the monkey are coded in motor error terms (Sparks and Mays 1983; Sparks and Porter 1983).

8.2.5 The Contralateral Representation of Auditory Spatial Location

One of the striking features of the auditory system to emerge from recent behavioral/lesion and electrophysiological studies is the finding that, at levels above the lower brainstem, the spatial location of a sound source is represented on the side of the brain contralateral to that half of space in which the sound is located. Unilateral lesions at levels above SOC result in profound deficits in localization of, or orientation to, contralaterally located sounds, but little or no deficit for ipsilateral sounds (Jenkins and Masterton 1982; Jenkins and Merzenich 1984; Thompson and Masterton 1978; see Sects. 5.3 and 7.6). Similar contralateral deficits have been reported in some clinical studies of humans with unilateral superior temporal (auditory cortex) lesions (e.g., Sanchez-Longo and Forster 1958; Wortis and Pfeffer 1948), but not in others (e.g., Shankweiler 1961; Walsh 1957; see Jenkins and Merzenich 1984; Phillips and Gates 1982 for discussion).

There can be little doubt that the contralateral spatial representation revealed in these studies reflects the characteristics of the IID and ITD sensitivity of binaural neurons. At levels above SOC, almost all neurons sensitive to IIDs have maximum response either at zero IID, corresponding to the median plane, or over some range of IIDs corresponding to contralateral azimuths (Sects. 7.2.4 and 7.5). Similarly, the delay sensitivity functions of the majority of IPD-sensitive neurons are characterized by maxima at delays corresponding to contralateral azimuths (Sect. 7.2.5). For those auditory midbrain and cortical neurons with restricted spatial receptive fields, the restriction is almost always to the contralateral half of space or to a region within it, and the form of the RFs is generally in accordance with what would be expected on the basis of IID or ITD sensitivity (Sect. 7.4). The best area map in MLD of the barn owl is similarly a map of contralateral space and reflects the IID and OTD sensitivity of binaural neurons (Sect. 7.3). Despite the remaining uncertainty (discussed in the preceding section) about the form of the spatial representation (or representations) in mammals, it seems clear that it is those neurons on each side of the brain that are sensitive to interaural disparities that provide a representation of the median plane and of the contralateral half of space.

The representation of contralateral spatial location in the auditory system is analogous to the representation of the contralateral visual field and body surface in the visual and somatosensory systems, respectively. In the latter systems, however, contralateral representation is achieved as a consequence of the fact that a given cerebral hemisphere receives projections only (or predominantly) from those portions of the receptor epithelia that are activated by contralaterally located stimuli. Thus, in the visual system, the partial decussation of optic nerve axons at the optic chiasm results in the projections to one side of the brain being derived from the ipsilateral temporal and contralateral nasal hemiretinae (on which the image of a contralateral stimulus falls). By contrast, auditory nuclei on each side of the brain receive projections from the entire length of the basilar membrane on each side and the great majority of neurons are binaurally influenced. In those neurons that are sensitive to spatial location, it is the *interaction* between the inputs from the two ears that results in the restriction of spatial sensitivity to the contralateral half of space. That is, contralateral spatial representation in the auditory system is derived from integrative neural processing, or computation, rather than from the sorting of projections.

Recently, Masterton and his colleagues (Glendenning and Masterton 1983; Masterton et al. 1981, 1982) have emphasized the importance of TB and SOC in the production of the contralateral representation of auditory spatial location. The fact that schematic illustrations of LSO's ascending projections suggest a chiasm-like arrangement (see Fig. 38C), and the analogy between the contralateral representations in the visual and auditory systems, have prompted them to suggest that TB – SOC serves as an "acoustic chiasm". Although this term provides a dramatic means of emphasizing contralateral representation in the auditory system, there are at least two respects in which it could obscure important differences between the auditory and visual systems. The first, and obvious danger, is that the use of an explicitly *structural* term might obscure the fact that, as elaborated previously, the auditory representation is based on neuronal integrative processes and not merely on the sorting of projections. That this is not a trivial danger is indicated by the uncertainty expressed in some of the papers of Masterton and his colleagues as to the contribution of "ipsilateral pathways" to sound localization (e.g., Masterton et al. 1982).

The second danger is rather more subtle, and relates to the question of precisely what it is that is represented contralaterally in the visual and auditory systems. In the visual system, the consequence of the decussation is that a stimulus in one half of space elicits activity *only* in (i.e., has its *sole* representation in) the contralateral geniculostriate system. That is, not only the spatial location of the stimulus, but also its form and other properties are directly represented (i.e., ignoring commissural connections) only in the contralateral hemisphere. This is not the situation in the auditory system. The properties of those binaural neurons that code for interaural disparities are such that the *spatial location* of an auditory stimulus is represented only in the contralateral half of the brain. In addition to these neurons, however, the auditory system contains other neurons that respond regardless of the spatial location of an adequate stimulus (i.e., have omnidirectional spatial RFs), presumably as a consequence of receiving EE or monaural input. An adequate auditory stimulus at *any* point in space will excite such neurons on both sides of the brain, and the occurrence of the stimulus and at least some of its spectral and other properties will therefore be represented bilaterally. This distinction between the auditory system's contralateral representation of the *spatial location* of a stimulus and the visual system's contralateral representation of *all* features of a stimulus is a fundamental one, and it is obscured by the analogy implicit in the term "acoustic chiasm".

Acknowledgments. Many people have contributed to the preparation of this review. Thanks are due to the many colleagues and friends who provided me with preprints of papers in press and/or in preparation and gave me permission to cite them. I am particularly grateful to those who read and made critical comments on an earlier version of the manuscript: Lindsay Aitkin, Mike Calford, Bill Rhode, Dave Ryugo, Don Sinex, Phil Smith, Lisa Wise. In other matters: Vladimir Kohout and Judy Sack labored mightily in the preparation of figures; Becky Morton, Elana Hartley, and Hazel Woodward gave invaluable bibliographic assistance; Craig Miller and Mike Durham helped me with the many difficulties with software and hardware that beset word processing on a laboratory computer. Ross Day, as Chairman of the Department of Psychology, has supported for many years what to some might seem a "nonpsychological" interest in the auditory system; neither my research nor this review would have been possible without his support and encouragement. Finally, I want to thank Dani for, among other virtues, a remarkable tolerance of her father's absorption in the auditory brainstem over the long gestation period of this review.

9 References

Abbas PJ (1978) Effects of stimulus frequency on two-tone suppression: a comparison of physiological and psychophysical results. J Acoust Soc Am 63:1878 – 1886

Abbas PJ, Sachs MB (1976) Two-tone suppression in auditory-nerve fibers: Extension of a stimulus-response relationship. J Acoust Soc Am 59:112 – 122

Adams JC (1976) Single unit studies on the dorsal and intermediate acoustic striae. J Comp Neurol 170:97 – 106

Adams JC (1979) Ascending projections to the inferior colliculus. J Comp Neurol 183:519 – 538

Adams JC (1983 a) Multipolar cells in the ventral cochlear nucleus project to the dorsal cochlear nucleus and the inferior colliculus. Neurosci Lett 37:205 – 208

Adams JC (1983 b) Cytology of periolivary cells and the organization of their projections in the cat. J Comp Neurol 215:275 – 289

Adams JC, Warr WB (1976) Origins of axons in the cat's acoustic striae determined by injection of horseradish peroxidase into severed tracts. J Comp Neurol 170: 107 – 122

Adamük E (1870) Über die Innervation der Augenbewegungen. Zentralbl Med Wiss 8:65 – 67

Aertsen AMH, Smolders JWT, Johannesma PIM (1979) Neural representation of the acoustic biotope: on the existence of stimulus-event relations for sensory neurons. Biol Cybern 32:175 – 185

Aitkin LM (1976) Tonotopic organization at higher levels of the auditory pathway. In: Porter R (ed) International review of physiology, Neurophysiology II, vol 10. University Park Press, Baltimore, pp 249 – 279

Aitkin LM (1979) The auditory midbrain. Trends Neurosci 2:308 – 310

Aitkin LM (1983) The pathway for low frequency sound localization in the cat hindbrain. In: Webster WR, Aitkin LM (eds) Mechanisms of hearing. Monash University Press, Clayton, p 100

Aitkin LM (1986) The auditory midbrain: Structure and function in the central auditory pathway. Humana, Clifton

Aitkin LM, Boyd J (1975) Responses of single units in cerebellar vermis of the cat to monaural and binaural stimuli. J Neurophysiol 38:418 – 429

Aitkin LM, Boyd J (1978) Acoustic input to the lateral pontine nuclei. Hear Res 1:67 – 77

Aitkin LM, Kenyon CE (1981) The auditory brain stem of a marsupial. Brain Behav Evol 19:126 – 143

Aitkin LM, Phillips SC (1984 a) Is the inferior colliculus an obligatory relay in the cat auditory system? Neurosci Lett 44:259 – 264

Aitkin LM, Phillips SC (1984 b) The interconnections of the inferior colliculi through their commissure. J Comp Neurol 228:210 – 216

Aitkin LM, Schuck DM (1985) Low frequency neurons in the lateral central nucleus of the cat inferior colliculus receive their input predominantly from the medial superior olive. Hear Res 17:87 – 93

Aitkin LM, Anderson DJ, Brugge JF (1970) Tonotopic organization and discharge characteristics of single neurons in nuclei of the lateral lemniscus of the cat. J Neurophysiol 33:421 – 440

Aitkin LM, Fryman S, Blake DW, Webster WR (1972 a) Responses of neurones in the rabbit inferior colliculus. I. Frequency-specificity and topographic arrangement. Brain Res 47:77 – 90

Aitkin LM, Blake DW, Fryman S, Bock GR (1972b) Responses of neurones in the rabbit inferior colliculus. II. Influence of binaural tonal stimulation. Brain Res 47:91 – 101
Aitkin LM, Webster WR, Veale JL, Crosby DC (1975) Inferior colliculus. I. Comparison of response properties of neurons in the central, pericentral, and external nuclei of adult cat. J Neurophysiol 38:1196 – 1207
Aitkin LM, Bush BMH, Gates GR (1978a) The auditory midbrain of a marsupial. The brush-tailed possum (*Trichosurus vulpecula*). Brain Res 150:29 – 44
Aitkin LM, Dickhaus H, Schult W, Zimmermann M (1978b) External nucleus of inferior colliculus: auditory and spinal somatosensory afferents and their interactions. J Neurophysiol 41:837 – 847
Aitkin LM, Kenyon CE, Philpott P (1981) The representation of the auditory and somatosensory systems in the external nucleus of the cat inferior colliculus. J Comp Neurol 196:25 – 40
Aitkin LM, Gates GR, Phillips SC (1984a) Responses of neurons in inferior colliculus to variations in sound-source azimuth. J Neurophysiol 52:1 – 17
Aitkin LM, Irvine DRF, Webster WR (1984b) Central neural mechanisms of hearing. In: Darian-Smith I (ed) Handbook of physiology, section 1: the nervous system, vol 3, sensory processes, chap 16. Am Physiol Soc, Bethesda, pp 675 – 737
Aitkin LM, Pettigrew JD, Calford MB, Phillips SC, Wise LZ (1985) Representation of stimulus azimuth by low-frequency neurons in inferior colliculus of the cat. J Neurophysiol 53:43 – 59
Allman JM, Kaas JH (1974) The organization of the second visual area (VII) in the owl monkey: a second order transformation of the visual hemifield. Brain Res 76:247 – 265
Allon N, Wollberg Z (1978) Responses of cells in the superior colliculus of the squirrel monkey to auditory stimuli. Brain Res 159:321 – 330
Altman JA (1968) Are there neurons detecting direction of sound source motion? Exp Neurol 22:13 – 25
Altman JA (1978) Sound localization – Neurophysiological mechanisms. In: Tonndorf J (ed) Translations of the Beltone Institute for Hearing Research. Beltone Inst Hearing Res, Chicago, no 30
Andersen RA, Roth GL, Aitkin LM, Merzenich MM (1980a) The efferent projections of the central nucleus and the pericentral nucleus of the inferior colliculus in the cat. J Comp Neurol 194:649 – 662
Andersen RA, Snyder RL, Merzenich MM (1980b) The tonotopic organization of corticocollicular projections from physiologically defined loci in the AI, AII, and anterior auditory cortical fields of the cat. J Comp Neurol 191:479 – 494
Anderson DJ (1973) Quantitative model for the effects of stimulus frequency upon synchronization of auditory nerve discharges. J Acoust Soc Am 54:361 – 364
Anderson DJ, Rose JE, Hind JE, Brugge JF (1971) Temporal position of discharges in single auditory nerve fibers within the cycle of a sine-wave stimulus: frequency and intensity effects. J Acoust Soc Am 49:1131 – 1139
Angell JR, Fite W (1901a) The monaural localization of sound. Psychol Rev 3:225 – 246
Angell JR, Fite W (1901b) Further observations on the monaural localization of sound. Psychol Rev 8:449 – 458
Arnesen AR, Osen KK (1978) The cochlear nerve in the cat: topography, cochleotopy and fiber spectrum. J Comp Neurol 178:661 – 678
Arthur RM (1976) Harmonic analysis of two-tone discharge patterns in cochlear nerve fibers. Biol Cybern 22:21 – 31
Arthur RM, Pfeiffer RR, Suga N (1971) Properties of "two-tone inhibition" in primary auditory neurones. J Physiol (Lond) 212:593 – 609
Auker CR, Meszler RM, Carpenter DO (1983) Apparent discrepancy between single-unit activity and [^{14}C] deoxyglucose labelling in optic tectum of the rattlesnake. J Neurophysiol 49:1504 – 1516

Batteau DW (1967) The role of the pinna in human localization. Proc R Soc Lond [Biol] Ser B 168:158–180
Behar A (1984) Intensity and sound pressure level. J Acoust Soc Am 76:632
Békésy von G (1930) Zur Theorie des Hörens: Über das Richtungshören bei einer Zeitdifferenz oder Lautstärkenungleichheit der beiderseitigen Schalleinwirkungen. Physik Zeits 31:824–835, 857–868. English translation in: Wever EG (ed) (1960) Experiments in hearing. McGraw-Hill, New York, pp 272–301
Belendiuk K, Butler RA (1975) Monaural localization of low-pass noise bands in the horizontal plane. J Acoust Soc Am 58:701–705
Belendiuk K, Butler RA (1977) Spectral cues which influence monaural localization in the horizontal plane. Percept Psychophysiol 22:353–358
Benevento LA, Coleman PD, Loe PR (1970) Responses of single cells in cat inferior colliculus to binaural click stimuli: combinations of intensity levels, time differences and intensity differences. Brain Res 17:387–405
Bengry MF, Silverman MS, Clopton BM (1977) Effects of lesioning the dorsal and intermediate acoustic striae on binaural interaction at the inferior colliculus. Exp Brain Res 28:211–219
Bergeijk van WA (1962) Variation on a theme of Békésy: a model of binaural interaction. J Acoust Soc Am 34:1431–1437
Berman AL (1968) The brainstem of the cat. A cytoarchitectonic atlas with stereotaxic coordinates. University Wisconsin Press, Madison
Beyerl BD (1978) Afferent projections to the central nucleus of the inferior colliculus in the rat. Brain Res 145:209–223
Blackstad TW, Osen KK, Mugnaini E (1984) Pyramidal neurones of the dorsal cochlear nucleus: a Golgi and computer reconstruction study in cat. Neuroscience 13:827–854
Blauert J (1969/1970) Sound localization in the median plane. Acustica 22:205–213
Blauert J (1980) Modelling of interaural time and intensity difference discrimination. In: Brink van den G, Bilsen FA (eds) Psychophysical, physiological and behavioural studies in hearing. University Press, Delft, pp 421–424
Blauert J (1982) Binaural localization. In: Pederson OJ, Poulsen T (eds) Binaural effects in normal and impaired hearing. Scand Audiol (Suppl) 15:7–26
Blauert J (1983) Spatial hearing. MIT, Cambridge
Bloom PJ (1977a) Determination of monaural sensitivity changes due to the pinna by use of minimum-audible-field measurements in the lateral vertical plane. J Acoust Soc Am 61:820–828
Bloom PJ (1977b) Creating source elevation illusions by spectral manipulation. J Audio Eng Soc 25:560–565
Bock GR, Webster WR (1974a) Spontaneous activity of single units in the inferior colliculus of anesthetized and unanesthetized cats. Brain Res 76:150–154
Bock GR, Webster WR (1974b) Coding of spatial location by single units in the inferior colliculus of the alert cat. Exp Brain Res 21:387–398
Bock GR, Webster WR, Aitkin LM (1972) Discharge patterns of single units in inferior colliculus of the alert cat. J Neurophysiol 35:265–277
Borg E (1973a) A neuroanatomical study of the brainstem auditory system of the rabbit. Part 1. Ascending connections. Acta Morphol Neerl Scand 11:31–48
Borg E (1973b) On the neuronal organization of the acoustic middle ear reflex. A physiological and anatomical study. Brain Res 49:101–123
Borg E, Zakrisson J-E (1975) The stapedius muscle and speech perception. Symp Zool Soc Lond 37:51–68
Bothe SJ, Elfner LF (1972) Monaural vs binaural auditory localization for noise bursts in the median vertical plane. J Aud Res 12:291–296
Boudreau JC, Tsuchitani C (1968) Binaural interaction in the cat superior olive S segment. J Neurophysiol 31:442–454

Boudreau JC, Tsuchitani C (1970) Cat superior olive S-segment cell discharge to tonal stimulation. In: Neff WD (ed) Contributions to sensory physiology, vol 4. Academic, New York, pp 143–213
Bourk TR (1976) Electrical responses of neural units in the anteroventral cochlear nucleus of the cat. PhD Thesis, MIT, Cambridge
Bourk TR (1977) Phase-locking in AVCN units of cat. J Acoust Soc Am 61: Suppl 1:S59
Bourk TR, Mielcarz JP, Norris BE (1981) Tonotopic organization of the anteroventral cochlear nucleus of the cat. Hear Res 4:215–241
Boyd J, Aitkin LM (1976) Responses of single units in the pontine nuclei of the cat to acoustic stimulation. Neurosci Lett 3:259–263
Brawer JR, Morest DK (1975) Relations between auditory nerve endings and cell types in the cat's anteroventral cochlear nucleus seen with the Golgi method and Nomarski optics. J Comp Neurol 160:491–506
Brawer JR, Morest DK, Kane EC (1974) The neuronal architecture of the cochlear nucleus of the cat. J Comp Neurol 155:251–299
Bregman AS (1978) The formation of auditory streams. In: Requin J (ed) Attention and performance, vol 7. Erlbaum, Hillsdale, pp 63–75
Britt RH (1976) Intracellular study of synaptic events related to phase-locking responses of cat cochlear nucleus cells to low frequency tones. Brain Res 112:313–327
Britt R, Starr A (1976a) Synaptic events and discharge patterns of cochlear nucleus cells. I. Steady-frequency tone bursts. J Neurophysiol 39:162–178
Britt R, Starr A (1976b) Synaptic events and discharge patterns of cochlear nucleus cells. II. Frequency-modulated tones. J Neurophysiol 39:179–194
Britt RH, Rossi GT, Morest DK (1983) Intracellular studies in cat cochlear nucleus: correlation of physiological responses and morphology of intracellularly labelled cat cochlear nucleus neurons. In: Webster WR, Aitkin LM (eds) Mechanisms of hearing. Monash University Press, Clayton, p 125
Brodal A (1981) Neurological anatomy in relation to clinical medicine (3rd ed). Oxford University Press, Oxford
Brown CH (1982) Primate auditory localization. In: Gatehouse RW (ed) Localization of sound: theory and applications. Amphora, Groton, pp 136–154
Brown CH, Beecher MD, Moody DB, Stebbins WC (1978) Localization of pure tones by Old World monkeys. J Acoust Soc Am 63:1484–1492
Brown CH, Beecher MD, Moody DB, Stebbins WC (1980) Localization of noise bands by Old World monkeys. J Acoust Soc Am 68:127–132
Brown CH, Schessler T, Moody D, Stebbins W (1982) Vertical and horizontal sound localization in primates. J Acoust Soc Am 72:1804–1811
Brown MC, Nuttall AL (1984) Efferent control of cochlear inner hair cell responses in the guinea-pig. J Physiol (Lond) 354:625–646
Brownell WE (1975) Organization of the cat trapezoid body and the discharge characteristics of its fibers. Brain Res 94:413–433
Brownell WE (1982) Cochlear transduction: an integrative model and review. Hear Res 6:335–360
Brownell WE (1983) Observations on a motile response in isolated outer hair cells. In: Webster WR, Aitkin LM (eds) Mechanisms of hearing. Monash University Press, Clayton, pp 5–10
Brownell WE, Manis PB, Ritz LA (1979) Ipsilateral inhibitory responses in the cat lateral superior olive. Brain Res 177:189–193
Brownell WE, Bader CR, Bertrand D, de Ribaupierre Y (1985) Evoked mechanical responses of isolated cochlear outer hair cells. Science 227:194–196
Browner RH, Webster DB (1975) Projections of the trapezoid body and the superior olivary complex of the kangaroo rat (*Dipodomys merriami*). Brain Behav Evol 11:322–354

Browner RH, Baruch A (1982) The cytoarchitecture of the dorsal cochlear nucleus in the 3-month- and 26-month-old C57BL/6 mouse: a Golgi impregnation study. J Comp Neurol 211:115–138
Brugge JF, Geisler CD (1978) Auditory mechanisms of the lower brain stem. Annu Rev Neurosci 1:363–394
Brugge JF, Merzenich MM (1973) Responses of neurons in auditory cortex of the macaque monkey to monaural and binaural stimulation. J Neurophysiol 36:1138–1158
Brugge JF, Anderson DJ, Hind JE, Rose JE (1969) Time structure of discharges in single auditory nerve fibers of the squirrel monkey in response to complex periodic sounds. J Neurophysiol 32:386–401
Brugge JF, Anderson DJ, Aitkin LM (1970) Responses of neurons in the dorsal nucleus of the lateral lemniscus of cat to binaural tonal stimulation. J Neurophysiol 33:441–458
Brunso-Bechtold JK, Thompson GC, Masterton RB (1981) HRP study of the organization of auditory afferents ascending to central nucleus of inferior colliculus in cat. J Comp Neurol 197:705–722
Buskirk RL van (1983) Subcortical auditory and somatosensory afferents to hamster superior colliculus. Brain Res Bull 10:583–587
Butler RA (1969) Monaural and binaural localization of noise bursts vertically in the median sagittal plane. J Aud Res 9:230–235
Butler RA (1971) The monaural localization of tonal stimuli. Percept Psychophys 9:99–101
Butler RA (1975) The influence of the external and middle ear on auditory discriminations. In: Keidel WD, Neff WD (eds) Handbook of sensory physiology, vol V. Auditory system, part 2. Springer, Berlin Heidelberg New York, pp 247–260
Butler RA, Belendiuk K (1977) Spectral cues utilized in the localization of sound in the median sagittal plane. J Acoust Soc Am 61:1264–1269
Butler RA, Planert N (1976) The influence of stimulus bandwidth on localization of sound in space. Percept Psychophys 19:103–108
Butler RA, Diamond IT, Neff WD (1957) Role of auditory cortex in discrimination of changes of frequency. J Neurophysiol 20:108–120
Caird D, Klinke R (1983) Processing of binaural stimuli by cat superior olivary complex neurons. Exp Brain Res 52:385–399
Caird D, Gottl K-H, Klinke R (1980) Interaural attenuation in the cat, measured with single fibre data. Hear Res 3:257–263
Calford MB (1983) The parcellation of the medial geniculate body of the cat defined by the auditory response properties of single units. J Neurosci 3:2350–2364
Calford MB, Aitkin LM (1983) Ascending projections to the medial geniculate body of the cat: evidence for multiple, parallel auditory pathways through the thalamus. J Neurosci 3:2365–2380
Calford MB, Pettigrew JD (1984) Frequency dependence of directional amplification at the cat's pinna. Hear Res 14:13–19
Calford MB, Webster WR, Semple MN (1983) Measurement of frequency selectivity of single neurons in the central auditory pathway. Hear Res 11:395–401
Calford MB, Moore DR, Hutchings ME (1986) Central and peripheral contributions to the coding of acoustic space by neurons in the inferior colliculus of the cat. J Neurophysiol 55:587–603
Campbell CBG, Hodos W (1970) The concept of homology and the evolution of the nervous system. Brain Behav Evol 3:353–367
Cant NB (1981) The fine structure of two types of stellate cells in the anterior division of the anteroventral cochlear nucleus of the cat. Neuroscience 6:2643–2655
Cant NB (1982) Identification of cell types in the anteroventral cochlear nucleus that project to the inferior colliculus. Neurosci Lett 32:241–246

Cant NB (1984) The fine structure of the lateral superior olivary nucleus of the cat. J Comp Neurol 227:63 – 77
Cant NB, Gaston KC (1982) Pathways connecting the right and left cochlear nuclei. J Comp Neurol 212:313 – 326
Cant NB, Morest DK (1978) Axons from non-cochlear sources in the anteroventral cochlear nucleus of the cat. A study with the rapid Golgi method. Neuroscience 3: 1003 – 1029
Cant NB, Morest DK (1979a) Organization of the neurons in the anterior division of the anteroventral cochlear nucleus of the cat. Light-microscopic observations. Neuroscience 4:1909 – 1923
Cant NB, Morest DK (1979b) The bushy cells in the anteroventral cochlear nucleus of the cat. A study with the electron microscope. Neuroscience 4:1925 – 1945
Carlyon RP, Moore BCJ (1984) Intensity discrimination: a severe departure from Weber's law. J Acoust Soc Am 76:1369 – 1376
Carpenter MB, Batton RB III, Peter P (1978) Transport of radioactivity from primary auditory neurons beyond the cochlear nuclei. J Comp Neurol 179:517 – 534
Carson KA, Mesulam MM (1982) Electron microscopic tracing of neural connections with horseradish peroxidase. In: Mesulam MM (ed) Tracing neural connections with horseradish peroxidase. Wiley, Chichester, pp 153 – 184
Casagrande VA, Diamond IT (1974) Ablation study of the superior colliculus in the tree shrew (*Tupaia glis*). J Comp Neurol 156:207 – 238
Caspary D (1972) Classification of subpopulations of neurons in the cochlear nuclei of the kangaroo rat. Exp Neurol 37:131 – 151
Casseday JH, Neff WD (1973) Localization of pure tones. J Acoust Soc Am 54:365 – 372
Casseday JH, Neff WD (1975) Auditory localization: role of auditory pathways in brainstem of the cat. J Neurophysiol 38:842 – 858
Casseday JH, Jones DR, Diamond IT (1979) Projections from cortex to tectum in the tree shrew, *Tupaia glis*. J Comp Neurol 185:253 – 292
Chalupa LM, Rhoades RW (1977) Responses of visual, somatosensory, and auditory neurones in the golden hamster's superior colliculus. J Physiol (Lond) 270:595 – 626
Chan JC, Yin TCT (1982) Topographical relationships along the isofrequency laminae of the cat inferior colliculus: correlation with the anatomical lamination and representation of binaural response properties. Soc Neurosci Abstr 8:348
Chan JCK, Verro P, Yin TCT (1985) Cross-correlation of binaural acoustic signals in the inferior colliculus of the cat. Assoc Res Otolaryngol 8:31
Clarey JC, Irvine DRF (1986) Auditory response properties of neurons in the anterior ectosylvian sulcus in the cat. Brain Res (in press)
Clark GM (1969) The ultrastructure of nerve endings in the medial superior olive of the cat. Brain Res 14:293 – 305
Clemo HR, Stein BE (1983) Organization of a fourth somatosensory area of cortex in cat. J Neurophysiol 50:910 – 925
Clopton BM, Winfield JA (1973) Tonotopic organization in the inferior colliculus of the rat. Brain Res 56:355 – 358
Clopton BM, Winfield JA, Flammino FJ (1974) Tonotopic organization: review and analysis. Brain Res 76:1 – 20
Cohen ES, Brawer JR, Morest DK (1972) Projections of the cochlea to the dorsal cochlear nucleus in the cat. Exp Neurol 35:470 – 479
Colburn HS (1973) Theory of binaural interaction based on auditory-nerve data. I. General strategy and preliminary results on interaural discrimination. J Acoust Soc Am 54:1458 – 1470
Colburn HS, Durlach NI (1978) Models of binaural interaction. In: Carterette EC, Friedman MP (eds) Handbook of perception, vol 4. Academic, New York, pp 467 – 518

Colburn HS, Latimer JS (1978) Theory of binaural interaction based on auditory-nerve data. III. Joint dependence on interaural time and amplitude differences in discrimination and detection. J Acoust Soc Am 64:95 – 106
Colburn HS, Moss PJ (1981) Binaural interaction models and mechanisms. In: Syka J, Aitkin L (eds) Neuronal mechanisms of hearing. Plenum, New York, pp 283 – 288
Coleman PD (1962) Failure to localize the source distance of an unfamiliar sound. J Acoust Soc Am 34:345 – 346
Coleman PD (1963) An analysis of cues to auditory depth perception in free space. Psychol Bull 60:302 – 315
Coles RB, Aitkin LM (1979) The response properties of auditory neurones in the midbrain of the domestic fowl (*Gallus gallus*) to monaural and binaural stimuli. J Comp Physiol A 134:241 – 251
Colonnier M (1968) Synaptic patterns on different cell types in the different laminae of the cat visual cortex. An electron microscope study. Brain Res 9:268 – 287
Colonnier M (1981) The electron-microscopic analysis of the neuronal organization of the cerebral cortex. In: Schmitt FO, Worden FG, Adelman G, Dennis SG (eds) The organization of the cerebral cortex. MIT, Cambridge, pp 125 – 152
Comis SD (1970) Centrifugal inhibitory processes affecting neurones in the cat cochlear nucleus. J Physiol (Lond) 210:751 – 760
Comis SD, Whitfield IC (1968) Influence of centrifugal pathways on unit activity in the cochlear nucleus. J Neurophysiol 31:62 – 68
Conlee JW, Kane ES (1982) Descending projections from the inferior colliculus to the dorsal cochlear nucleus in the cat: an autoradiographic study. Neuroscience 7:161 – 178
Costalupes JA, Young ED, Gibson DJ (1984) Effects of continuous noise backgrounds on rate response of auditory nerve fibers in cat. J Neurophysiol 51:1326 – 1344
Crommelinck M, Guitton D, Roucoux A (1977) Retinotopic versus spatial coding of saccades: clues obtained by stimulating deep layers of cat's superior colliculus. In: Baker R, Berthoz A (eds) Control of gaze by brain stem neurons. Developments in neuroscience. Vol 1. Elsevier, Amsterdam, pp 425 – 435
Crow G, Rupert AL, Moushegian G (1978) Phase locking in monaural and binaural medullary neurons: implications for binaural phenomena. J Acoust Soc Am 64:493 – 501
Crow G, Langford TL, Moushegian G (1980) Coding of interaural time differences by some high-frequency neurons of the inferior colliculus: responses to noise bands and two-tone complexes. Hear Res 3:147 – 153
Czihak E, Santibáñez-K M, Klimann M, Santibáñez-H G (1983) Audio-visual targeting reaction after unilateral lesions of the superior colliculus in cats. Acta Neurobiol Exp (Warsz) 43:15 – 25
Dallos P (1973) The auditory periphery. Biophysics and physiology. Academic, New York
Dallos P (1981) Cochlear physiology. Annu Rev Psychol 32:153 – 190
Dallos P, Harris D (1978) Properties of auditory nerve responses in absence of outer hair cells. J Neurophysiol 41:365 – 383
Davis M, Gendelman DS, Tischler MD, Gendelman PM (1982) A primary acoustic startle circuit: lesion and stimulation studies. J Neurosci 2:791 – 805
Deatherage BH, Hirsh IJ (1959) Auditory localization of clicks. J Acoust Soc Am 31:486 – 492
Delgutte B (1980) Representation of speech-like sounds in the discharge patterns of auditory-nerve fibers. J Acoust Soc Am 68:843 – 857
Delgutte B (1984) Speech coding in the auditory nerve: II. Processing schemes for vowel-like sounds. J Acoust Soc Am 75:879 – 886
Delgutte B, Kiang NYS (1984a) Speech coding in the auditory nerve: I. Vowel-like sounds. J Acoust Soc Am 75:866 – 878
Delgutte B, Kiang NYS (1984b) Speech coding in the auditory nerve: III. Voiceless fricative consonants. J Acoust Soc Am 75:887 – 896

Delgutte B, Kiang NYS (1984c) Speech coding in the auditory nerve: IV. Sounds with consonant-like dynamic characteristics. J Acoust Soc Am 75:897–907
Delgutte B, Kiang NYS (1984d) Speech coding in the auditory nerve: V. Vowels in background noise. J Acoust Soc Am 75:908–918
Diamond IT, Jones EG, Powell TPS (1969) The projection of the auditory cortex upon the diencephalon and brain stem of the cat. Brain Res 15:305–340
Disterhoft JF, Perkins RE, Evans S (1980) Neuronal morphology of the rabbit cochlear nucleus. J Comp Neurol 192:687–702
Domnitz R (1973) The interaural time jnd as a simultaneous function of interaural time and interaural amplitude. J Acoust Soc Am 53:1549–1552
Domnitz RH, Colburn HS (1977) Lateral position and interaural discrimination. J Acoust Soc Am 61:1586–1598
Don M, Starr A (1972) Lateralization performance of squirrel monkey (*Saimiri sciureus*) to binaural click signals. J Neurophysiol 35:493–500
Dräger UC, Hubel DH (1975a) Physiology of visual cells in mouse superior colliculus and correlation with somatosensory and auditory input. Nature 253:203–204
Dräger UC, Hubel DH (1975b) Responses to visual stimulation and relationship between visual, auditory, and somatosensory inputs in mouse superior colliculus. J Neurophysiol 38:690–713
Druga R, Syka J (1984a) Ascending and descending projections to the inferior colliculus in the rat. Physiol Bohemoslov 33:31–42
Druga R, Syka J (1984b) Neocortical projections to the inferior colliculus in the rat. (An experimental study using anterograde degeneration techniques). Physiol Bohemoslov 33:251–253
Druga R, Syka J (1984c) Projections from auditory structures to the superior colliculus in the rat. Neurosci Lett 45:247–252
Durlach NI, Colburn HS (1978) Binaural phenomena. In: Carterette EC, Friedman MP (eds) Handbook of perception, vol IV. Academic, New York, pp 365–466
Eccles JC (1964) The physiology of synapses. Springer, Berlin Göttingen Heidelberg New York
Edwards SB (1980) The deep cell layers of the superior colliculus: their reticular characteristics and structural organization. In: Hobson JA, Brazier MAB (eds) The reticular formation revisited. Raven, New York, pp 193–209
Edwards SB, Henkel CK (1978) Superior colliculus connections with the extraocular motor nuclei in the cat. J Comp Neurol 179:451–468
Edwards SB, Ginsburgh CL, Henkel CK, Stein BE (1979) Sources of subcortical projections to the superior colliculus in the cat. J Comp Neurol 184:309–329
Elverland HH (1977) Descending connections between the superior olivary and cochlear nucleus complexes in the cat studied by autoradiographic and horseradish peroxidase methods. Exp Brain Res 27:397–412
Elverland HH (1978) Ascending and intrinsic projections of the superior olivary complex in the cat. Exp Brain Res 32:117–134
Emmerich DS, Brown WS, Fantini DA, Navarro NC (1983) Frequency discrimination and signal detection in band-reject noise. J Acoust Soc Am 74:1702–1708
Erulkar SD (1959) The responses of single units of the inferior colliculus of the cat to acoustic stimulation. Proc R Soc Lond [Biol] Ser B 150:336–355
Erulkar SD (1972) Comparative aspects of spatial localization of sound. Physiol Rev 52:237–360
Erulkar SD (1975) Physiological studies of the inferior colliculus and medial geniculate complex. In: Keidel WD, Neff WD (eds) Handbook of sensory physiology. vol V. Auditory system. part 2. Springer, Berlin Heidelberg New York, pp 145–198
Erulkar SD, Butler RA, Gerstein GL (1968a) Excitation and inhibition in cochlear nucleus. II. Frequency-modulated tones. J Neurophysiol 31:537–548

Erulkar SD, Nelson PG, Bryan JS (1968b) Experimental and theoretical approaches to neural processing in the central auditory pathway. In: Neff WD (ed) Contributions to sensory physiology, vol 3. Academic, New York, pp 149–189
Evans EF (1974) Neural processes for the detection of acoustic patterns and for sound localization. In: Schmitt FO, Worden FG (eds) The neurosciences – third study program. MIT, Cambridge, pp 131–145
Evans EF (1975) Cochlear nerve and cochlear nucleus. In: Keidel WD, Neff WD (eds) Handbook of sensory physiology, vol V. Auditory system, part 2. Springer, Berlin Heidelberg New York, pp 1–108
Evans EF (1978) Place and time coding of frequency in the peripheral auditory system: some physiological pros and cons. Audiology 17:369–420
Evans EF (1980) "Phase-locking" of cochlear fibres and the problem of dynamic range. In: Van den Brink G, Bilsen FA (eds) Psychophysical, physiological and behavioural studies in hearing. Delft University Press, Delft, pp 300–309
Evans EF (1981) The dynamic range problem: place and time coding at the level of the cochlear nerve and nucleus. In: Syka J, Aitkin L (eds) Neuronal mechanisms of hearing. Plenum, New York, pp 69–85
Evans EF (1983) Theoretical and pragmatic approaches to coding for cochlear implant prostheses. In: Webster WR, Aitkin LM (eds) Mechanisms of hearing. Monash University Press, Clayton, pp 168–172
Evans EF, Nelson PG (1973a) The responses of single neurons in the cochlear nucleus of the cat as a function of their location and anaesthetic state. Exp Brain Res 17:402–427
Evans EF, Nelson PG (1973b) On the functional relationship between the dorsal and ventral divisions of the cochlear nucleus of the cat. Exp Brain Res 17:428–442
Evans EF, Palmer AR (1980) Relationship between the dynamic range of cochlear nerve fibres and their spontaneous activity. Exp Brain Res 40:115–118
Faye-Lund H, Osen KK (1985) Anatomy of the inferior colliculus in rat. Anat Embryol (Berl) 171:1–20
Fekete DM, Rouiller EM, Liberman MC, Ryugo DK (1984) The central projections of intracellularly labeled auditory nerve fibers in cats. J Comp Neurol 229:432–450
Feng AS, Capranica RR (1978) Sound localization in Anurans. II. Binaural interaction in superior olivary nucleus of the green tree fog (*Hyla cinerea*). J Neurophysiol 41:43–54
Feng AS, Vater M (1985) Functional organization of the cochlear nucleus of rufous horseshoe bats (*Rhinolophus rouxi*): frequencies and internal connections are arranged in slabs. J Comp Neurol 235:529–554
Fernald RD, Gerstein GL (1972) Response of cat cochlear nucleus neurons to frequency and amplitude modulated tones. Brain Res 45:417–435
Fernandez C, Karapas F (1967) The course and termination of the striae of Monakow and Held in the cat. J Comp Neurol 131:371–386
Ferster D (1981) A comparison of binocular depth mechanisms in areas 17 and 18 of the cat visual cortex. J Physiol (Lond) 311:623–655
Fitzpatrick KA (1975) Cellular architecture and topographic organization of the inferior colliculus of the squirrel monkey. J Comp Neurol 164:185–208
Fitzpatrick KA, Imig TJ (1978) Projections of auditory cortex upon the thalamus and midbrain in the owl monkey. J Comp Neurol 177:537–556
Flammino F, Clopton BM (1975) Neural responses in the inferior colliculus of albino rat to binaural stimuli. J Acoust Soc Am 57:692–695
Flynn WE, Elliott DN (1965) Role of the pinna in hearing. J Acoust Soc Am 38:104–105
Freedman SJ, Fisher HG (1968) The role of the pinna in auditory localization. In: Freedman SJ (ed) The neuropsychology of spatially oriented behavior. Dorsey, Homewood, pp 135–152
Fuzessery ZM, Pollak GD (1984) Neural mechanisms of sound localization in an echolocating bat. Science 225:725–728

Galambos R, Schwartzkopff J, Rupert A (1959) Microelectrode study of superior olivary nuclei. Am J Physiol 197:527–536

Gardner MB (1969) Lateral localization of 0°- or near-0°-oriented speech signals in anechoic space. J Acoust Soc Am 44:797–802

Gardner MB (1973) Some monaural and binaural facets of median plane localization. J Acoust Soc Am 54:1489–1495

Gardner MB, Gardner RS (1973) Problem of localization in the median plane: effect of pinnae cavity occlusion. J Acoust Soc Am 53:400–408

Geisler CD, Sinex DG (1980) Responses of primary auditory fibers to combined noise and tonal stimuli. Hear Res 3:317–334

Geisler CD, Rhode WS, Hazelton DW (1969) Responses of inferior colliculus neurons in the cat to binaural acoustic stimuli having wide-band spectra. J Neurophysiol 32:960–974

Geisler CD, Deng L, Greenberg S (1985) Thresholds for primary auditory fibers using statistically defined criteria. J Acoust Soc Am 77:1102–1109

Geniec P, Morest DK (1971) The neuronal architecture of the human posterior colliculus: a study with the Golgi method. Acta Otolaryngol (Stockh) Suppl 295:1–33

Gerstein GL, Butler RA, Erulkar SD (1968) Excitation and inhibition in cochlear nucleus. I. Tone-burst stimulation. J Neurophysiol 31:526–536

Gibson DJ (1982) Interaural crosstalk in the cat. Hear Res 7:325–333

Gisbergen van JAM, Grashuis JL, Johannesma PIM, Vendrik AJH (1975a) Spectral and temporal characteristics of activation and suppression of units in the cochlear nuclei of the anaesthetized cat. Exp Brain Res 23:367–386

Gisbergen van JAM, Grashuis JL, Johannesma PIM, Vendrik AJH (1975b) Neurons in the cochlear nucleus investigated with tone and noise stimuli. Exp Brain Res 23:387–406

Gisbergen van JAM, Grashuis JL, Johannesma PIM, Vendrik AJH (1975c) Statistical analysis and interpretation of the initial response of cochlear nucleus neurons to tone bursts. Exp Brain Res 23:407–423

Glattke TJ (1969) Unit responses of the cat cochlear nucleus to amplitude-modulated stimuli. J Acoust Soc Am 45:419–425

Glendenning KK, Masterton RB (1983) Acoustic chiasm: efferent projections of the lateral superior olive. J Neurosci 3:1521–1537

Glendenning KK, Brunso-Bechtold JK, Thompson GC, Masterton RB (1981) Ascending auditory afferents to the nuclei of the lateral lemniscus. J Comp Neurol 197:673–703

Glendenning KK, Hutson KA, Nudo RJ, Masterton RB (1985) Acoustic chiasm II: anatomical basis of binaurality in lateral superior olive of cat. J Comp Neurol 232: 261–285

Godfrey DA, Kiang NYS, Norris BE (1975a) Single unit activity in the posteroventral cochlear nucleus of the cat. J Comp Neurol 162:247–268

Godfrey DA, Kiang NYS, Norris BE (1975b) Single unit activity in the dorsal cochlear nucleus of the cat. J Comp Neurol 162:269–284

Goldberg JM (1975) Physiological studies of the auditory nuclei of the pons. In: Keidel WD, Neff WD (eds) Handbook of sensory physiology, vol V. Auditory system, part 2. Springer, Berlin Heidelberg, New York, pp 109–144

Goldberg JM, Brown PB (1968) Functional organization of the dog superior olivary complex: an anatomical and electrophysiological study. J Neurophysiol 31:639–656

Goldberg JM, Brown PB (1969) Response of binaural neurons of dog superior olivary complex to dichotic tonal stimuli: some physiological mechanisms of sound localization. J Neurophysiol 32:613–636

Goldberg JM, Brownell WE (1973) Discharge characteristics of neurons in anteroventral and dorsal cochlear nuclei of cat. Brain Res 64:35–54

Goldberg JM, Greenwood DD (1966) Response of neurons of the dorsal and posteroventral cochlear nuclei of the cat to acoustic stimuli of long duration. J Neurophysiol 29:72–93

Goldberg JM, Moore RY (1967) Ascending projections of the lateral lemniscus in the cat and monkey. J Comp Neurol 129:143–156

Goldberg JM, Neff WD (1961a) Frequency discrimination after bilateral ablation of cortical auditory areas. J Neurophysiol 24:119–128

Goldberg JM, Neff WD (1961b) Frequency discrimination after bilateral section of the brachium of the inferior colliculus. J Comp Neurol 116:265–290

Goldberg JM, Smith FD, Adrian HO (1963) Response of single units of superior olivary complex of the cat to acoustic stimuli: laterality of afferent projections. Anat Rec 145:232

Goldberg ME, Robinson DL (1978) Visual system: superior colliculus. In: Masterton RB (ed) Handbook of behavioural neurobiology, vol I. Sensory integration. Plenum, New York, pp 119–184

Goldstein JL (1978) Mechanisms of signal analysis and pattern perception in periodicity pitch. Audiology 17:421–445

Goldstein JL (1980) On the signal processing potential of high threshold auditory nerve fibers. In: van den Brink G, Bilsen FA (eds) Psychophysical, physiological and behavioral studies in hearing. Delft University Press, Delft, pp 293–299

Goldstein JL, Srulovicz P (1977) Auditory-nerve spike intervals as an adequate basis for aural frequency measurement. In: Evans EF, Wilson JP (eds) Psychophysics and physiology of hearing. Academic, New York, pp 337–346

Gordon B (1973) Receptive fields in deep layers of cat superior colliculus. J Neurophysiol 36:157–178

Gordon B (1975) Superior colliculus: structure, physiology and possible functions. In: Hunt CC (ed) Physiology, ser 1, vol 3: Neurophysiology. Butterworths, London, pp 185–230

Gottlieb DI, Cowan WM (1972) On the distribution of axonal terminals containing spheroidal and flattened synaptic vesicles in the hippocampus and dentate gyrus of the cat and rat. Z Zellforsch Mikrosk Anat 129:413–429

Gourevitch G (1980) Directional hearing in terrestrial mammals. In: Popper AN, Fay RR (eds) Comparative Studies of Hearing in Vertebrates. Springer, Berlin Heidelberg New York, pp 357–373

Graham J (1977) An autoradiographic study of the efferent connections of the superior colliculus in the cat. J Comp Neurol 173:629–654

Graham J, Pearson HE, Berman N, Murphy EH (1981) Laminar organization of superior colliculus in the rabbit: a study of receptive-field properties of single units. J Neurophysiol 45:915–932

Gray EG (1959) Axo-somatic and axo-dendritic synapses of the cerebral cortex: an electron microscope study. J Anat 93:420–433

Gray EG (1969) Electron microscopy of excitatory and inhibitory synapses: a brief review. Prog Brain Res 31:141–155

Graybiel AM (1972) Some fiber pathways related to the posterior thalamic region in the cat. Brain Behav Evol 6:363–393

Graybiel AM (1973) The thalamo-cortical projection of the so-called posterior nuclear group: a study with anterograde degeneration methods in the cat. Brain Res 49:229–244

Greenwood DD (1974) Critical bandwidth in man and in some other species in relation to the traveling wave envelope. In: Moskowitz HR, Scharf B, Stevens JC (eds) Sensation and measurement. Dordrecht, Reidel, pp 231–239

Greenwood DD, Goldberg JM (1970) Response of neurons in the cochlear nuclei to variations in noise bandwidth and to tone-noise combinations. J Acoust Soc Am 47:1022–1040

Greenwood DD, Maruyama N (1965) Excitatory and inhibitory response areas of auditory neurons in the cochlear nucleus. J Neurophysiol 28:863 – 892
Greenwood DD, Merzenich MM, Roth GL (1976) Some preliminary observations on the interrelations between two-tone suppression and combination-tone driving in the anteroventral cochlear nucleus of the cat. J Acoust Soc Am 59:607 – 633
Grinnell AD (1963 a) The neurophysiology of audition in bats: intensity and frequency parameters. J Physiol (Lond) 167:38 – 66
Grinnell AD (1963 b) The neurophysiology of audition in bats: directional localization and binaural interaction. J Physiol (Lond) 167:97 – 113
Grinnell AD, Grinnell VS (1965) Neural correlates of vertical localization by echo-locating bats. J Physiol (Lond) 181:830 – 851
Grofová I, Ottersen OP, Rinvik E (1978) Mesencephalic and diencephalic afferents to the superior colliculus and periaqueductal grey substance demonstrated by retrograde axonal transport of horseradish peroxidase in the cat. Brain Res 146:205 – 220
Guinan JJ Jr, Guinan SS, Norris BE (1972 a) Single auditory units in the superior olivary complex. I. Response to sounds and classifications based on physiological properties. Int J Neurosci 4:101 – 120
Guinan JJ Jr, Norris BE, Guinan SS (1972 b) Single auditory units in the superior olivary complex. II: Locations of unit categories and tonotopic organization. Int J Neurosci 4:147 – 166
Guinan JJ Jr, Warr WB, Norris BE (1983) Differential olivocochlear projections from lateral versus medial zones of the superior olivary complex. J Comp Neurol 221:358 – 370
Guinan JJ Jr, Warr WB, Norris BE (1984) Topographic organization of the olivocochlear projections from the lateral and medial zones of the superior olivary complex. J Comp Neurol 226:21 – 27
Hafter ER, Carrier SC (1972) Binaural interaction in low-frequency stimuli: the inability to trade time and intensity completely. J Acoust Soc Am 51:1852 – 1862
Hall JL (1965) Binaural interaction in the accessory superior-olivary nucleus of the cat. J Acoust Soc Am 37:814 – 823
Hand PJ, van Winkle T (1977) The efferent connections of the feline nucleus cuneatus. J Comp Neurol 171:83 – 110
Harnischfeger G (1978) Single unit study in the inferior colliculus of the house mouse (*Mus musculus*). Neurosci Lett 9:279 – 284
Harnischfeger G (1980) Brainstem units of echolocating bats code binaural time differences in the microsecond range. Naturwissenschaften 67:314 – 315
Harnischfeger G, Neuweiler G, Schlegel P (1985) Interaural time and intensity coding in superior olivary complex and inferior colliculus of the echolocating bat *Molossus ater*. J Neurophysiol 53:89 – 109
Harris JD (1972) A florilegium of experiments on directional hearing. Acta Otolaryngol (Stockh) Suppl. 298:1 – 26
Harris LR (1980) The superior colliculus and movements of the head and eyes in cats. J Physiol (Lond) 300:367 – 391
Harris LR, Blakemore C, Donaghy M (1980) Integration of visual and auditory space in the mammalian superior colliculus. Nature 288:56 – 59
Harrison JM (1978) Functional properties of the auditory system of the brain stem. In: Masterton RB (ed) Handbook of behavioural neurobiology, vol 1. Sensory integration. Plenum, New York, pp 409 – 458
Harrison JM, Downey P (1970) Intensity changes at the ear as a function of the azimuth of a tone source: a comparative study. J Acoust Soc Am 47:1509 – 1518
Harrison JM, Feldman ML (1970) Anatomical aspects of the cochlear nucleus and superior olivary complex. In: Neff WD (ed) Contributions to sensory physiology, vol 4. Academic, New York, pp 95 – 142

Harrison JM, Howe ME (1974a) Anatomy of the afferent auditory nervous system of mammals. In: Keidel WD, Neff WD (eds) Handbook of sensory physiology, vol V. Auditory system, part 1. Springer, Berlin Heidelberg New York, pp 284–336
Harrison JM, Howe ME (1974b) Anatomy of the descending auditory system (mammalian). In: Keidel WD, Neff WD (eds) Handbook of sensory physiology, vol V. Auditory system, part 1. Springer, Berlin Heidelberg New York, pp 365–388
Harrison JM, Irving R (1965) The anterior ventral cochlear nucleus. J Comp Neurol 124:15–42
Harrison JM, Irving R (1966a) Ascending connections of the anterior ventral cochlear nucleus in the rat. J Comp Neurol 126:51–64
Harrison JM, Irving R (1966b) The organization of the posterior ventral cochlear nucleus in the rat. J Comp Neurol 126:391–402
Harrison JM, Irving R (1966c) Visual and nonvisual auditory systems in mammals. Science 154:738–743
Harrison JM, Warr WB (1962) A study of the cochlear nuclei and ascending auditory pathways of the medulla. J Comp Neurol 119:341–380
Hashikawa T (1983) The inferior colliculopontine neurons of the cat in relation to other collicular descending neurons. J Comp Neurol 219:241–249
Hashikawa T, Kawamura K (1983) Retrograde labeling of ascending and descending neurons in the inferior colliculus. A fluorescent double labeling study in the cat. Exp Brain Res 49:457–461
Hebrank J, Wright D (1974a) Are two ears necessary for localization of sound sources on the median plane? J Acoust Soc Am 56:935–938
Hebrank J, Wright D (1974b) Spectral cues used in the localization of sound sources on the median plane. J Acoust Soc Am 56:1829–1834
Heffner H, Masterton B (1980) Hearing in glires: domestic rabbit, cotton rat, feral house mouse and kangaroo rat. J Acoust Soc Am 68:1584–1599
Heffner RS, Heffner HE (1982) Hearing in the elephant (*Elephas maximus*): absolute sensitivity, frequency discrimination and sound localization. J Comp Physiol Psychol 96:926–944
Held H (1893) Die centrale Gehörleitung. Archiv Anat Entwicklungsgesch 1g:201–248
Held R (1968) Dissociation of visual functions by deprivation and rearrangement. Psychol Forsch 31:338–348
Held R, Ingle D, Schneider G, Trevarthen C (1967) Locating and identifying: two modes of visual processing. Psychol Forsch 31:41–42
Henkel CK (1981) Afferent sources of a lateral midbrain tegmental zone associated with the pinnae in the cat as mapped by retrograde transport of horseradish peroxidase. J Comp Neurol 203:213–226
Henkel CK (1983) Evidence of sub-collicular auditory projections to the medial geniculate nucleus in the cat: an autoradiographic and horseradish peroxidase study. Brain Res 259:21–30
Henkel CK, Edwards SB (1978) The superior colliculus control of pinna movements in the cat: possible anatomical connections. J Comp Neurol 182:763–776
Henkel CK, Spangler KM (1983) Organization of the efferent projections of the medial superior olivary nucleus in the cat as revealed by HRP and autoradiographic tracing methods. J Comp Neurol 221:416–428
Henning GB (1974) Detectability of interaural delay in high-frequency complex waveforms. J Acoust Soc Am 55:84–90
Hershkowitz RM, Durlach NI (1969) Interaural time and amplitude jnds for a 500-Hz tone. J Acoust Soc Am 46:1464–1467
Hind JE (1972) Physiological correlates of auditory stimulus periodicity. Audiology 11:42–57

Hind JE, Anderson DJ, Brugge JF, Rose JE (1967) Coding of information pertaining to paired low-frequency tones in single auditory nerve fibers of the squirrel monkey. J Neurophysiol 30:794–816

Hind JE, Goldberg JM, Greenwood DD, Rose JE (1963) Some discharge characteristics of single neurons in the inferior colliculus of the cat. II. Timing of the discharges and observations on binaural stimulation. J Neurophysiol 26:321–341

Hirsch HR (1968) Perception of the range of a sound source of unknown strength. J Acoust Soc Am 43:373–374

Hirsch HR, Morton Gibson M (1976) Responses of single units in the cat cochlear nucleus to sinusoidal amplitude modulation of tones and noise: linearity and relation to speech perception. J Neurosci Res 2:337–356

Hirsch JA, Chan JCK, Yin TCT (1985) Responses of neurons in the cat's superior colliculus to acoustic stimuli. I. Monaural and binaural response properties. J Neurophysiol 53:726–745

Holt RE, Thurlow WR (1969) Subject orientation and judgement of distance of a sound source. J Acoust Soc Am 46:1584–1585

Horn G, Hill RM (1966) Responsiveness to sensory stimulation of units in the superior colliculus and subjacent tectotegmental regions of the rabbit. Exp Neurol 14:199–223

Houben D, Gourevitch G (1979) Auditory lateralization in monkeys: an examination of two cues serving directional hearing. J Acoust Soc Am 66:1057–1063

Houtsma AJM, Goldstein JL (1972) The central origin of the pitch of complex tones: evidence from musical interval recognition. J Acoust Soc Am 51:520–529

Huerta MF, Harting JK (1984) The mammalian superior colliculus: studies of its morphology and connections. In: Vanegas H (ed) The comparative neurology of the optic tectum. Plenum, New York, pp 687–773

Hughes A (1979) A rose by any other name ... On 'Naming of Neurones' by Rowe and Stone. Brain Behav Evol 16:52–64

Hui GS, Disterhoft JF (1980) Cochlear nucleus unit responses to pure tones in the unanesthetized rabbit. Exp Neurol 69:576–588

Ibata Y, Pappas GD (1976) The fine structure of synapses in relation to the large spherical neurons in the anterior ventral cochlear nucleus of the cat. J Neurocytol 5:395–406

Imig TJ, Adrian HO (1977) Binaural columns in the primary field (AI) of cat auditory cortex. Brain Res 138:241–257

Imig TJ, Morel A (1983) Organization of the thalamocortical auditory system in the cat. Annu Rev Neurosci 6:95–120

Inbody SB, Feng AS (1981) Binaural response characteristics of single neurons in the medial superior olivary nucleus of the albino rat. Brain Res 210:361–366

Irvine DRF (1976) Effects of reflex middle-ear muscle contractions on cochlear responses to bone-conducted sound. Audiology 15:433–444

Irvine DRF, Gago G (1985) Sensitivity of high-frequency neurones in inferior colliculus of the cat to interaural intensity differences: effects of variation in average binaural intensity. Proc Aust Physiol Pharmacol Soc 16:214P

Irvine DRF, Jackson GD (1983) Auditory response properties of neurons in mesencephalic and rostral pontine reticular formation of the cat. J Neurophysiol 49:1319–1343

Irvine DRF, Wise LZ (1983a) Topographic organization of interaural-intensity-difference sensitivity and the representation of auditory azimuthal location in the deep layers of the superior colliculus. In: Webster WR, Aitkin LM (eds) Mechanisms of hearing. Monash University Press, Clayton, pp 101–106

Irvine DRF, Wise LZ (1983b) The neural representation of auditory space. Proc Aust Physiol Pharmacol Soc 14:84–96

Irving R, Harrison JM (1967) The superior olivary complex and audition: a comparative study. J Comp Neurol 130:77–86

Jane JA, Masterton RB, Diamond IT (1965) The function of the tectum for attention to auditory stimuli in the cat. J Comp Neurol 125:165–191
Javel E (1981) Suppression of auditory nerve responses I: temporal analysis, intensity effects and suppression contours. J Acoust Soc Am 69:1735–1745
Javel E, Geisler CD, Ravindran A (1978) Two-tone suppression in auditory nerve of the cat: rate-intensity and temporal analyses. J Acoust Soc Am 63:1093–1104
Jay M, Sparks DL (1984) Auditory receptive fields in primate superior colliculus shift with changes in eye position. Nature 309:345–347
Jean-Baptiste M, Morest DK (1975) Transneuronal changes of synaptic endings and nuclear chromatin in the trapezoid body following cochlear ablations in cats. J Comp Neurol 162:111–134
Jeffress LA (1948) A place theory of sound localization. J Comp Psychol 41:35–39
Jeffress LA (1975) Localization of sound. In: Keidel WD, Neff WD (eds) Handbook of sensory physiology, vol V. Auditory system, part 2. Springer, Berlin Heidelberg New York, pp 449–459
Jen PH-S (1980) Coding of directional information by single neurones in the S-segment of the FM bat, *Myotis lucifugus*. J Exp Biol 87:203–216
Jen PH-S, Schlegel PA (1982) Auditory physiological properties of the neurones in the inferior colliculus of the big brown bat, *Eptesicus fuscus*. J Comp Physiol A 147:351–363
Jen PH, Sun X (1984) Pinna orientation determines the maximal directional sensitivity of bat auditory neurons. Brain Res 301:157–161
Jen PH, Sun X, Kamada T, Zhang S, Shimozawa T (1984) Auditory response properties and spatial response areas of superior colliculus neurons of the FM bat, *Eptesicus fuscus*. J Comp Physiol A 154:407–413
Jenkins WM, Masterton RB (1982) Sound localization: effects of unilateral lesions in central auditory system. J Neurophysiol 47:987–1016
Jenkins WM, Merzenich MM (1984) Role of cat primary auditory cortex for sound localization behaviour. J Neurophysiol 52:819–847
Johnson DH (1980) The relationship between spike rate and synchrony in responses of auditory-nerve fibers to single tones. J Acoust Soc Am 68:1115–1122
Johnstone BM, Taylor KJ (1969) Use of probe microphones to measure sound pressures in the ear. J Acoust Soc Am 46:1404–1405
Jones DR, Casseday JH (1979a) Projections of auditory nerve in the cat as seen by anterograde transport methods. Neuroscience 4:1299–1313
Jones DR, Casseday JH (1979b) Projections to laminae in dorsal cochlear nucleus in the tree shrew, *Tupaia glis*. Brain Res 160:131–133
Jones DR, Morest DK, Oliver DL, Potashner SJ (1984) Transganglionic transport of D-aspartate from cochlear nucleus to cochlea – a quantitative autoradiographic study. Hear Res 15:197–213
Jones EG (1981) Functional subdivision and synaptic organization of the mammalian thalamus. In: Porter R (ed) International review of physiology. Neurophysiology IV, vol 25, chapter 5. University Park Press, Baltimore, pp 173–245
Jones EG, Burton H (1976) Areal differences in the laminar distribution of thalamic afferents in cortical fields of the insular, parietal and temporal regions of primates. J Comp Neurol 168:197–248
Jones EG, Powell TPS (1970) Electron microscopy of the somatic sensory cortex of the cat. I. Cell types and synaptic organization. Philos Trans R Soc Lond [Biol] Ser B 257:1–11
Jones LS, Disterhoft JF (1983) The effect of auditory stimulus rate on [^{14}C] 2-deoxyglucose uptake in rabbit inferior colliculus. Brain Res 279:85–91
Kaas JH (1977) Sensory representation in mammals. In: Stent GS (ed) Function and formation of neural systems. Dahlem Konferenzen, Berlin, pp 65–80

Kanaseki T, Sprague JM (1974) Anatomical organization of pretectal nuclei and tectal laminae in the cat. J Comp Neurol 158:319–338

Kane EC (1973) Octopus cells in the cochlear nucleus of the cat: heterotypic synapses on homeotypic neurons. Int J Neurosci 5:251–279

Kane EC (1974a) Patterns of degeneration in the caudal cochlear nucleus of the cat after cochlear ablation. Anat Rec 179:67–91

Kane EC (1974b) Synaptic organization in the dorsal cochlear nucleus of the cat: a light and electron microscopic study. J Comp Neurol 155:301–329

Kane ES (1976a) Descending projections to specific regions of cat cochlear nucleus: a light microscopic study. Exp Neurol 52:372–388

Kane ES (1976b) Descending inputs to caudal cochlear nucleus in cats: a horseradish peroxidase (HRP) study. Am J Anat 146:433–441

Kane ES (1977a) Descending inputs to the octopus cell area of the cat cochlear nucleus: an electron microscopic study. J Comp Neurol 173:337–354

Kane ES (1977b) Autoradiographic evidence of primary projections to the caudal cochlear nucleus in cats. Am J Anat 150:641–652

Kane ES (1977c) Descending inputs to the cat dorsal cochlear nucleus: an electron microscopic study. J Neurocytol 6:583–605

Kane ES, Barone LM (1980) The dorsal nucleus of the lateral lemniscus in the cat: neuronal types and their distributions. J Comp Neurol 192:797–826

Kane ES, Conlee JW (1979) Descending inputs to the caudal cochlear nucleus of the cat: degeneration and autoradiographic studies. J Comp Neurol 187:759–784

Kane ES, Finn RC (1977) Descending and intrinsic inputs to dorsal cochlear nucleus of cats: a horseradish peroxidase study. Neuroscience 2:897–912

Karten HJ (1967) The organization of the ascending auditory pathway in the pigeon (*Columba livia*). I. Diencephalic projection of the inferior colliculus (*nucleus mesencephali lateralis, pars dorsalis*). Brain Res 6:409–427

Katsuki Y (1961) Neural mechanism of auditory sensation in cats. In: Rosenblith WA (ed) Sensory communication. MIT, Cambridge, pp 561–583

Kawamura K (1975) The pontine projection from the inferior colliculus in the cat. An experimental anatomical study. Brain Res 95:309–322

Kawamura K, Konno T (1979) Various types of corticotectal neurons of cats as demonstrated by means of retrograde axonal transport of horseradish peroxidase. Exp Brain Res 35:161–175

Khanna SM, Stinson MR (1985) Specification of the acoustical input to the ear at high frequencies. J Acoust Soc Am 77:577–589

Kiang NY-S (1965) Stimulus coding in the auditory nerve and cochlear nucleus. Acta Otolaryngol (Stockh) 59:186–200

Kiang NY-S (1968) A survey of recent developments in the study of auditory physiology. Ann Otol Rhinol Laryngol 77:656–675

Kiang NY-S (1975) Stimulus representation in the discharge patterns of auditory neurons. In: Tower DB (ed) The nervous system. Human communication and its disorders, vol 3. Raven, New York, pp 81–96

Kiang NYS (1984) Peripheral neural processing of auditory information. In: Darian-Smith I (ed) Handbook of physiology, section 1. The nervous system, vol III. Sensory processes, chapter 15. Am Physiol Soc, Bethesda, pp 639–674

Kiang NYS, Moxon EC (1972) Physiological considerations in artificial stimulation of the inner ear. Ann Otol Rhinol Laryngol 81:714–730

Kiang NYS, Moxon EC (1974) Tails of tuning curves of auditory-nerve fibers. J Acoust Soc Am 55:620–630

Kiang NYS, Watanabe T, Thomas EC, Clark LF (1965a) Discharge patterns of single fibers in the cat's auditory nerve. MIT, Cambridge

Kiang NYS, Pfeiffer RR, Warr WB, Backus ASN (1965b) Stimulus coding in the cochlear nucleus. Ann Otol Rhinol Laryngol 74:463–485

Kiang NYS, Morest DK, Godfrey DA, Guinan JJ Jr, Kane EC (1973) Stimulus coding at caudal levels of the cat's auditory nervous system. I. Response characteristics of single neurons. In: Møller AR (ed) Basic mechanisms in hearing. Academic, New York, pp 455–478

Kiang NYS, Godfrey DA, Norris BE, Moxon SE (1975) A block model of the cat cochlear nucleus. J Comp Neurol 162:221–246

Kiang NYS, Liberman MC, Levine RA (1976) Auditory-nerve activity in cats exposed to ototoxic drugs and high-intensity sounds. Ann Otol Rhinol Laryngol 85:752–768

Kiang NYS, Eddington DK, Delgutte B (1979) Fundamental considerations in designing auditory implants. Acta Otolaryngol 87:204–218

Kiang NYS, Rho JM, Northrop CC, Liberman MC, Ryugo DK (1982) Hair-cell innervation by spiral ganglion cells in adult cats. Science 217:175–177

Kim DO, Molnar CE (1979) A population study of cochlear nerve fibers: comparison of spatial distributions of average-rate and phase-locking measures of responses to single tones. J Neurophysiol 42:16–30

King AJ, Palmer AR (1983) Cells responsive to free-field auditory stimuli in guinea-pig superior colliculus: distribution and response properties. J Physiol (Lond) 342:361–381

Kiss A, Majorossy K (1983) Neuron morphology and synaptic architecture in the medial superior olivary nucleus. Exp Brain Res 52:315–327

Kitzes LM, Morton Gibson M, Rose JE, Hind JE (1978) Initial discharge latency and threshold considerations for some neurons in cochlear nuclear complex of the cat. J Neurophysiol 41:1165–1182

Kitzes LM, Wrege KS, Cassady JM (1980) Patterns of responses of cortical cells to binaural stimulation. J Comp Neurol 192:455–472

Klinke R, Galley N (1974) Efferent innervation of vestibular and auditory receptors. Physiol Rev 54:316–357

Klinke R, Boerger G, Gruber J (1969) Studies on the functional significance of efferent innervation in the auditory system: afferent neuronal activity as influenced by contralaterally applied sound. Pflügers Arch 306:165–175

Klumpp RG, Eady HR (1956) Some measurements of interaural time difference thresholds. J Acoust Soc Am 28:859–860

Knudsen EI (1982) Auditory and visual maps of space in the optic tectum of the owl. J Neurosci 2:1177–1194

Knudsen EI (1983a) Subdivisions of the inferior colliculus in the barn owl (*Tyto alba*). J Comp Neurol 218:174–186

Knudsen EI (1983b) Early auditory experience aligns the auditory map of space in the optic tectum of the barn owl. Science 222:939–942

Knudsen EI (1983c) Space coding in the vertebrate auditory system. In: Lewis B (ed) Bioacoustics: a comparative approach. Academic, London, pp 311–344

Knudsen EI (1984a) Synthesis of a neural map of auditory space in the owl. In: Edelman GM, Gall WE, Cowan WM (eds) Dynamic aspects of neocortical function. Wiley, New York, pp 375–396

Knudsen EI (1984b) Auditory properties of space-tuned units in owl's optic tectum. J Neurophysiol 52:709–723

Knudsen EI (1984c) The role of auditory experience in the development and maintenance of sound localization. Trends Neurosci 7:326–330

Knudsen EI, Knudsen PF (1983) Space-mapped auditory projections from the inferior colliculus to the optic tectum in the barn owl (*Tyto alba*). J Comp Neurol 218:187–196

Knudsen EI, Konishi M (1978a) A neural map of auditory space in the owl. Science 200:795–797

Knudsen EI, Konishi M (1978b) Space and frequency are represented separately in auditory midbrain of the owl. J Neurophysiol 41:870–884
Knudsen EI, Konishi M (1978c) Center-surround organization of auditory receptive fields in the owl. Science 202:778–780
Knudsen EI, Konishi M (1979) Mechanisms of sound localization in the barn owl (*Tyto alba*). J Comp Physiol A 133:13–21
Knudsen EI, Konishi M (1980) Monaural occlusion shifts receptive-field locations of auditory midbrain units in the owl. J Neurophysiol 44:687–695
Knudsen EI, Konishi M, Pettigrew JD (1977) Receptive fields of auditory neurons in the owl. Science 198:1278–1280
Knudsen EI, Blasdel GG, Konishi M (1979) Sound localization by the barn owl (*Tyto alba*) measured with the search coil technique. J Comp Physiol A 133:1–11
Koerber KC, Pfeiffer RR, Warr WB, Kiang NYS (1966) Spontaneous spike discharges from single units in the cochlear nucleus after destruction of the cochlea. Exp Neurol 16:119–130
Konishi M (1973a) Locatable and non-locatable acoustic signals for barn owls. Am Nat 107:775–785
Konishi M (1973b) How the owl tracks its prey. Am Sci 61:414–427
Konishi M (1978) Ethological aspects of auditory pattern recognition. In: Held H, Leibowitz W, Teuber H-L (eds) Handbook of sensory physiology, vol VIII, Perception. Springer, Berlin Heidelberg New York, pp 289–309
Konishi M (1983) Neuroethology of acoustic prey localization in the barn owl. In: Huber F, Markl H (eds) Neuroethology and behavioral physiology. Springer, Berlin Heidelberg New York, pp 303–317
Kudo M (1981) Projections of the nuclei of the lateral lemniscus in the cat: An autoradiographic study. Brain Res 221:57–69
Kudo M, Niimi K (1978) Ascending projections of the inferior colliculus onto the medial geniculate body in the cat studied by anterograde and retrograde tracing techniques. Brain Res 155:113–117
Kudo M, Niimi K (1980) Ascending projections of the inferior colliculus in the cat. An autoradiographic study. J Comp Neurol 191:545–556
Kudo M, Tashiro T, Higo S, Matsuyama T, Kawamura S (1984) Ascending projections from the nucleus of the brachium of the inferior colliculus in the cat. Exp Brain Res 54:203–211
Kuhl PK (1978) Predispositions for the perception of speech-sound categories: a species-specific phenomenon? In: Minifie FD, Lloyd LL (eds) Communication and cognitive abilities – early behavioral assessment. University Park Press, Baltimore, pp 229–255
Kuhl PK (1981) Discrimination of speech by nonhuman animals: basic auditory sensitivities conducive to the perception of speech-sound categories. J Acoust Soc Am 70:340–349
Kuhl PK, Miller JD (1978) Speech perception by the chinchilla: identification functions for synthetic VOT stimuli. J Acoust Soc Am 63:905–917
Kuhn GF (1977) Model for the interaural time differences in the azimuthal plane. J Acoust Soc Am 62:157–167
Kuwada S, Yin TCT (1983) Binaural interaction in low-frequency neurons in the inferior colliculus of the cat. I. Effects of long interaural delays, intensity, and repetition rate on interaural delay function. J Neurophysiol 50:981–999
Kuwada S, Yin TCT, Wickesberg RE (1979) Response of cat inferior colliculus neurons to binaural beat stimuli: possible mechanisms for sound localization. Science 206:586–588
Kuwada S, Yin TCT, Haberly LB, Wickesberg RE (1980) Binaural interaction in the cat inferior colliculus: physiology and anatomy. In: van den Brink G, Bilsen FA (eds) Psychophysical, physiological and behavioral studies in hearing. Delft University Press, Delft, pp 40–50

Kuwada S, Yin TCT, Syka J, Buunen TJF, Wickesberg RE (1984) Binaural interaction in low-frequency neurons in inferior colliculus of the cat. IV. Comparison of monaural and binaural response properties. J Neurophysiol 51:1306 – 1325
Langford TL (1984) Responses elicited from medial superior olivary neurons by stimuli associated with binaural masking and unmasking. Hear Res 15:39 – 50
Langner G (1981) Neuronal mechanisms for pitch analysis in the time domain. Exp Brain Res 44:450 – 454
Langner G (1983) Evidence for neuronal periodicity detection in the auditory system of the guinea fowl: implications for pitch analysis in the time domain. Exp Brain Res 52:333 – 355
Lavine RA (1971) Phase-locking in response of single neurons in cochlear nuclear complex of the cat to low-frequency tonal stimuli. J Neurophysiol 34:467 – 483
Leake-Jones PA, Snyder RL (1982) Uptake and transport of horseradish peroxidase by cochlear spiral ganglion neurons. Hear Res 8:199 – 223
Leiman AL, Hafter ER (1972) Responses of inferior colliculus neurons to free field auditory stimuli. Exp Neurol 35:431 – 449
Lenn NJ, Reese TS (1966) The fine structure of nerve endings in the nucleus of the trapezoid body and the ventral cochlear nucleus. Am J Anat 118:375 – 389
LeVay S (1973) Synaptic patterns in the visual cortex of the cat and monkey. Electron microscopy of Golgi preparations. J Comp Neurol 150:53 – 86
Lewis B (1983) Directional cues for auditory localization. In: Lewis B (ed) Bioacoustics. A comparative approach. Academic, London, pp 233 – 257
Li RYS, Guinan JJ (1971) Antidromic and orthodromic stimulation of neurons receiving calyces of Held. Quarterly progress report, research laboratory of electronics. MIT 100:227 – 234
Liberman AM, Cooper FS, Shankweiler DP, Studdert-Kennedy M (1967) Perception of the speech code. Psychol Rev 74:431 – 461
Liberman MC (1978) Auditory-nerve response from cats raised in a low-noise chamber. J Acoust Soc Am 63:442 – 455
Liberman MC (1980) Morphological differences among radial afferent fibers in the cat cochlea: an electron-microscopic study of serial sections. Hear Res 3:45 – 63
Liberman MC (1982a) The cochlear frequency map for the cat: labeling auditory-nerve fibers of known characteristic frequency. J Acoust Soc Am 72:1441 – 1449
Liberman MC (1982b) Single-neuron labeling in the cat auditory nerve. Science 216:1239 – 1241
Liberman MC, Dodds LW (1984) Single-neuron labeling and chronic cochlear pathology. III. Stereocilia damage and alterations of threshold tuning curves. Hear Res 16:55 – 74
Liberman MC, Kiang NYS (1984) Single-neuron labeling and chronic cochlear pathology. IV. Stereocilia damage and alterations in rate– and phase–level functions. Hear Res 16:75 – 90
Liberman MC, Oliver ME (1984) Morphometry of intracellularly labeled neurons of the auditory nerve: correlations with functional properties. J Comp Neurol 223:163 – 176
Licklider JCR (1951) A duplex theory of pitch perception. Experientia 7:128 – 134
Lindsey BG (1975) Fine structure and distribution of axon terminals from the cochlear nucleus on neurons in the medial superior olivary nucleus of the cat. J Comp Neurol 160:81 – 103
Lippe WR, Steward O, Rubel EW (1980) The effect of unilateral basilar papilla removal upon nuclei laminaris and magnocellularis of the chick examined with [^{3}H]2-deoxy-D-glucose autoradiography. Brain Res 196:43 – 58
Loeb GE, White WM, Merzenich MM (1983) Spatial cross-correlation: a proposed mechanism for acoustic pitch perception. Biol Cybern 47:149 – 163
Lorente de Nó R (1933a) Anatomy of the eighth nerve. The central projection of the nerve endings of the internal ear. Laryngoscope 43:1 – 38

Lorente de Nó R (1933b) Anatomy of the eighth nerve. III. General plan of structure of the primary cochlear nuclei. Laryngoscope 43:327–350

Lorente de Nó R (1979) Central representation of the eighth nerve. In: Goodhill V (ed) Ear diseases, deafness and dizziness. Harper and Row, Hagerstown, pp 64–83

Lorente de Nó R (1981) The primary acoustic nuclei. Raven, New York

Luk GD, Morest DK, McKenna NM (1974) Origins of the crossed olivocochlear bundle shown by an acid phosphatase method in the cat. Ann Otol Rhinol Laryngol 83:382–392

Manis PB, Brownell WE (1983) Synaptic organization of eighth nerve afferents to cat dorsal cochlear nucleus. J Neurophysiol 50:1156–1181

Manley G (1983) Auditory nerve fibre activity in mammals. In: Lewis B (ed) Bioacoustics: a comparative approach. Academic, London, pp 207–232

Martin MR (1981) Morphology of the cochlear nucleus of the normal and reeler mutant mouse. J Comp Neurol 197:141–152

Mast TE (1970) Binaural interaction and contralateral inhibition in dorsal cochlear nucleus of the chinchilla. J Neurophysiol 33:108–115

Mast TE (1973) Dorsal cochlear nucleus of the chinchilla: excitation by contralateral sound. Brain Res 62:61–70

Mast TE, Chung DY (1973) Binaural interaction in the superior colliculus of the chinchilla. Brain Res 62:227–230

Masterton B, Jane JA, Diamond IT (1967) Role of brainstem auditory structures in sound localization. I. Trapezoid body, superior olive, and lateral lemniscus. J Neurophysiol 30:341–359

Masterton B, Heffner H, Ravizza R (1969) The evolution of human hearing. J Acoust Soc Am 45:966–985

Masterton B, Thompson GC, Bechtold JK, RoBards MJ (1975) Neuroanatomical basis of binaural phase-difference analysis for sound localization: a comparative study. J Comp Physiol Psychol 89:379–386

Masterton RB (1974) Adaptation for sound localization in the ear and brainstem of mammals. Fed Proc 33:1904–1910

Masterton RB, Diamond IT (1967) Medial superior olive and sound localization. Science 155:1696–1697

Masterton RB, Imig TJ (1984) Neural mechanisms for sound localization. Annu Rev Physiol 46:275–287

Masterton RB, Jane JA, Diamond IT (1968) Role of brain-stem auditory structures in sound localization. II. Inferior colliculus and its brachium. J Neurophysiol 31:96–108

Masterton RB, Glendenning KK, Nudo RJ (1981) Anatomical-behavioral analyses of hindbrain sound localization mechanisms. In: Syka J, Aitkin L (eds) Neuronal mechanisms of hearing. Plenum, New York, pp 263–275

Masterton RB, Glendenning KK, Nudo RJ (1982) Anatomical pathways subserving the contralateral representation of a sound source. In: Gatehouse RW (ed) Localization of sound: theory and applications. Amphora, Groton, pp 113–125

Mays LE, Sparks DL (1980) Saccades are spatially, not retinocentrically, coded. Science 208:1163–1165

McDonald DM, Rasmussen GL (1971) Ultrastructural characteristics of synaptic endings in the cochlear nucleus having acetylcholinesterase activity. Brain Res 28:1–18

McFadden D (1981) The problem of different interaural time differences at different frequencies. J Acoust Soc Am 69:1836–1837

McFadden D, Pasanen EG (1976) Lateralization at high frequencies based on interaural time differences. J Acoust Soc Am 59:634–639

McHaffie JG, Stein BE (1982) Eye movements evoked by electrical stimulation in the superior colliculus of rats and hamsters. Brain Res 247:243–253

McIlwain JT (1975) Visual receptive fields and their images in superior colliculus of the cat. J Neurophysiol 38:219–230
McIlwain JT (1976) Large receptive fields and spatial transformations in the visual system. In: Porter R (ed) International review of physiology. Neurophysiology II, vol 10. University Park Press, Baltimore, pp 223–248
McIlwain JT (1982) Lateral spread of neural excitation during microstimulation in intermediate gray layer of cat's superior colliculus. J Neurophysiol 47:167–178
Melzer P (1984) The central auditory pathway of the gerbil *Psammomys obesus*: a deoxyglucose study. Hear Res 15:187–195
Meredith MA, Stein BE (1983) Interactions among converging sensory inputs in the superior colliculus. Science 221:389–391
Mershon DH, King LE (1975) Intensity and reverberation as factors in the auditory perception of egocentric distance. Percept Psychophys 18:409–415
Merzenich MM (1983) Auditory nerve array representation of complex electrical and sound stimuli. In: Webster WR, Aitkin LM (eds) Mechanisms of hearing. Monash University Press, Clayton, pp 163–167
Merzenich MM, Reid MD (1974) Representation of the cochlea within the inferior colliculus of the cat. Brain Res 77:397–415
Merzenich MM, Roth GL, Andersen RA, Knight PL, Colwell SA (1977) Some basic features of organization of the central auditory nervous system. In: Evans EF, Wilson JP (eds) Psychophysics and physiology of hearing. Academic, London, pp 485–497
Merzenich MM, Andersen RA, Middlebrooks JH (1979) Functional and topographic organization of the auditory cortex. In: Creutzfeldt O, Scheich H, Schreiner C (eds) Hearing mechanisms and speech. Springer, Berlin Heidelberg New York, pp 61–75
Merzenich MM, Jenkins WM, Middlebrooks JC (1984) Observations and hypotheses on special organizational features of the central auditory nervous system. In: Edelman GM, Gall WE, Cowan WM (eds) Dynamic aspects of neocortical function. Wiley, New York, pp 397–424
Middlebrooks JC, Dykes RW, Merzenich MM (1980) Binaural response-specific bands in primary auditory cortex (AI) of the cat: topographical organization orthogonal to isofrequency contours. Brain Res 181:31–48
Middlebrooks JC, Knudsen EI (1984) A neural code for auditory space in the cat's superior colliculus. J Neurosci 4:2621–2634
Middlebrooks JC, Pettigrew JD (1981) Functional classes of neurons in primary auditory cortex of the cat distinguished by sensitivity to sound location. J Neurosci 1: 107–120
Miller JD (1970) Audibility curve of the chinchilla. J Acoust Soc Am 48:513–523
Miller JD (1977) Perception of speech sounds in animals: evidence for speech processing by mammalian auditory mechanisms. In: Bullock TH (ed) Recognition of complex acoustic signals. Abakon Verlagsgesellschaft, Berlin, pp 49–58
Miller MI, Sachs MB (1983) Representation of stop consonants in the discharge patterns of auditory-nerve fibers. J Acoust Soc Am 74:502–517
Mills AW (1958) On the minimum audible angle. J Acoust Soc Am 30:237–246
Mills AW (1960) Lateralization of high frequency tones. J Acoust Soc Am 32:132–134
Mills AW (1972) Auditory localization. In: Tobias JV (ed) Foundations of modern auditory theory, vol II. Academic, New York, pp 303–348
Mogus MA (1972) Single unit responses to frequency-modulated tones and possible relationship to inhibitory effect of two-tone stimuli. Brain Res 43:668–671
Mohler CW, Wurtz RH (1977) Role of striate cortex and superior colliculus in visual guidance of saccadic eye movements in monkeys. J Neurophysiol 40:74–94
Moiseff A, Konishi M (1981) Neuronal and behavioral sensitivity to binaural time differences in the owl. J Neurosci 1:40–48

Moiseff A, Konishi M (1983a) The neural mechanisms of sound localization in the barn owl. In: Webster WR, Aitkin LM (eds) Mechanisms of hearing. Monash University Press, Clayton, pp 107–110

Moiseff A, Konishi M (1983b) Binaural characteristics of units in the owl's brainstem auditory pathway: precursors of restricted spatial receptive fields. J Neurosci 3:2553–2562

Molino J (1973) Perceiving the range of a sound source when the direction is known. J Acoust Soc Am 53:1301–1304

Møller AR (1969a) Unit responses in the rat cochlear nucleus to repetitive, transient sounds. Acta Physiol Scand 75:542–551

Møller AR (1969b) Unit responses in the cochlear nucleus of the rat to sweep tones. Acta Physiol Scand 76:503–512

Møller AR (1970) Periodicity coding in the peripheral auditory system. In: Andersen P, Jansen JKS (eds) Excitatory synaptic mechanisms. Scandinavian University Book, Oslo, pp 287–293

Møller AR (1971) Unit responses in the rat cochlear nucleus to tones of rapidly varying frequency and amplitude. Acta Physiol Scand 81:540–556

Møller AR (1972a) Coding of amplitude and frequency modulated sounds in the cochlear nucleus of the rat. Acta Physiol Scand 86:223–238

Møller AR (1972b) Coding of sounds in lower levels of the auditory system. Q Rev Biophys 5:59–155

Møller AR (1974a) The acoustic middle ear muscle reflex. In: Keidel WD, Neff WD (eds) Handbook of sensory physiology, vol V. Auditory system, part 1. Springer, Berlin Heidelberg New York, pp 519–548

Møller AR (1974b) Responses of units in the cochlear nucleus to sinusoidally amplitude-modulated tones. Exp Neurol 45:104–117

Møller AR (1974c) Coding of sounds with rapidly varying spectrum in the cochlear nucleus. J Acoust Soc Am 55:631–640

Møller AR (1977) Coding of time-varying sounds in the cochlear nucleus. Audiology 17:446–468

Møller AR (1983) Auditory physiology. Academic, New York

Möller J, Neuweiler G, Zöller H (1978) Response characteristics of inferior colliculus neurons of the awake CF-FM bat *Rhinolophus ferrumequinum*. I. Single tone stimulation. J Comp Physiol A 125:217–225

Molnar CE, Pfeiffer RR (1968) Interpretation of spontaneous spike discharge patterns of neurons in the cochlear nucleus. Proc IEEE 56:993–1004

Moore BCJ (1973) Frequency difference limens for short-duration tones. J Acoust Soc Am 54:610–619

Moore BCJ (1982) An introduction to the psychology of hearing. 2nd edn. Academic, London

Moore BCJ, Raab DH (1974) Pure-tone intensity discrimination: some experiments relating to the "near-miss" to Weber's law. J Acoust Soc Am 55:1049–1054

Moore CN, Casseday JH, Neff WD (1974) Sound localization: the role of the commissural pathways of the auditory system of the cat. Brain Res 82:13–26

Moore DR, Irvine DRF (1979) A developmental study of the sound pressure transformation by the head of the cat. Acta Otolaryngol (Stockh) 87:434–440

Moore DR, Irvine DRF (1980) Development of binaural input, response patterns, and discharge rate in single units of the cat inferior colliculus. Exp Brain Res 38:103–108

Moore DR, Irvine DRF (1981) Development of responses to acoustic interaural intensity differences in the cat inferior colliculus. Exp Brain Res 41:301–309

Moore DR, Hutchings ME, Addison PD, Semple MN, Aitkin LM (1984a) Properties of spatial receptive fields in the central nucleus of the cat inferior colliculus. II. Stimulus intensity effects. Hear Res 13:175–188

Moore DR, Semple MN, Addison PD, Aitkin LM (1984b) Properties of spatial receptive fields in the central nucleus of the cat inferior colliculus. I. Responses to tones of low intensity. Hear Res 13:159–174
Moore JK (1980) The primate cochlear nuclei: loss of lamination as a phylogenetic process. J Comp Neurol 193:609–629
Moore JK, Moore RY (1971) A comparative study of the superior olivary complex in the primate brain. Folia Primatol (Basel) 16:35–51
Moore JK, Osen KK (1979) The cochlear nuclei in man. Am J Anat 154:393–418
Moore TJ, Cashin JL Jr (1974) Response patterns of cochlear nucleus neurons to excerpts from sustained vowels. J Acoust Soc Am 56:1565–1576
Moore TJ, Cashin JL Jr (1976) Response of cochlear-nucleus neurons to synthetic speech. J Acoust Soc Am 59:1443–1449
Moore RY, Goldberg JM (1963) Ascending projections of the inferior colliculus in the cat. J Comp Neurol 121:109–136
Morest DK (1964a) The laminar structure of the inferior colliculus of the cat. Anat Rec 148:314
Morest DK (1964b) The neuronal architecture of the medial geniculate body of the cat. J Anat 98:611–630
Morest DK (1964c) The probable significance of synaptic and dendritic patterns of the thalamic and midbrain auditory system. Anat Rec 148:390–391
Morest DK (1965a) The laminar structure of the medial geniculate body of the cat. J Anat 99:143–160
Morest DK (1965b) The lateral tegmental system of the midbrain and the medial geniculate body: a study with Golgi and Nauta methods in cats. J Anat 99:611–634
Morest DK (1966) The cortical structure of the inferior quadrigeminal lamina of the cat. Anat Rec 154:389–390
Morest DK (1968a) The collateral system of the medial nucleus of the trapezoid body of the cat, its neuronal architecture and relation to the olivo-cochlear bundle. Brain Res 9:288–311
Morest DK (1968b) The growth of synaptic endings in the mammalian brain: a study of the calyces of the trapezoid body. Z Anat Entwicklungsgesch 127:201–220
Morest DK (1973) Auditory neurons of the brain stem. Adv Oto-rhino-laryngol 20:337–356
Morest DK (1981) The Golgi methods. In: Heym ChH, Forssmann WG (eds) Techniques in neuroanatomical research. Springer, Berlin Heidelberg New York, pp 124–138
Morest DK, Bohne BA (1983) Noise-induced degeneration in the brain and representation of inner and outer hair cells. Hear Res 9:145–151
Morest DK, Oliver DL (1984) The neuronal architecture of the inferior colliculus in the cat: defining the functional anatomy of the auditory midbrain. J Comp Neurol 222:209–236
Morest DK, Kiang NYS, Kane EC, Guinan JJ Jr, Godfrey DA (1973) Stimulus coding at caudal levels of the cat's auditory nervous system: II. Patterns of synaptic organization. In: Møller AR (ed) Basic mechanisms in hearing. Academic, New York, pp 479–509
Moskowitz N, Liu JC (1972) Central projections of the spiral ganglion of the squirrel monkey. J Comp Neurol 144:335–344
Mountain DC (1980) Changes in endolymphatic potential and crossed olivocochlear bundle stimulation alter cochlear mechanics. Science 210:71–72
Mountcastle VB (1978) An organizing principle for cerebral function: the unit module and the distributed system. In: Edelman GM, Mountcastle VB (eds) The mindful brain. Cortical organization and the group-selective theory of higher brain function. MIT, Cambridge, pp 7–50

Moushegian G, Rupert A, Galambos R (1962) Microelectrode study of ventral cochlear nucleus of the cat. J Neurophysiol 25:515 – 529
Moushegian G, Rupert A, Whitcomb MA (1964) Medial superior-olivary-unit response patterns to monaural and binaural clicks. J Acoust Soc Am 36:196 – 202
Moushegian G, Rupert AL, Langford TL (1967) Stimulus coding by medial superior olivary neurons. J Neurophysiol 30:1239 – 1261
Moushegian G, Rupert AL, Gidda JS (1975) Functional characteristics of superior olivary neurons to binaural stimuli. J Neurophysiol 38:1037 – 1048
Mucke L, Norita M, Benedek G, Creutzfeldt O (1982) Physiologic and anatomic investigation of a visual cortical area situated in the ventral bank of the anterior ectosylvian sulcus of the cat. Exp Brain Res 46:1 – 11
Mugnaini E, Warr WB, Osen KK (1980a) Distribution and light microscopic features of granule cells in the cochlear nuclei of cat, rat, and mouse. J Comp Neurol 191:581 – 606
Mugnaini E, Osen KK, Dahl A-L, Friedrich VL Jr, Korte G (1980b) Fine structure of granule cells and related interneurons (termed Golgi cells) in the cochlear nucleus complex of cat, rat and mouse. J Neurocytol 9:537 – 570
Musicant AD, Butler RA (1984) The influence of pinnae-based spectral cues on sound localization. J Acoust Soc Am 75:1195 – 1200
Nakajima Y (1971) Fine structure of the medial nucleus of the trapezoid body of the bat with special reference to two types of synaptic endings. J Cell Biol 50:121 – 134
Neff WD (1962) Neural structures concerned in localization of sound in space. Psychol Beitr 6:492 – 500
Neff WD (1968) Localization and lateralization of sound in space. In: de Reuck AVS, Knight J (eds) Hearing mechanism in vertebrates. Churchill, London, pp 207 – 233
Neff WD, Casseday JH (1977) Effects of unilateral ablation of auditory cortex on monaural cat's ability to localize sound. J Neurophysiol 40:44 – 52
Neff WD, Fisher JF, Diamond IT, Yela M (1956) Role of auditory cortex in discrimination requiring localization of sound in space. J Neurophysiol 19:500 – 512
Neff WD, Diamond IT, Casseday JH (1975) Behavioral studies of auditory discrimination: central nervous system. In: Keidel WD, Neff WD (eds) Handbook of sensory physiology. Vol V. Auditory system, part 2. Springer, Berlin Heidelberg New York, pp 307 – 400
Nelson PG, Erulkar SD (1963) Synaptic mechanisms of excitation and inhibition in the central auditory pathway. J Neurophysiol 26:908 – 923
Nelson PG, Erulkar SD, Bryan JS (1966) Responses of units of the inferior colliculus to time-varying acoustic stimuli. J Neurophysiol 29:834 – 860
Neuweiler G, Vater M (1977) Response patterns to pure tones of cochlear nucleus units in the CF-FM bat, *Rhinolophus ferrumequinum*. J Comp Physiol A 115:119 – 133
Niimi K, Naito FD (1974) Cortical projections of the medial geniculate body in the cat. Exp Brain Res 19:326 – 342
Noda Y, Pirsig W (1974) Anatomical projection of the cochlea to the cochlear nuclei of the guinea pig. Arch Otorhinolaryngol 208:107 – 120
Noort van J (1969) The structure and connections of the inferior colliculus. An investigation of the lower auditory system. Van Gorcum, Assen
Nordeen KW, Killackey HP, Kitzes LM (1983) Ascending auditory projections to the inferior colliculus in the adult gerbil, *Meriones unguiculatus*. J Comp Neurol 214:131 – 143
Norita M (1980) Neurons and synaptic patterns in the deep layers of the superior colliculus of the cat. A Golgi and electron microscopic study. J Comp Neurol 190:29 – 48
Nudo RJ, Masterton RB (1984) 2-Deoxyglucose studies of stimulus coding in the brainstem auditory system of the cat. In: Neff WD (ed) Contributions to sensory physiology, vol 8. Academic, London, pp 79 – 97
Nuetzel JM, Hafter ER (1976) Lateralization of complex waveforms: effects of fine structure, amplitude, and duration. J Acoust Soc Am 60:1339 – 1346

Oertel D (1983) Synaptic responses and electrical properties of cells in brain slices of the mouse anteroventral cochlear nucleus. J Neurosci 3:2043 – 2053

Oldfield SR, Parker SPA (1984a) Acuity of sound localization: a topography of auditory space, I. Normal hearing conditions. Perception 13:581 – 600

Oldfield SR, Parker SPA (1984b) Acuity of sound localization: a topography of auditory space, II. Pinna cues absent. Perception 13:601 – 617

Oliver DL (1984a) Neuron types in the central nucleus of the inferior colliculus that project to the medial geniculate body. Neuroscience 11:409 – 424

Oliver DL (1984b) Dorsal cochlear nucleus projections to the inferior colliculus in the cat: a light and electron microscopic study. J Comp Neurol 224:155 – 172

Oliver DL, Hall WC (1978) The medial geniculate body of the tree shrew, *Tupaia glis*. I. Cytoarchitecture and midbrain connections. J Comp Neurol 182:423 – 458

Oliver DL, Morest DK (1979) Cochlear nucleus projections to the inferior colliculus of the cat studied with light and electron microscopic autoradiography. Soc Neurosci Abstr 5:27

Oliver DL, Morest DK (1984) The central nucleus of the inferior colliculus in the cat. J Comp Neurol 222:237 – 264

Oliver DL, Potashner SJ, Jones DR, Morest DK (1983) Selective labeling of spiral ganglion and granule cells with D-aspartate in the auditory system of cat and guinea pig. J Neurosci 3:455 – 472

Ollo C, Schwartz IR (1979) The superior olivary complex in C57BL/6 mice. Am J Anat 155:349 – 374

Osen KK (1969a) Cytoarchitecture of the cochlear nuclei in the cat. J Comp Neurol 136:453 – 484

Osen KK (1969b) The intrinsic organization of the cochlear nuclei in the cat. Acta Otolaryngol (Stockh) 67:352 – 359

Osen KK (1970) Course and termination of the primary afferents in the cochlear nuclei of the cat. An experimental anatomical study. Arch Ital Biol 108:21 – 51

Osen KK (1972) Projection of the cochlear nuclei on the inferior colliculus in the cat. J Comp Neurol 144:355 – 372

Osen KK (1983) Orientation of dendritic arbors studied in Golgi sections of the cat dorsal cochlear nucleus. In: Webster WR, Aitkin LM (eds) Mechanisms of hearing. Monash University Press, Clayton, pp 83 – 89

Osen KK, Jansen J (1965) The cochlear nuclei in the common porpoise, *Phocaena phocaena*. J Comp Neurol 125:223 – 258

Osen KK, Mugnaini E (1981) Neuronal circuits in the dorsal cochlear nucleus. In: Syka J, Aitkin L (eds) Neuronal mechanisms of hearing. Plenum, New York, pp 119 – 125

Osen KK, Roth K (1969) Histochemical localization of cholinesterases in the cochlear nuclei of the cat. With notes on the origin of acetylcholinesterase-positive afferents and the superior olive. Brain Res 16:165 – 185

Osen KK, Mugnaini E, Dahl AL, Christiansen AH (1984) Histochemical localization of acetylcholinesterase in the cochlear and superior olivary nuclei. A reappraisal with emphasis on the cochlear granule cell system. Arch Ital Biol 122:169 – 212

Palay SL (1967) Principles of cellular organization in the nervous system. In: Quarton GC, Melnechuk T, Schmitt FO (eds) The Neurosciences – a study program. Rockefeller University Press, New York, pp 24 – 31

Palmer AR, Evans EF (1979) On the peripheral coding of the level of individual frequency components of complex sounds at high sound levels. In: Creutzfeldt O, Scheich H, Schreiner C (eds) Hearing mechanisms and speech. Springer, Berlin Heidelberg New York, pp 19 – 26

Palmer AR, Evans EF (1982) Intensity coding in the auditory periphery of the cat: responses of cochlear nerve and cochlear nucleus neurons to signals in the presence of bandstop masking noise. Hear Res 7:305 – 323

Palmer AR, King AJ (1982) The representation of auditory space in the mammalian superior colliculus. Nature 299:248–249
Palmer AR, King AJ (1983) Monaural and binaural contributions to an auditory space map in the guinea-pig superior colliculus. In: Klinke R, Hartmann R (eds) Hearing: physiological bases and psychophysics. Springer, Berlin Heidelberg New York, pp 230–236
Paula-Barbosa MM, Sousa-Pinto A (1973) Auditory cortical projections to the superior colliculus in the cat. Brain Res 50:47–61
Payne RS (1971) Acoustic location of prey by barn owls (*Tyto alba*). J Exp Biol 54:535–573
Perkins RE (1973) An electron microscopic study of synaptic organization in the medial superior olive of normal and experimental chinchillas. J Comp Neurol 148:387–416
Perry DR (1974) Acoustic responses of cells in deep superior colliculus of rabbit. Proc Aust Physiol Pharmacol Soc 5:237–238
Perry DR, Webster WR (1981) Neuronal organization of the rabbit cochlear nucleus: some anatomical and electrophysiological observations. J Comp Neurol 197:623–638
Peters A, Palay SL, Webster H de F (1976) The fine structure of the nervous system: the neurons and supporting cells. Saunders, Philadelphia
Pfalz R (1973) Efferent crossed inhibition in the ventral cochlear nuclei. In: Møller AR (ed) Basic mechanisms in hearing. Academic, New York, pp 773–784
Pfalz RKJ (1962) Centrifugal inhibition of afferent secondary neurons in the cochlear nucleus by sound. J Acoust Soc Am 34:1472–1477
Pfeiffer RR (1966a) Classification of response patterns of spike discharges for units in the cochlear nucleus: tone-burst stimulation. Exp Brain Res 1:220–235
Pfeiffer RR (1966b) Anteroventral cochlear nucleus: waveforms of extracellularly recorded spike potentials. Science 154:667–668
Pfeiffer RR, Kiang NYS (1965) Spike discharge patterns of spontaneous and continuously stimulated activity in the cochlear nucleus of anesthetized cats. Biophys J 5:301–316
Pfeiffer RR, Kim DO (1972) Response patterns of single cochlear nerve fibers to click stimuli: descriptions for cat. J Acoust Soc Am 52:1669–1677
Pfeiffer RR, Kim DO (1975) Cochlear nerve fiber responses: distribution along the cochlear partition. J Acoust Soc Am 58:867–869
Phillips DP, Brugge JF (1985) Progress in neurophysiology of sound localization. Annu Rev Psychol 36:245–274
Phillips DP, Gates GR (1982) Representation of the two ears in the auditory cortex: a reexamination. Int J Neurosci 16:41–46
Phillips DP, Irvine DRF (1979) Methodological considerations in mapping auditory cortex: binaural columns in AI of cat. Brain Res 161:342–346
Phillips DP, Irvine DRF (1981a) Responses of single neurons in physiologically defined primary auditory cortex (AI) of the cat: frequency tuning and responses to intensity. J Neurophysiol 45:48–58
Phillips DP, Irvine DRF (1981b) Responses of single neurons in physiologically defined area AI of cat cerebral cortex: sensitivity to interaural intensity differences. Hear Res 4:299–307
Phillips DP, Irvine DRF (1983) Some features of binaural input to single neurons in physiologically defined area AI of cat cerebral cortex. J Neurophysiol 49:383–395
Phillips DP, Orman SS (1984) Responses of single neurons in posterior field of cat auditory cortex to tonal stimulation. J Neurophysiol 51:147–163
Phillips DP, Calford MB, Pettigrew JD, Aitkin LM, Semple MN (1982) Directionality of sound pressure transformation at the cat's pinna. Hear Res 8:13–28
Plomp R (1964) The ear as a frequency analyzer. J Acoust Soc Am 36:1628–1636
Poggio GF (1979) Mechanisms of stereopsis in monkey visual cortex. Trends Neurosci 2:199–201

Poggio GF, Fischer B (1977) Binocular interaction and depth sensitivity in striate and prestriate cortex of behaving rhesus monkey. J Neurophysiol 40:1392–1405

Poggio GF, Poggio T (1984) The analysis of stereopsis. Annu Rev Neurosci 7:379–412

Poggio GF, Talbot WH (1981) Mechanisms of static and dynamic stereopsis in foveal cortex of the rhesus monkey. J Physiol (Lond) 315:469–492

Poljak S (1926) The connections of the acoustic nerve. J Anat 60:465–469

Pollack GD (1980) Organizational and encoding features of single neurons in the inferior colliculus of bats. In: Busnel R-G, Fis JF (eds) Animal sonar systems. Plenum, New York, pp 549–587

Pollack GD, Marsh DS, Bodenhamer R, Souther A (1978) A single-unit analysis of inferior colliculus in unanesthetized bats: response patterns and spike-count functions generated by constant-frequency and frequency-modulated sounds. J Neurophysiol 41:677–691

Popelář J, Syka J (1982) Response properties of neurons in the inferior colliculus of the guinea pig. Acta Neurobiol Exp (Warsz) 42:299–310

Pöppel E (1973) Comment on "Visual system's view of acoustic space". Nature 243:231

Poussin C, Schlegel P (1984) Directional sensitivity of auditory neurons in the superior colliculus of the bat, *Eptesicus fuscus,* using free field sound stimulation. J Comp Physiol A 154:253–261

Powell EW, Hatton JB (1969) Projections of the inferior colliculus in cat. J Comp Neurol 136:183–192

Powell TPS, Cowan WM (1962) An experimental study of the projection of the cochlea. J Anat 96:269–284

Rall W (1964) Theoretical significance of dendritic trees for neuronal input-output relations. In: Reiss RF (ed) Neural theory and modeling. Proceedings of the 1962 Ojai symposium. Stanford University Press, Stanford, pp 73–97

Ramon y Cajal S (1909) Histologie du système nerveux de l'homme et des vertebres. Maloine, Paris

Rassmussen GL (1960) Efferent fibers of the cochlear nerve and cochlear nucleus. In: Rasmussen GL, Windle WE (eds) Neural mechanisms of the auditory and vestibular systems. Thomas, Springfield, pp 105–115

Rasmussen GL (1964) Anatomic relationships of the ascending and descending auditory systems. In: Fields WS, Alford BR (eds) Neurological aspects of auditory and vestibular disorders. Thomas, Springfield, pp 1–19

Rasmussen GL (1967) Efferent connections of the cochlear nucleus. In: Graham AB (ed) Sensorineural hearing processes and disorders. Little Brown, Boston, pp 61–75

Rauschecker JP, Harris LR (1983) Auditory compensation of the effects of visual deprivation in the cat's superior colliculus. Exp Brain Res 50:69–83

Reale RA, Geisler CD (1980) Auditory-nerve fiber encoding of two-tone approximations to steady-state vowels. J Acoust Soc Am 67:891–902

Regan D (1982) Visual information channeling in normal and disordered vision. Psychol Rev 89:407–444

Regan D, Tansley BW (1979) Selective adaptation to frequency-modulated tones: evidence for an information-processing channel selectively sensitive to frequency changes. J Acoust Soc Am 65:1249–1257

Rhode WS (1976) A digital system for auditory neurophysiological research. In: Brown P (ed) Current computer technology in neurobiology. Hemisphere, Washington, pp 543–567

Rhode WS, Geisler CD, Kennedy DT (1978) Auditory nerve fiber responses to wide-band noise and tone combinations. J Neurophysiol 41:692–704

Rhode WS, Kettner R (1986) Physiological study of neurons in the dorsal and posteroventral cochlear nucleus of the unanesthetized cat. J Neurophysiol (in press)

Rhode WS, Smith PH (1985) Characteristics of tone-pip response patterns in relationship to spontaneous rate in cat auditory nerve fibers. Hear Res 18:159 – 168
Rhode WS, Smith PH (1986a) Encoding timing and intensity in the ventral cochlear nucleus of the cat. J Neurophysiol (in press)
Rhode WS, Smith PH (1986b) Physiological studies of neurons in the dorsal cochlear nucleus of cat. J Neurophysiol (in press)
Rhode WS, Smith PH, Oertel D (1983a) Physiological response properties of cells labeled intracellularly with horseradish peroxidase in cat dorsal cochlear nucleus. J Comp Neurol 213:426 – 447
Rhode WS, Oertel D, Smith PH (1983b) Physiological response properties of cells labeled intracellularly with horseradish peroxidase in cat ventral cochlear nucleus. J Comp Neurol 213:448 – 463
Ribaupierre F de, Rouiller E, Toros A, Ribaupierre Y de (1980) Transmission delay of phase-locked cells in the medial geniculate body. Hear Res 3:65 – 77
Richards W (1970) Stereopsis and stereoblindness. Exp Brain Res 10:380 – 388
Richards W (1971) Anomalous stereoscopic depth perception. J Opt Soc Am 61:410 – 414
Ritz LA, Brownell WE (1982) Single unit analysis of the posteroventral cochlear nucleus of the decerebrate cat. Neuroscience 7:1995 – 2010
RoBards MJ (1979) Somatic neurons in the brain stem and neocortex projecting to the external nucleus of the inferior colliculus: an anatomical study in the opossum. J Comp Neurol 184:547 – 566
RoBards MJ, Watkins DW III, Masterton RB (1976) An anatomical study of some somesthetic afferents to the intercollicular terminal zone of the midbrain of the opossum. J Comp Neurol 170:499 – 524
Robertson D (1984) Horseradish peroxidase injection of physiologically characterized afferent and efferent neurones in the guinea pig spiral ganglion. Hear Res 15:113 – 121
Robertson D, Cody AR, Bredberg G, Johnstone BM (1980) Response properties of spiral ganglion neurons in cochleas damaged by direct mechanical trauma. J Acoust Soc Am 67:1295 – 1303
Robinson D (1972) Eye movements evoked by collicular stimulation in the alert monkey. Vision Res 12:1795 – 1808
Rockel AJ, Jones EG (1973a) The neuronal organization of the inferior colliculus of the adult cat. I. The central nucleus. J Comp Neurol 147:11 – 60
Rockel AJ, Jones EG (1973b) Observations on the fine structure of the central nucleus of the inferior colliculus of the cat. J Comp Neurol 147:61 – 92
Rockel AJ, Jones EG (1973c) The neuronal organization of the inferior colliculus of the adult cat. II. The pericentral nucleus. J Comp Neurol 149:301 – 334
Rodieck RW, Brening RK (1983) Retinal ganglion cells: properties, types, genera, pathways and trans-species comparisons. Brain Behav Evol 23:121 – 164
Rodieck RW, Kiang NYS, Gerstein GL (1962) Some quantitative methods for the study of spontaneous activity of single neurons. Biophys J 2:351 – 368
Roffler SK, Butler RA (1968a) Factors that influence the localization of sound in the vertical plane. J Acoust Soc Am 43:1255 – 1259
Roffler SK, Butler RA (1968b) Localization of tonal stimuli in the vertical plane. J Acoust Soc Am 43:1260 – 1266
Romand R (1978) Survey of intracellular recording in the cochlear nucleus of the cat. Brain Res 148:43 – 65
Romand R (1979) Intracellular recording of "chopper responses" in the cochlear nucleus of the cat. Hear Res 1:95 – 99
Rose JE (1960) Organization of frequency sensitive neurons in the cochlear nuclear complex of the cat. In: Rasmussen GL, Windle WF (eds) Neural mechanisms of the auditory and vestibular systems. Thomas, Springfield, pp 116 – 136

Rose JE, Galambos R, Hughes JR (1959) Microelectrode studies of the cochlear nuclei of the cat. Bull Johns Hopkins Hosp 104:211 – 251
Rose JE, Greenwood DD, Goldberg JM, Hind JE (1963) Some discharge characteristics of single neurons in the inferior colliculus of the cat. I. Tonotopical organization, relation of spike-counts to tone intensity, and firing patterns of single elements. J Neurophysiol 26:294 – 320
Rose JE, Gross NB, Geisler CD, Hind JE (1966) Some neural mechanisms in the inferior colliculus of the cat which may be relevant to localization of a sound source. J Neurophysiol 29:288 – 314
Rose JE, Brugge JF, Anderson DJ, Hind JE (1967) Phase-locked response to low-frequency tones in single auditory nerve fibers of the squirrel monkey. J Neurophysiol 30:769 – 793
Rose JE, Hind JE, Anderson DJ, Brugge JF (1971) Some effects of stimulus intensity on response of auditory nerve fibers in the squirrel monkey. J Neurophysiol 34:685 – 699
Rose JE, Kitzes LM, Morton Gibson M, Hind JE (1974) Observations on phase-sensitive neurons of anteroventral cochlear nucleus of the cat: nonlinearity of cochlear output. J Neurophysiol 37:218 – 253
Roth GL, Aitkin LM, Andersen RA, Merzenich MM (1978) Some features of the spatial organization of the central nucleus of the inferior colliculus of the cat. J Comp Neurol 182:661 – 680
Roth GL, Kochhar RK, Hind JE (1980) Interaural time differences: implications regarding the neurophysiology of sound localization. J Acoust Soc Am 68:1643 – 1651
Roucoux A, Crommelinck M (1976) Eye movements evoked by superior colliculus stimulation in the alert cat. Brain Res 106:349 – 363
Rouiller EM, Ryugo DK (1984) Intracellular marking of physiologically characterized cells in the ventral cochlear nucleus of the cat. J Comp Neurol 225:167 – 186
Rowe MH, Stone J (1977) Naming of neurones. Classification and naming of cat retinal ganglion cells. Brain Behav Evol 14:185 – 216
Rowe MH, Stone J (1979) The importance of knowing our own presuppositions. Brain Behav Evol 16:65 – 80
Rowe MH, Stone J (1980) Parametric and feature extraction analyses of the receptive fields of visual neurones. Two streams of thought in the study of a sensory pathway. Brain Behav Evol 17:103 – 122
Ruggero M, Santi PA, Rich NC (1982) Type II cochlear ganglion cells in the chinchilla. Hear Res 8:339 – 356
Ruggero MA (1973) Response to noise of auditory nerve fibers in the squirrel monkey. J Neurophysiol 36:569 – 587
Rupert AL, Moushegian G (1970) Neuronal responses of kangaroo rat ventral cochlear nucleus to low-frequency tones. Exp Neurol 26:84 – 102
Ryan A, Miller J (1977) Effects of behavioral performance on single-unit firing patterns in inferior colliculus of the rhesus monkey. J Neurophysiol 40:943 – 956
Ryan A, Miller J (1978) Single unit responses in the inferior colliculus of the awake and performing rhesus monkey. Exp Brain Res 32:389 – 407
Ryan AF, Woolf NK, Sharp FR (1982) Tonotopic organization in the central auditory pathway of the Mongolian gerbil: a 2-deoxyglucose study. J Comp Neurol 207:369 – 380
Ryugo DK, Fekete DM (1982) Morphology of primary axosomatic endings in the anteroventral cochlear nucleus of the cat: a study of the endbulbs of Held. J Comp Neurol 210:239 – 257
Ryugo D, Willard FH, Fekete DM (1981) Differential afferent projections to the inferior colliculus from the cochlear nucleus in the albino mouse. Brain Res 210:342 – 349
Sachs MB (1969) Stimulus-response relation for auditory-nerve fibers: two-tone stimuli. J Acoust Soc Am 45:1025 – 1036

Sachs MB, Abbas PJ (1974) Rate versus level functions for auditory-nerve fibers in cats: tone-burst stimuli. J Acoust Soc Am 56:1835–1847
Sachs MB, Kiang NYS (1968) Two-tone inhibition in auditory-nerve fibers. J Acoust Soc Am 43:1120–1128
Sachs MB, Young ED (1979) Encoding of steady-state vowels in the auditory nerve: representation in terms of discharge rate. J Acoust Soc Am 66:470–479
Sachs MB, Young ED, Schalk TB, Bernardin CP (1980) Suppression effects in the responses of auditory-nerve fibers to broadband stimuli. In: van den Brink G, Bilsen FA (eds) Psychophysical, physiological and behavioural studies in hearing. Delft University Press, Delft, pp 284–291
Sachs MB, Voigt HF, Young ED (1983) Auditory nerve representation of vowels in background noise. J Neurophysiol 50:27–45
Sanchez-Longo LP, Forster FM (1958) Clinical significance of impairment of sound localization. Neurology (NY) 8:119–125
Sandel TT, Teas DC, Feddersen WE, Jeffress LA (1955) Localization of sound from single and paired sources. J Acoust Soc Am 27:842–852
Sando I (1965) The anatomical interrelationships of the cochlear nerve fibers. Acta Otolaryngol (Stockh) 59:417–436
Sayers BMcA, Cherry EC (1957) Mechanism of binaural fusion in the hearing of speech. J Acoust Soc Am 29:973–987
Schaefer KP (1970) Unit analysis and electrical stimulation in the optic tectum of rabbits and cats. Brain Behav Evol 3:222–240
Schaefer KP, Schneider H (1968) Reizversuche im Tectum opticum des Kaninchens. Ein experimenteller Beitrag zur sensomotorischen Koordination des Hirnstammes. Archiv Psychiat u Zeits ges Neurol 211:118–137
Schalk TB, Sachs MB (1980) Nonlinearities in auditory-nerve fiber responses to bandlimited noise. J Acoust Soc Am 67:903–913
Shankweiler DP (1961) Performance of brain-damaged patients on two tests of sound localization. J Comp Physiol Psychol 54:375–381
Scheibel ME, Scheibel AB (1974) Neuropil organization in the superior olive of the cat. Exp Neurol 43:339–348
Schiller PH, Sandell JH (1983) Interactions between visually and electrically elicited saccades before and after superior colliculus and frontal eye field ablations in the rhesus monkey. Exp Brain Res 49:381–392
Schiller PH, Stryker M (1972) Single-unit recording and stimulation in superior colliculus of the alert rhesus monkey. J Neurophysiol 35:915–924
Schiller PH, True SD, Conway JL (1980) Deficits in eye movements following frontal eye-field and superior colliculus ablations. J Neurophysiol 44:1175–1189
Schlegel P (1977) Directional coding by binaural brainstem units of the CF-FM bat, *Rhinolophus ferrumequinum*. J Comp Physiol A 118:327–352
Schmiedt RA (1982) Boundaries of two-tone rate suppression of cochlear-nerve activity. Hear Res 7:335–351
Schneider GE (1967) Contrasting visuomotor functions of tectum and cortex in the golden hamster. Psychol Forsch 31:52–62
Schneider GE (1969) Two visual systems. Science 163:895–902
Schneider GE (1970) Mechanisms of functional recovery following lesions of visual cortex or superior colliculus in neonate and adult hamsters. Brain Behav Evol 3:295–323
Schouten MEH (1980) The case against a speech mode of perception. Acta Psychol (Amst) 44:71–98
Schwartz IR (1972) Axonal endings in the cat medial superior olive: coated vesicles and intercellular substance. Brain Res 46:187–202
Schwartz IR (1977) Dendritic arrangements in the cat medial superior olive. Neuroscience 2:81–101

Schwartz IR (1980) The differential distribution of synaptic terminal on marginal and central cells in the cat medial superior olivary nucleus. Am J Anat 159:25 – 31

Schwartz IR (1984) Axonal organization in the cat medial superior olivary nucleus. In: Neff WD (ed) Contributions to sensory physiology, vol 8. Academic, New York, pp 99 – 129

Schwartz WJ, Smith CB, Davidsen L, Savaki H, Sokoloff L, Mata M, Fink DJ, Gainer H (1979) Metabolic mapping of functional activity in the hypothalamo-neurohypophysial system of the rat. Science 205:723 – 725

Schweitzer H (1981) The connections of the inferior colliculus and the organization of the brainstem auditory system in the greater horseshoe bat (*Rhinolophus ferrumequinum*). J Comp Neurol 201:25 – 49

Searle CL, Braida LD, Cuddy DR, Davis MF (1975) Binaural pinna disparity: Another auditory localization cue. J Acoust Soc Am 57:448 – 455

Searle CL, Braida LD, Davis MF, Colburn HS (1976) Model for auditory localization. J Acoust Soc Am 60:1164 – 1175

Segal MM (1983) Specification of synaptic action. Trends Neurosci 6:118 – 121

Semple MN, Aitkin LM (1979) Representation of sound frequency and laterality by units in central nucleus of cat inferior colliculus. J Neurophysiol 42:1626 – 1639

Semple MN, Aitkin LM (1980) Physiology of pathway from dorsal cochlear nucleus to inferior colliculus revealed by electrical and auditory stimulation. Exp Brain Res 41:19 – 28

Semple MN, Aitkin LM (1981) Integration and segregation of input to the cat inferior colliculus. In: Syka J, Aitkin L (eds) Neuronal mechanisms of hearing. Plenum, New York, pp 155 – 161

Semple MN, Kitzes LM (1984) Sensitivity of single units in the gerbil inferior colliculus to interaural intensity differences at different average binaural levels. Proc Neurosci Abstr 10:1148

Semple MN, Kitzes LM (1985) Single-unit responses in the inferior colliculus: different consequences of contralateral and ipsilateral auditory stimulation. J Neurophysiol 53:1467 – 1482

Semple MN, Aitkin LM, Calford MB, Pettigrew JD, Phillips DP (1983) Spatial receptive fields in the cat inferior colliculus. Hear Res 10:203 – 215

Servière J, Webster WR (1981) A combined electrophysiological and [^{14}C] 2-deoxyglucose study of the frequency organization of the inferior colliculus of the cat. Neurosci Lett 27:113 – 118

Servière J, Webster WR, Calford MB (1984) Iso-frequency labelling revealed by a combined [^{14}C]-2-deoxyglucose, electrophysiological, and horseradish peroxidase study of the inferior colliculus of the cat. J Comp Neurol 228:463 – 477

Shaw EAG (1974a) Transformation of sound pressure level from the free field to the eardrum in the horizontal plane. J Acoust Soc Am 56:1848 – 1861

Shaw EAG (1974b) The external ear. In: Keidel WD, Neff WD (eds) Handbook of sensory physiology, vol V. Auditory system, part 1. Springer, Berlin Heidelberg New York, pp 455 – 490

Shaw EAG (1982) External ear response and sound localization. In: Gatehouse RW (ed) Localization of sound: theory and applications. Amphora, Groton, pp 30 – 41

Shaw EAG, Teranishi R (1968) Sound pressure generated in an external-ear replica and real human ears by a nearby point source. J Acoust Soc Am 44:240 – 249

Shimozawa T, Suga N, Hendler P, Schuetze S (1974) Directional sensitivity of echolocation system in bats producing frequency modulated signals. J Exp Biol 60:53 – 69

Shimozawa T, Sun X, Jen PHS (1984) Auditory space representation in the superior colliculus of the big brown bat, *Eptesicus fuscus*. Brain Res 311:289 – 296

Siebert WM (1965) Some implications of the stochastic behavior of primary auditory neurons. Kybernetik 2:206 – 215

Siebert WM (1968) Stimulus transformations in the peripheral auditory system. In: Kolers PA, Eden M (eds) Recognizing patterns. MIT, Cambridge, pp 104–133
Siebert WM (1970) Frequency discrimination in the auditory system: place or periodicity mechanisms? Proc IEEE 58:723–730
Siegel JH, Kim DO (1982) Efferent neural control of cochlear mechanics? Olivocochlear bundle stimulation affects cochlear biomechanical nonlinearity. Hear Res 6:171–182
Silverman MS, Clopton BM (1977) Plasticity of binaural interaction. I. Effect of early auditory deprivation. J Neurophysiol 40:1266–1274
Sinex DG, Geisler CD (1981) Auditory-nerve fiber responses to frequency-modulated tones. Hear Res 4:127–148
Sinex DG, Geisler CD (1983) Responses of auditory-nerve fibers to consonant-vowel syllables. J Acoust Soc Am 73:602–615
Sinex DG, Geisler CD (1984) Comparison of the responses of auditory nerve fibers to consonant-vowel syllables with predictions from linear models. J Acoust Soc Am 76:116–121
Smith PH, Rhode WS (1985) Electron microscopic features of physiologically characterized, HRP-labeled fusiform cells in the cat dorsal cochlear nucleus. J Comp Neurol 237:127–143
Smith RL, Brachman ML (1980a) Dynamic responses of single auditory-nerve fibers: some effects of intensity and time. In: van den Brink G, Bilsen FA (eds) Psychophysical, physiological and behavioural studies in hearing. Delft University Press, Delft, pp 312–319
Smith RL, Brachman ML (1980b) Operating range and maximum response of single auditory nerve fibres. Brain Res 184:499–505
Smolders JWT, Aertsen AMH, Johannesma PIM (1979) Neural representation of the acoustic biotope. Biol Cybernetics 35:11–20
Smoorenburg GF, Morton Gibson M, Kitzes LM, Rose JE, Hind JE (1976) Correlates of combination tones observed in the response of neurons in the anteroventral cochlear nucleus of the cat. J Acoust Soc Am 59:945–962
Sparks DL, Holland R, Guthrie BL (1976) Size and distribution of movement fields in the monkey superior colliculus. Brain Res 113:21–34
Sparks DL, Mays LE (1981) The role of the monkey superior colliculus in the control of saccadic eye movements: a current perspective. In: Fuchs A, Becker W (eds) Progress in oculomotor research. Elsevier, New York, pp 137–144
Sparks DL, Mays LE (1983) Spatial localization of saccade targets. I. Compensation for stimulus-induced perturbations in eye position. J Neurophysiol 49:45–63
Sparks DL, Pollack JG (1977) The neural control of saccadic eye movements: the role of the superior colliculus. In: Brooks BA, Bajandos FJ (eds) Eye movements – ARVO Symposium 1976. Plenum, New York, pp 179–219
Sparks DL, Porter JD (1983) Spatial localization of saccade targets. II. Activity of superior colliculus neurons preceding compensatory saccades. J Neurophysiol 49:64–74
Spoendlin H (1969) Innervation patterns in the organ of Corti of the cat. Acta Otolaryngol (Stockh) 67:239–254
Spoendlin H (1971) Degeneration behaviour of the cochlear nerve. Arch klin exp Ohr-Nas-u Kehlk Heilk 200:275–291
Spoendlin H (1978) The afferent innervation of the cochlea. In: Naunton RF, Fernandez C (eds) Evoked electrical activity in the auditory nervous system. Academic, New York, pp 21–39
Spoendlin H (1979) Neural connections of the outer haircell system. Acta Otolaryngol (Stockh) 87:381–387
Sprague JM, Meikle TH (1965) The role of the superior colliculus in visually guided behavior. Exp Neurol 11:115–146

Sprague JM, Levitt M, Robson K, Liu CN, Stellar E, Chambers WW (1963) A neuroanatomical and behavioral analysis of the syndromes resulting from midbrain lemniscal and reticular lesions in the cat. Arch Ital Biol 101:225–295

Srulovicz P, Goldstein JL (1983) A central spectrum model: a synthesis of auditory-nerve timing and place cues in monaural communication of frequency spectrum. J Acoust Soc Am 73:1266–1276

Starr A, Britt R (1970) Intracellular recordings from cat cochlear nucleus during tone stimulation. J Neurophysiol 33:137–147

Starr A, Wernick JS (1968) Olivocochlear bundle stimulation: effects on spontaneous and tone-evoked activities of single units in cat cochlear nucleus. J Neurophysiol 31:549–564

Stein BE, Arigbede MO (1972) Unimodal and multimodal response properties of neurons in the cat's superior colliculus. Exp Neurol 36:179–196

Stein BE, Clamann HP (1981) Control of pinna movements and sensorimotor register in cat superior colliculus. Brain Behav Evol 19:180–192

Stein BE, Dixon JP (1979) Properties of superior colliculus neurons in the golden hamster. J Comp Neurol 2:269–284

Stein BE, Goldberg SJ, Clamann HP (1976) The control of eye movements by the superior colliculus in the alert cat. Brain Res 118:469–474

Stein BE, Spencer RF, Edwards SB (1983) Corticotectal and corticothalamic efferent projections of SIV somatosensory cortex in cat. J Neurophysiol 50:896–909

Stern RM Jr, Colburn HS (1978) Theory of binaural interaction based on auditory-nerve data. IV. A model for subjective lateral position. J Acoust Soc Am 64:127–140

Stern RM Jr, Colburn HS (1985) Lateral-position-based models of interaural discrimination. J Acoust Soc Am 77:753–755

Stevens KN (1980) Acoustic correlates of some phonetic categories. J Acoust Soc Am 68:836–842

Stevens KN (1981) Constraints imposed by the auditory system on the properties used to classify speech sounds: data from phonology, acoustics, and psychoacoustics. In: Myers T, Laver J, Anderson J (eds) The cognitive representation of speech. Elsevier, Amsterdam, pp 61–74

Stevens KN, House AS (1972) Speech perception. In: Tobias JV (ed) Foundations of modern auditory theory, vol II. Academic, New York, pp 3–62

Stevens SS, Newman EB (1936) The localization of actual sources of sound. Am J Psychol 48:297–306

Stillman RD (1971 a) Characteristic delay neurons in the inferior colliculus of the kangaroo rat. Exp Neurol 32:404–412

Stillman RD (1971 b) Pattern responses of low-frequency inferior colliculus neurons. Exp Neurol 33:432–440

Stillman RD (1972) Responses of high-frequency inferior colliculus neurons to interaural intensity differences. Exp Neurol 36:118–126

Stone J (1983) Parallel processing in the visual system. The classification of retinal ganglion cells and its impact on the neurobiology of vision. Plenum, New York

Stopp PE (1983) The distribution of the olivo-cochlear bundle and its possible role in frequency/intensity coding. In: Klinke R, Hartmann R (eds) Hearing – physiological bases and psychophysics. Springer, Berlin Heidelberg New York Tokyo, pp 176–180

Stotler WA (1953) An experimental study of the cells and connections of the superior olivary complex of the cat. J Comp Neurol 98:401–431

Straschill M, Rieger P (1973) Eye movements evoked by focal stimulation of the cat's superior colliculus. Brain Res 59:211–227

Strominger NL (1978) The anatomical organization of the primate auditory pathway. In: Noback CR (ed) Sensory systems of primates. Plenum, New York, pp 53–91

Strominger NL, Oesterreich RE (1970) Localization of sound after section of the brachium of the inferior colliculus. J Comp Neurol 138:1 – 18
Strominger NL, Strominger AI (1971) Ascending brain stem projections of the anteroventral cochlear nucleus in the rhesus monkey. J Comp Neurol 143:217 – 242
Strutz J (1981) Efferent innervation of the cochlea. Ann Otol 90:158 – 160
Strutz J, Bielenberg K (1984) Efferent acoustic neurons within the lateral superior olivary nucleus of the guinea pig. Brain Res 299:174 – 177
Strutz J, Spatz WB (1980) Superior olivary and extraolivary origin of centrifugal innervation of the cochlea in the guinea pig. A horseradish peroxidase study. Neurosci Lett 17:227 – 230
Strybel TZ, Perrott DR (1984) Discrimination of relative distance in the auditory modality: the success and failure of the loudness discrimination hypothesis. J Acoust Soc Am 76:318 – 320
Stryker M, Blakemore C (1972) Saccadic and disjunctive eye movements in cats. Vision Res 12:2005 – 2013
Suga N (1964) Single unit activity in cochlear nucleus and inferior colliculus of echo-locating bats. J Physiol (Lond) 172:449 – 474
Suga N (1968) Analysis of frequency-modulated and complex sounds by single auditory neurones of bats. J Physiol (Lond) 198:51 – 80
Suga N (1971) Feature detection in the cochlear nucleus, inferior colliculus, and auditory cortex. In: Sachs MB (ed) Physiology of the auditory system. NEC, Baltimore, pp 197 – 206
Suga N (1977) Amplitude spectrum representation in the Doppler-shifted-CF processing area of the auditory cortex of the mustache bat. Science 196:64 – 67
Suga N, Schlegel P (1973) Coding and processing in the auditory systems of FM-signal-producing bats. J Acoust Soc Am 54:174 – 190
Sullivan WE (1985) Classification of response patterns in cochlear nucleus of barn owl: correlation with functional response properties. J Neurophysiol 53:201 – 216
Sullivan WE, Konishi M (1984) Segregation of stimulus phase and intensity coding in the cochlear nucleus of the barn owl. J Neurosci 4:1787 – 1799
Syka J, Radil-Weiss T (1971) Electrical stimulation of the tectum in freely moving cats. Brain Res 28:567 – 572
Syka J, Straschill M (1970) Activation of superior colliculus neurons and motor responses after electrical stimulation of the inferior colliculus. Exp Neurol 28:384 – 392
Syka J, Radionova EA, Popelář J (1981) Discharge characteristics of neuronal pairs in the rabbit inferior colliculus. Exp Brain Res 44:11 – 18
Syka J, Popelář J, Druga R (1983) Structure and function of crossed and uncrossed pathways to the inferior colliculus in the rat. In: Klinke R, Hartmann R (eds) Hearing – physiological bases and psychophysics. Springer, Berlin Heidelberg New York, pp 224 – 229
Szentagothai J, Arbib MA (1975) Conceptual models of neural organization. MIT, Cambridge
Taber E (1961) The cytoarchitecture of the brain stem of the cat. J Comp Neurol 116:27 – 70
Takahashi T, Moiseff A, Konishi M (1984) Time and intensity cues are processed independently in the auditory system of the owl. J Neurosci 4:1781 – 1786
Terhune JM (1974) Sound localization abilities of untrained humans using complex and sinusoidal sounds. Scand Audiol 3:115 – 120
Theurich M, Müller CM, Scheich H (1984) 2-Deoxyglucose accumulation parallels extracellularly recorded spike activity in the avian auditory neostriatum. Brain Res 322:157 – 161
Thompson GC, Masterton RB (1978) Brain stem auditory pathways involved in reflexive head orientation to sound. J Neurophysiol 41:1183 – 1202

Thompson RF (1960) Function of auditory cortex of cat in frequency discrimination. J Neurophysiol 23:321 – 334

Thurlow WR, Mangels JW, Runge PS (1967) Head movements during sound localization. J Acoust Soc Am 42:489 – 493

Thurlow WR, Runge PS (1967) Effect of induced head movements on localization of direction of sounds. J Acoust Soc Am 42:480 – 488

Tiao YC, Blakemore C (1976) Functional organization in the superior colliculus of the golden hamster. J Comp Neurol 168:483 – 503

Tobias JV, Zerlin S (1959) Lateralization threshold as a function of stimulus duration. J Acoust Soc Am 31:1591 – 1594

Tolbert LP, Morest DK (1982a) The neuronal architecture of the anteroventral cochlear nucleus of the cat in the region of the cochlear nerve root: Golgi and Nissl methods. Neuroscience 7:3013 – 3030

Tolbert LP, Morest DK (1982b) The neuronal architecture of the anteroventral cochlear nucleus of the cat in the region of the cochlear nerve root: electron microscopy. Neuroscience 7:3053 – 3067

Tolbert LP, Morest DK, Yurgelun-Todd DA (1982) The neuronal architecture of the anteroventral cochlear nucleus of the cat in the region of the cochlear nerve root: horseradish peroxidase labelling of identified cell types. Neuroscience 7:3031 – 3052

Tortelly A, Reinoso-Suarez F, Llamas A (1980) Projections from non-visual cortical areas to the superior colliculus demonstrated by retrograde transport of HRP in the cat. Brain Res 188:543 – 549

Tsuchitani C (1977) Functional organization of lateral cell groups of cat superior olivary complex. J Neurophysiol 40:296 – 318

Tsuchitani C (1978) Lower auditory brainstem structures of the cat. In: Naunton RF, Fernandez C (eds) Evoked electrical activity in the auditory nervous system. Academic, New York, pp 373 – 401

Tsuchitani C (1982) Discharge patterns of cat lateral superior olivary units to ipsilateral tone-burst stimuli. J Neurophysiol 47:479 – 500

Tsuchitani C, Boudreau JC (1964) Wave activity in the superior olivary complex of the cat. J Neurophysiol 27:814 – 827

Tsuchitani C, Boudreau JC (1966) Single unit analysis of cat superior olive S segment with tonal stimuli. J Neurophysiol 29:684 – 697

Tsuchitani C, Boudreau JC (1967) Encoding of stimulus frequency and intensity by cat superior olive S-segment cells. J Acoust Soc Am 42:794 – 805

Tsuchitani C, Boudreau JC (1969) Stimulus level of dichotically presented tones and cat superior olive S-segment cell discharge. J Acoust Soc Am 46:979 – 988

Tsuchitani C, Johnson DH (1985) The effects of ipsilateral tone burst stimulus level on the discharge patterns of cat lateral superior olivary units. J Acoust Soc Am 77: 1484 – 1496

Tunkl JE (1980) Location of auditory and visual stimuli in cats with superior colliculus ablations. Exp Neurol 68:395 – 402

Tyner CF (1975) The naming of neurons: applications of taxonomic theory to the study of cellular populations. Brain Behav Evol 12:75 – 96

Uchizono K (1965) Characteristics of excitatory and inhibitory synapses in the central nervous system of the cat. Nature 207:642 – 643

Uchizono K (1967) Synaptic organization of the Purkinje cells in the cerebellum of the cat. Exp Brain Res 4:97 – 113

Updyke BV (1974) Characteristics of unit responses in superior colliculus of the *Cebus* monkey. J Neurophysiol 37:896 – 909

Vater M (1982) Single unit responses in cochlear nucleus of horseshoe bats to sinusoidal frequency and amplitude modulated signals. J Comp Physiol A 149:369 – 388

Vater M, Schlegel P, Zöller H (1979) Comparative auditory neurophysiology of the inferior colliculus of two molossid bats, *Molossus ater* and *Molossus molossus*. I. Gross evoked potentials and single unit responses to pure tones. J Comp Physiol A 131:137–145

Viemeister NF (1974) Intensity discrimination of noise in the presence of band-reject noise. J Acoust Soc Am 56:1594–1600

Viemeister NF (1983) Auditory intensity discrimination at high frequencies in the presence of noise. Science 221:1206–1208

Voigt HF, Young ED (1980) Evidence of inhibitory interactions between neurons in dorsal cochlear nucleus. J Neurophysiol 44:76–96

Wakeford OS, Robinson DE (1974) Lateralization of tonal stimuli by the cat. J Acoust Soc Am 55:649–652

Walberg F (1965) An electron microscopic study of terminal degeneration in the inferior olive of the cat. J Comp Neurol 125:205–222

Wallach H (1939) On sound localization. J Acoust Soc Am 10:270–274

Walsh BT, Miller JB, Gacek RR, Kiang NYS (1972) Spontaneous activity in the eighth cranial nerve of the cat. Int J Neurosci 3:221–236

Walsh EG (1957) An investigation of sound localization in patients with neurological abnormalities. Brain 80:222–250

Warr WB (1966) Fiber degeneration following lesions in the anterior ventral cochlear nucleus of the cat. Exp Neurol 14:453–474

Warr WB (1969) Fiber degeneration following lesions in the posteroventral cochlear nucleus of the cat. Exp Neurol 23:140–155

Warr WB (1972) Fiber degeneration following lesions in the multipolar and globular cell areas in the ventral cochlear nucleus of the cat. Brain Res 40:247–270

Warr WB (1975) Olivocochlear and vestibular efferent neurons of the feline brain stem: their location, morphology and number determined by retrograde axonal transport and acetylcholinesterase histochemistry. J Comp Neurol 161:159–182

Warr WB (1982) Parallel ascending pathways from the cochlear nucleus: neuroanatomical evidence of functional specialization. In: Neff WD (ed) Contributions to sensory physiology, vol 7. Academic, New York, pp 1–38

Warr WB, Guinan JJ Jr (1979) Efferent innervation of the organ of Corti: two separate systems. Brain Res 173:152–155

Watanabe T, Ohgushi K (1968) FM sensitive auditory neuron. Proc Japan Acad 44:968–973

Watanabe T, Liao T-T, Katsuki Y (1968) Neuronal response patterns in the superior olivary complex of the cat to sound stimulation. Jpn J Physiol 18:267–287

Watkins AJ (1978) Psychoacoustical aspects of synthesized vertical locale cues. J Acoust Soc Am 63:1152–1165

Watkins AJ (1982) The monaural perception of azimuth: a synthesis approach. In: Gatehouse RW (ed) Localization of sound: theory and applications. Amphora, Groton, pp 194–206

Webster DB, Trune DR (1982) Cochlear nucleus complex of mice. Am J Anat 163:103–130

Webster WR (1977) Chopper units recorded in the cochlear nucleus of the awake cat. Neurosci Lett 7:261–265

Webster WR, Aitkin LM (1975) Central auditory processing. In: Gazzaniga MS, Blakemore C (eds) Handbook of psychobiology. Academic, New York, pp 325–364

Webster WR, Servière J, Batini C, Laplante S (1978) Autoradiographic demonstration with [^{14}C]-2-deoxyglucose of frequency selectivity in the auditory system of cats under conditions of functional activity. Neurosci Lett 10:43–48

Webster WR, Servière J, Brown M (1984a) Inhibitory contours in the inferior colliculus as revealed by the 2-deoxyglucose method. Exp Brain Res 56:577–581

Webster WR, Servière J, Crewther D, Crewther S (1984b) Iso-frequency 2-DG contours in the inferior colliculus of the awake monkey. Exp Brain Res 56:427–437
Webster WR, Servière J, Martin R, Brown M (1985) Uncrossed and crossed inhibition in the inferior colliculus of the cat: A combined 2-deoxyglucose and electrophysiological study. J Neurosci 5:1820–1832
Wenstrup JJ, Ross LS, Pollack GD (1985) A functional organization of binaural responses in the inferior colliculus. Hear Res 17:191–195
Westerman LA, Smith RL (1984) Rapid and short-term adaptation in auditory nerve responses. Hear Res 15:249–260
Wettschureck RG (1973) Die absoluten Unterschiedsschwellen der Richtungswahrnehmung in der Medianebene beim natürlichen Hören, sowie beim Hören über ein Kunstkopf-Übertragungssystem. Acustica 28:197–208
Wever EG (1949) Theory of hearing. Wiley, New York
White JS, Warr WB (1983) The dual origins of the olivocochlear bundle in the albino rat. J Comp Neurol 219:203–214
Whitfield IC (1967) The auditory pathway. Arnold, London
Whitfield IC (1970) Central nervous processing in relation to spatio-temporal discrimination of auditory patterns. In: Plomp R, Smoorenberg GF (eds) Frequency analysis and periodicity detection in hearing. Sijthoff, Leiden, pp 136–152
Whitley JM, Henkel CK (1984) Topographical organization of the inferior collicular projection and other connections of the ventral nucleus of the lateral lemniscus in the cat. J Comp Neurol 229:257–270
Whittington DA, Hepp-Reymond M-C, Flood W (1981) Eye and head movements to auditory targets. Exp Brain Res 41:358–363
Wickelgren BG (1971) Superior colliculus: some receptive field properties of bimodally responsive cells. Science 173:69–71
Wiener FM (1947) On the diffraction of a progressive sound wave by the human head. J Acoust Soc Am 19:143–146
Wiener FM, Pfeiffer RR, Backus ASN (1966) On the sound pressure transformation by the head and auditory meatus of the cat. Acta Otolaryngol (Stockh) 61:255–269
Wier CC, Jesteadt W, Green DM (1977) Frequency discrimination as a function of frequency and sensation level. J Acoust Soc Am 61:178–184
Willard FH, Martin GF (1983) The auditory brainstem nuclei and some of their projections to the inferior colliculus in the North American opossum. Neuroscience 10:1203–1232
Willard FH, Martin GF (1984) Collateral innervation of the inferior colliculus in the North American opossum: a study using fluorescent markers in a double-labeling paradigm. Brain Res 303:171–182
Willard FH, Ryugo DK (1979) External nucleus of the inferior colliculus: a site of overlap for ascending auditory and somatosensory projections in the mouse. Soc Neurosci Abstr 5:33
Willard FH, Ryugo DK (1983) Anatomy of the central auditory system. In: Willott JF (ed) The auditory psychobiology of the mouse. Thomas, Springfield, pp 201–304
Willott JF (1983) Central nervous system physiology. In: Willott JF (ed) The auditory psychobiology of the mouse. Thomas, Springfield, pp 305–338
Willott JF, Urban GP (1978) Response properties of neurons in nuclei of the mouse inferior colliculus. J Comp Physiol 127:175–184
Willott JF, Chalupa LM, Henry KR (1977) Responses of single units in the inferior colliculus of the mouse (*Mus musculus*) as a function of tone intensity. Exp Brain Res 28:443–448
Wilson ME, Cragg BG (1969) Projections from the medial geniculate body to the cerebral cortex in the cat. Brain Res 13:462–475

Wise LZ, Irvine DRF (1983) Auditory response properties of neurons in deep layers of cat superior colliculus. J Neurophysiol 49:674–685

Wise LZ, Irvine DRF (1984) Interaural intensity difference sensitivity based on facilitatory binaural interaction in cat superior colliculus. Hear Res 16:181–187

Wise LZ, Irvine DRF (1985) Topographic organization of interaural intensity difference sensitivity in deep layers of cat superior colliculus: implications for auditory spatial representation. J Neurophysiol 54:185–211

Wise LZ, Irvine DRF, Pettigrew JD, Calford MB (1982) Auditory spatial receptive field properties of neurones in intermediate and deep layers of cat superior colliculus. Neurosci Letts Suppl 8:S88

Wong D (1984) Spatial tuning of auditory neurons in the superior colliculus of the echolocating bat, *Myotis lucifugus*. Hear Res 16:261–270

Woodworth RS (1938) Experimental psychology. Holt, New York

Worden FG, Galambos R (1972) Auditory processing of biologically significant sounds. Neurosci Res Prog Bull 10:1–119

Wortis SB, Pfeffer AZ (1948) Unilateral auditory-spatial agnosia. J Nerv Ment Dis 108:181–186

Wouterlood FG, Mugnaini E (1984) Cartwheel neurons of the dorsal cochlear nucleus: a Golgi-electron microscopic study in rat. J Comp Neurol 227:136–157

Wouterlood FG, Mugnaini E, Osen KK, Dahl AL (1984) Stellate neurons in rat dorsal cochlear nucleus studied with combined Golgi impregnation and electron microscopy: synaptic connections and mutual coupling by gap junctions. J Neurocytol 13:639–664

Wright D, Hebrank JH, Wilson B (1974) Pinna reflections as cues for localization. J Acoust Soc Am 56:957–962

Wu SH, Oertel D (1984) Intracellular injection with horseradish peroxidase of physiologically characterized stellate and bushy cells in slices of mouse anteroventral cochlear nucleus. J Neurosci 4:1577–1588

Yin TCT, Kuwada S (1983a) Binaural interaction in low-frequency neurons in inferior colliculus of the cat. II. Effects of changing rate and direction of interaural phase. J Neurophysiol 50:1000–1019

Yin TCT, Kuwada S (1983b) Binaural interaction in low-frequency neurons in inferior colliculus of the cat. III. Effects of changing frequency. J Neurophysiol 50:1020–1042

Yin TCT, Kuwada S (1984) Neuronal mechanisms of binaural interaction. In: Edelman GM, Gall WE, Cowan WM (eds) Dynamic aspects of neocortical function. Wiley, New York, pp 263–313

Yin TCT, Chan JCK, Kuwada S (1983) Characteristic delays and their topographical distribution in the inferior colliculus of the cat. In: Webster WR, Aitkin LM (eds) Mechanisms of hearing. Monash University Press, Clayton, pp 94–99

Yin TCT, Kuwada S, Sujaku Y (1984) Interaural time sensitivity of high frequency neurons in the inferior colliculus. J Acoust Soc Am 76:1401–1410

Yin TCT, Chan JCK, Irvine DRF (1986) Effects of interaural time delays of noise stimuli on low-frequency cells in the cat's inferior colliculus. I. Responses to wide-band noise. J Neurophysiol 55:280–300

Yin TCT, Hirsch JA, Chan JCK (1985) Responses of neurons in the cat's superior colliculus to acoustic stimuli. II. A model of interaural intensity sensitivity. J Neurophysiol 53:746–758

Yost WA (1974) Discrimination of interaural phase differences. J Acoust Soc Am 55:1299–1303

Young ED (1980) Identification of response properties of ascending axons from dorsal cochlear nucleus. Brain Res 200:23–37

Young ED, Brownell WE (1976) Responses to tones and noise of single cells in dorsal cochlear nucleus of unanesthetized cats. J Neurophysiol 39:282–300

Young ED, Sachs MB (1979) Representation of steady-state vowels in the temporal aspects of the discharge patterns of populations of auditory-nerve fibers. J Acoust Soc Am 66:1381–1403

Young ED, Sachs MB (1981) Processing of speech in the peripheral auditory system. In: Myers T, Laver J, Anderson J (eds) The cognitive representation of speech. North-Holland, Amsterdam, pp 75–93

Young ED, Voigt HF (1982) Response properties of type II and type III units in dorsal cochlear nucleus. Hear Res 6:153–169

Zahn JR, Abel LA, Dell'Osso LF (1978) Audio-ocular response characteristics. Sens Proc 2:32–37

Zambarbieri D, Schmid R, Magenes G, Prablanc C (1982) Saccadic responses evoked by presentation of visual and auditory targets. Exp Brain Res 47:417–427

Zambarbieri D, Schmid R, Prablanc C, Magenes G (1981) Characteristics of eye movements evoked by the presentation of acoustic targets. In: Fuchs A, Becker W (eds) Progress in oculomotor research. Elsevier, New York, pp 559–566

Zook JM, Casseday JH (1980) Identification of auditory centers in lower brain stem of echolocating bats: Evidence from injection of horseradish peroxidase into inferior colliculus. In: Wilson DE, Gardner AL (eds) Proceedings fifth international bat conference. Texas Tech, Lubbock, pp 51–59

Zook JM, Casseday JH (1982a) Cytoarchitecture of auditory system in lower brainstem of the mustache bat, *Pteronotus parnelli.* J Comp Neurol 207:1–13

Zook JM, Casseday JH (1982b) Origin of the ascending projections to the inferior colliculus in the mustache bat, *Pteronotus parnelli.* J Comp Neurol 207:14–28

Zook JM, Winer JA, Pollak GD, Bodenhamer RD (1985) Topology of the central nucleus of the mustache bat's inferior colliculus: correlation of single unit properties and neuronal architecture. J Comp Neurol 231:530–546

Zvorykin VP (1964) Morphological substrate of ultrasonic and locational capacities in the dolphin. Fed Proc 23:T647–T654

Zwicker E (1970) Masking and psychological excitation as consequences of the ear's frequency analysis. In: Plomp R, Smoorenburg GF (eds) Frequency analysis and periodicity detection in hearing. Sijthoff, Leiden, pp 376–396

Zwislocki J, Feldman RS (1956) Just-noticeable differences in dichotic phase. J Acoust Soc Am 28:860–864

10 Subject Index

MIX
Papier aus verantwortungsvollen Quellen
Paper from responsible sources
FSC® C105338

If you have any concerns about our products,
you can contact us on
ProductSafety@springernature.com

In case Publisher is established outside the EU,
the EU authorized representative is:
Springer Nature Customer Service Center GmbH
Europaplatz 3, 69115 Heidelberg, Germany

Printed by Libri Plureos GmbH
in Hamburg, Germany